Urban Water Security: Managing Risks

T0186391

Urban Water Series – UNESCO-IHP

ISSN 1749-0790

Series Editors:

Čedo Maksimović

Department of Civil and Environmental Engineering
Imperial College
London, United Kingdom

Alberto Tejada-Guibert

International Hydrological Programme (IHP)
United Nations Educational, Scientific and Cultural Organization (UNESCO)
Paris, France

Urban Water Security:
Managing Risks

Edited by

Blanca Jiménez and Joan Rose

UNESCO Publishing

United Nations
Educational, Scientific and
Cultural Organization

CRC Press
Taylor & Francis Group
Boca Raton London New York Leiden

CRC Press is an imprint of the
Taylor & Francis Group, an **informa** business

A BALKEMA BOOK

Cover illustration

Drainage tunnel 'Emisor Central' of the sewer system of Mexico City, conveying from 50 m³/s (dry season) up to 270 m³/s (rainy season) of wastewater – Universidad Nacional Autónoma de México, Mexico.

Published jointly by

The United Nations Educational, Scientific and Cultural Organization (UNESCO)
7, place de Fontenoy
75007 Paris, France
www.unesco.org/publishing

and

Taylor & Francis The Netherlands
P.O. Box 447
2300 AK Leiden, The Netherlands
www.taylorandfrancis.com – www.balkema.nl – www.crcpress.com
Taylor & Francis is an imprint of the Taylor & Francis Group, an informa business, London, United Kingdom.

Typeset by Macmillan Publishing Solutions, Chennai, India

ISBN UNESCO, paperback: 978-92-3-104063-4
ISBN Taylor & Francis, hardback: 978-0-415-48566-1
ISBN Taylor & Francis, paperback: 978-0-415-48567-1
ISBN Taylor & Francis e-book: 978-0-203-88162-0

Urban Water Series: ISSN 1749-0790

Volume 5

The designations employed and the presentation of material throughout this publication do not imply the expression of any opinion whatsoever on the part of UNESCO or Taylor & Francis concerning the legal status of any country, territory, city or area or of its authorities, or the delimitation of its frontiers or boundaries.
The authors are responsible for the choice and the presentation of the facts contained in this book and for the opinions expressed therein, which are not necessarily those of UNESCO nor those of Taylor & Francis and do not commit the Organization.

British Library Cataloguing in Publication Data
A catalogue record for this book is available from the British Library

Library of Congress Cataloging-in-Publication Data

Urban water security: managing risks / Edited by Blanca Jiménez and Joan Rose.
      p. cm. – (Urban water series ; v.5)
   Includes bibliographical references and index.
   ISBN 978-0-415-48566-1 (hardcover : alk. paper) – ISBN 978-0-415-48567-8 (pbk. : alk. paper) –
ISBN 978-0-203-88162-0 (e-book)   1. Municipal water supply–Management–Case studies.   2. Water quality management–Case studies.   3. Water resources development–Case studies.   4. Sewerage–Case studies.
5. Drinking water–Contamination–Case studies.   6. Bacterial pollution of water–Case studies.   7. Viral pollution of water–Case studies.   I. Jiménez P., Blanca M.   II. Rose, Joan B.   III. Title.   IV. Series.

TD220.2.U73 2009
363.6'1091732–dc22   2009006431

Foreword

Urban water security is a complex concept, involving water availability, security of water supply, public health risks and water hazards.

The disposal of inadequately treated or untreated wastewater into urban waters constitutes a serious risk to human health and represents a major problem across the world. The uncertainties and risks posed by climate change and climate variability with increasingly frequent and intense water-related hazards, and other global drivers such as population growth and urbanization are likely to have significant impacts on the availability and quality of water resources. Managing risks is therefore of crucial importance. Approaches to mitigating these risks must be based on principles of closing the loop of the urban water cycle and adapting to the uncertainties posed by global changes.

In a context of growing concern over water security, this publication examines the risks associated with urban water systems and services. It addresses the concept of urban water security in relation to the health risks that arise from chemical and microbiological pollution of urban waters, as well as the impact of climate change on urban water management. It presents approaches to controlling and managing these risks, along with case studies that look at specific issues from different parts of the world.

The book springs from UNESCO's project on "Urban Water Security, Human Health and Disaster Prevention", implemented during the Sixth Phase of UNESCO's International Hydrological Programme (2002–2007). It includes the findings of workshops held in June 2004 in Sao Paulo, Brazil, and in November 2005 in Guanajuato, Mexico, as well as ensuing efforts by experts and collaborators. Case studies from around the world have enriched the publication. The role of Ms Blanca Jiménez (Universidad Nacional Autónoma de México) as the lead editor of the book was significant and is gratefully acknowledged. The publication, which is part of the UNESCO-IHP Urban Water Series, was prepared under the responsibility and coordination of Mr J. Alberto Tejada-Guibert, Deputy-Secretary of IHP and Responsible Officer for the Urban Water Management Programme of IHP, and Ms Sarantuyaa Zandaryaa, Programme Specialist in urban water management and water quality at UNESCO-IHP.

We extend our thanks to all the contributors for their remarkable effort and are confident that the conclusions, recommendations and case studies presented here will be of value to urban water management practitioners, policy-makers and educators alike, throughout the world.

András Szöllösi-Nagy
Secretary of UNESCO's International Hydrological Programme (IHP)
Director of UNESCO's Division of Water Sciences
Deputy Assistant Director-General for the Natural Sciences Sector of UNESCO

Contents

List of Figures

List of Tables

List of Boxes

Acronyms

ASR	Aquifer storage recovery
BAC	Biological activated carbon
BDCM	Bromodichloromethane
BOD	Biological oxygen demand
BTEX	Benzene, Toluene, Ethylbenzene, and Xylene
CCP	Critical control point
COD	Chemical oxygen demand
CPWF	Challenge Program on Water and Food
CWA	Clean Water Act
DAF	Dissolved air flotation
DBP	Disinfection by-product
DDT	Dichlorodiphenyltrichloroethane
DEHP	Diethylhexyl phthalate
DO	Dissolved oxygen
EDC	Endocrine disruptive compounds
ERP	Enforcement Response Plan
EU	European Union
FAO	Food and Agriculture Organization
IDRC	International Development Research Centre
CGIAR	Consultative Group on International Research
FBI	Federal Bureau of Investigation
FC	Fecal coliform
GAC	Granular activated carbon
GHG	Greenhouse gas
GWRP	Goreangab Water Reclamation Plant
HAA	Haloacetic acid
HACCP	Hazard analysis and critical control point
HAV	Hepatitis A virus
HCB	Hexachlorobenzene
HEV	Hepatitis E virus
HMO	Hazardous material ordinance
HMPC	Hazardous Materials Spills Control Plan
HPC	Heterotrophic plate count
IPCC	Intergovernmental Panel on Climate Change
IU	Industrial users

IWA	International Water Association
IWMI	International Water Management Institute
IWRM	Integrated water resource management
LAS	Linear alkylbenzene sulphonate
LEL	Lower explosive limit
MCL	Maximum contaminant level
MDG	Millennium Development Goals of the United Nations
MRA	Microbiological risk assessment
MSD	Metropolitan sewer district
MTBE	Methyl ter-butyl ether
NAPL	Non-aqueous phase liquid
NAWQA	USGS National Water Quality Assessment Programme
NGO	Non-governmental organization
NOM	Natural organic matter
NOV	Notice of violation
NPE	Nonylphenol ethoxylate
NRC	National Research Council
OPI	Organic pollution index
OWC	Organic wastewater contaminant
PAC	Powder activated carbon
PAH	Polynuclear aromatic hydrocarbon
PCB	Polychlorinated biphenyl
PCCP	Personal care and pharmaceutical products
PCDDF	Polychlorinated dibenzofurans
PCE	Perchloroethene
PCN	Polychlorinated naphthalene
PCR	Polymerase chain reaction
POP	Persistent organic pollutant
PT	Pre-treatment
PTEs	Potentially toxic elements
PVC	Polyvinyl chloride
QMRA	Quantitative microbial risk assessment
RUAF	Resource Centers on Urban Agriculture and Food Security
SAT	Soil-aquifer treatment
SIC	Standard industrial classification
SIU	Significant industrial users
SNC	Significant non-compliance
SSA	Sub-Saharan Africa
SUVA	Specific ultraviolet absorbance
TC	Total coliform
TCE	Trichloroethene
TDS	Total Dissolved Solids
TEQ	Toxicity equivalent quotient
THM	Trihalomethane
TRC	Technical review criteria
UDR	Unusual discharge request
UNDP	United Nations Development Programme

UNEP	United Nations Environmental Programme
UNFCCC	United Nations Framework Convention on Climate Change
US EPA	United States Environmental Protection Agency
UV	Ultraviolet (radiation)
VBNC	Viable but not culturable (state)
VOC	Volatile Organic Compound
WDR	Wastewater discharge regulation
WHO	World Health Organization
WWTW	Wastewater treatment works

Glossary

Adenoviruses   Viruses most commonly cause respiratory illness; however, they may also cause various other illnesses, such as gastroenteritis, conjunctivitis, cystitis and rash illness. Adenoviruses are unusually stable to chemical or physical agents and adverse pH conditions, allowing for prolonged survival outside the body.

Aerobic   Condition of water which contains dissolved oxygen sufficient to support aerobic bacteria.

Anaerobic digestion   The degradation of organic matter by microorganisms in the absence of oxygen, it can be used to treat wastewater or sludge. Anaerobic digestion plants produce methane that sometimes is used as a heating fuel.

Anoxic   Condition in which the concentration of dissolved oxygen is so low that certain groups of microorganisms prefer oxidized forms of nitrogen, sulphur or carbon as an electron acceptor.

Aqueous speciation   The formation of new and distinct species in the course of evolution in water.

Ataxic   A disturbance of the natural animal functions; irregular.

Bedrock lithology   Nature and composition of the solid rock underlying alluvial and other superficial formations.

Bioaccumulation   The increase in concentration of a chemical in organisms that reside in environments contaminated with low concentrations of various organic compounds. Also used to describe the progressive increase in the amount of a chemical in an organism resulting from rates of absorption of a substance in excess of its metabolism and excretion.

Biochemical Oxygen Demand (BOD)   The amount of dissolved oxygen consumed by microorganisms as they decompose organic material in polluted water; a water quality indicator; BOD_5 is the biochemical oxygen demand over a five-day period.

Chloramines   Compounds formed by the reaction of hypochlorous acid (or aqueous chlorine) with ammonia.

Class A biosolids   Biosolids in which the pathogens (including enteric viruses, pathogenic bacteria and viable Helminth ova) are reduced below current detectable levels.

Class B biosolids   Biosolids in which the pathogens are reduced to levels that are unlikely to pose a threat to public health and the environment under specific use conditions.

Coagulation process   Process of adding a chemical (the coagulant) which causes the destabilization and aggregation of dispersed colloidal material into flocs.

Colloidal suspension   Suspension containing particles, often electrically charged, which do not settle but may be removed by coagulation.

Cryptosporidium   A microorganism commonly found in lakes and rivers, which is highly resistant to disinfection. Cryptosporidium has caused several large outbreaks of gastrointestinal illness, with symptoms that include diarrhoea, nausea and/or stomach cramps.

Cyanotoxin   Toxin produced by certain cyanobacteria (blue-green algae). When produced during algal blooms ('red tide'), cyanotoxins can poison and even kill animals and humans. Cyanotoxins can also accumulate in other animals such as fish and shellfish, and cause poisonings such as shellfish poisoning.

Cyanosis   A dusky bluish or purplish discoloration of the skin or mucous membranes due to insufficient oxygen in the blood, which could be caused by excessive nitrates in drinking water and methemoglobinemia in infants.

Cysts   The sac enclosing a hydatid, or larval form of a species of Tænia or tapeworm, found parasitic in man and various other animals.

Cytopathic   Of, pertaining to, or producing damage to cells.

Deltaic   Alluvial deposit at the mouth of a river; the geographical and geomorphological unit which results from it.

Disability Adjusted Life Years (DALY)   A measure of the overall disease burden. Originally developed by the World Health Organization, it is becoming increasingly common in the field of public health and health impact assessment (HIA). It is designed to quantify the impact of premature death and disability on a population by combining them into a single, comparable measure. In so doing, mortality and morbidity are combined into a single, common metric.

Disinfection by-products (DBPs)   A compound formed by the reaction of a disinfectant such as chlorine with organic material in the water supply.

Endocrine Disrupters Compounds (EDC)   Endocrine disruptors are chemicals that interfere with the normal function of hormones and the way hormones control growth, metabolism and body functions.

Electroplating   The process of coating an article with metal by means of electrolysis.

Endemic   Something peculiar to a particular people or locality, such as a disease which is always present in the population.

Enteric diseases   Diseases of or relating to the small intestine.

Eutrophication   Enrichment of water by nutrients, especially compounds of nitrogen and phosphorus, which increases productivity of ecosystems, leading usually to lower water quality and several adverse ecological and social effects (e.g., secondary pollution due to accelerated growth of algae and toxic cyanobacteria, depletion of oxygen).

Fecal coliform   Bacteria present in the intestines or feces of warm-blooded animals. They often are used as indicators of the sanitary quality of the water. In the laboratory, they are defined as all organisms that produce blue colonies within 24 hours when incubated at 44.5°C plus or minus 0.2°C on M-FC medium (nutrient medium for bacterial growth). Their concentrations are expressed as number of colonies per 100 ml. of sample.)

Flocculation process   Process by which clumps of solids in water or sewage aggregate through biological or chemical action so they can be separated from water or sewage.

Fluorosis   An abnormal condition caused by excessive intake of fluorine, characterized chiefly by mottling of the teeth.

Fresh weight   The weight, including the water content, of a specimen. It is also called
wet weight.

Giardia   Flagellate protozoan which is shed during its cyst stage into the feces of man
and animals. When water containing these cysts is ingested, the protozoan causes
a severe gastrointestinal disease, known as Giardiasis.

Grade   The inclination or slope of a pipeline, conduit, stream channel, or natural
ground surface; usually expressed in terms of the ratio or percentage of number of
units of vertical rise or fall per unit of horizontal distance. A 0.5% grade would be
a drop of one-half foot per hundred feet of pipe.

Helminths   Parasitic worms. There are three different kinds of helminths:
(a) Plathelminths or flat worms, (b) Nemathelminths (Aschelminths) or non-
segmented round worms, and (c) Annelida or segmented round worms. Those
infecting humans through wastewater, sludge or fecal sludge belong only to the
first two groups.

Hypolimnion   Water below the thermocline in a stratified body of water. It is remote
from surface influences and has a relatively small temperature gradient.

Leacheate   Water that collects contaminants as it trickles through wastes, pesticides
or fertilizers. Leaching may occur in farming areas, feedlots and landfills, and may
result in hazardous substances entering surface water, groundwater, or soil.
(USEPA)

Lixiviation   The action or process of separating a soluble substance from one that is
insoluble by The percolation of water, as salts from wood ashes.

Maximum Contaminant Levels (MCLs)   The MCLs are primary standard that cannot be exceeded for any waters delivered to any user of a U.S. public water system.
These standards went into effect in 1977 with the 'Safe Drinking Water Act of
1977', Public Law 93-523. In practice it is also used to define the maximum
amount of pollution that a water body or any other environment compartment
can accept without displaying negative effects.

Monte Carlo simulation   A method for obtaining the probability distribution of an
output, given the probability distribution of the input. Thus, in Monte Carlo simulation studies, three steps are usually required, namely, determining the input,
transforming the input into the output, and then analyzing the output.

Nitrification   The process whereby ammonia in wastewater is oxidized to nitrite and
then to nitrate by bacterial or chemical reactions.

Non intentional or incidental water recharge   The amount of water which percolates
down to the water table after it is used, excluding water that is added to an aquifer
pursuant to the underground storage, savings and replenishment programme.

Non-Point source   Pollution sources which are diffuse and do not have a single point
of origin or are not introduced into a receiving stream from a specific outlet.
The pollutants are generally carried off the land by stormwater runoff.
The commonly used categories for non-point sources are: agriculture, forestry,
urban, mining, construction, dams and channels, land disposal and saltwater
intrusion.

Oocysts   A resistant structure that forms from a zygote and develops into infective
sporozoites.

Organochlorines   Any chemicals that contain carbon and chlorine. Organochlorine compounds that are important in investigations of water, sediment and biological quality include certain pesticides and industrial compounds.

Organophosphates   Any organic compound whose molecule contains one or more phosphate ester groups, especially any of a large class of pesticides of this kind.

Ozonation The application of ozone to water for disinfection or for taste and odour control.

Persistent Organic Pollutant (POP)   The resistance to degradation of an organic compound as measured by the period of time required for complete decomposition of a material.

Pharmaceuticals and Personal Care Products (PPCPs)   Any product used by individuals for personal health or cosmetic reasons or used by agribusiness to enhance growth or health of livestock. PPCPs comprise a diverse collection of thousands of chemical substances, including prescription and over-the-counter therapeutic drugs, veterinary drugs, fragrances and cosmetics.

Phenolic compounds   Organic compounds that are derivatives of benzene.

Point source A stationary location or fixed facility from which pollutants are discharged; any single identifiable source of pollution; e.g., a pipe, ditch, ship, ore pit, factory smokestack.

Polychlorinated biphenyls (PCBs)   Mixtures of more than 200 different organochlorine chemical substances (cogeners) that are known for toxicity and persistence.

Protozoa   One-celled animals larger and more complex than bacteria that may cause disease.

Quantitative microbial risk assessment (QMRA) An important discipline for addressing complex food and water safety problems, it combines existing laboratory and surveillance databases with computational techniques to yield models that predict public health outcomes.

Raw water Water which has received no treatment whatsoever, or water entering a plant for further treatment.

Recalcitrant compounds   Non-degraded organic microorganisms.

Rotavirus   Any member of the genus Rotavirus of wheel-shaped reoviruses which are pathogens of a wide range of mammals and birds, typically causing severe diarrhoeal illness, esp. in the young.

Runoff   That part of precipitation that appears as streamflow.

Septic system   An on-site system designed to treat and dispose of domestic sewage. A typical septic system consists of tank that receives waste from a residence or business and a system of tile lines or a pit for disposal of the liquid effluent (sludge) that remains after decomposition of the solids by bacteria in the tank and must be pumped out periodically.

Sewer   An underground system of conduits (pipes and/or tunnels) that collect and transport wastewaters and/or runoff; gravity sewers carry free-flowing water and wastes; pressurized sewers carry pumped wastewaters under pressure.

Sewerage   The entire system of sewage collection, treatment and disposal.

Sorption process   The attachment of a portion of the dissolved phase of a substance to a solid by electrochemical or thermodynamic processes such as cation exchange or by adsorption.

South-South   A term indicating collaboration or technology sharing between low income and/or middle income developing countries.

Stochastic hydrology   Hydrological processes and phenomena which are described and analyzed by the methods of probability theory.

Stormwater   Runoff from buildings and land surfaces resulting from storm precipitation.

Superfund   Federal law which authorizes the US EPA to manage the clean-up of abandoned or uncontrolled hazardous waste sites.

Turbidity   Condition of a liquid due to fine, visible material in suspension, which impedes the passage of light through the liquid.

Unsaturated zone   That portion of the lithosphere in which the interstices are filled partly with air and partly with water.

Vector   A person, animal, or plant which carries a pathogenic agent and acts as a potential source of infection for members of another species.

Volatile Organic Compounds (VOCs)   Organic compounds that turn into vapour at relatively low temperatures.

Xenobiota   Biota displaced from its normal habitat; a chemical foreign to a biological system.

Zoonotic pathogens   Microorganisms that cause disease in animals and can be transmitted to humans.

Zooplankton   Small, usually microscopic animals (such as protozoans), found in lakes and reservoirs.

Zygotes   A body of living protoplasm, such as a cell or cell-nucleus, formed by the conjugation or fusion of two such bodies in reproduction.

List of Contributors

- Tuula Tuhkanen, Tampere University of Technology, Institute of Environmental Engineering and Biotechnology, Tampere, Finland
- Joan B. Rose, Homer Nowlin Chair in Water Research, Michigan State University, USA
- Samuel R. Farrah, Department of Microbiology and Cell Science, University of Florida, Gainesville, USA
- Inés Navarro, Instituto de Ingeniería, Universidad Nacional Autónoma de México, Mexico
- Francisco J. Zagmutt, DVM, MPVM, Vose Consulting, Boulder, Colorado, USA
- Claudia Sheinbaum Pardo, Instituto de Ingeniería, Universidad Nacional Autónoma de México, Mexico
- Francisco Cubillo, Research Development and Innovation Department, Canal de Isabel II, Madrid, Spain
- Blanca Jimenéz, Instituto de Ingeniería, Universidad Nacional Autónoma de México, Mexico
- Stephen R. Smith, Centre of Environmental Control and Waste Management, Imperial College, London, United Kingdom
- Ben van der Merwe, Environmental Engineering Services, Ausspannplatz, Windhoek, Namibia
- Pay Drechsel, International Water Management Institute (IWMI)-Africa Office, Accra, Ghana
- Liqa Raschid-Sally, International Water Management Institute (IWMI)-Africa Office, Accra, Ghana
- Robert Abaidoo, Kwame Nkrumah University of Science and Technology (KNUST), Kumassi, Ghana
- Gordon Garner, PE, Vice President, Water Business Group, CH2MHILL, United Kingdom
- Darren Saywell, Development Director, International Water Association, London, United Kingdom
- J.M. Anderson, Afton Water Solutions P/L, Berowra, Australia

Acknowledgments for Chapter 9

Chapter 9 is a condensed version of a paper 'Water Reuse in Windhoek, Namibia: 40 Years and still the only Case of Direct Water Reuse for Human Consumption' written by Petrus du Pisani, Jürgen Menge, Erich König and Ben van der Merwe as part of an International Water Association publication on water reuse.

Appreciation is recorded for the numerous individuals and organizations who have been involved with the Windhoek Water Reclamation Plant since the early sixties; for their vision, perseverance and sterling work. The authors acknowledge the enormous contribution made during the past years by diligent individuals who performed innumerable analyses, who operated pilot plants under trying conditions and who analyzed vast quantities of data. These efforts have provided the scientific base necessary for the critical selection process for the new process design.

The German KfW (Kreditanstalt für Wiederaufbau) and the European Investment Bank financed the construction of the NGWRP through favourable loan conditions. The City of Windhoek made all information available for the preparation of this paper.

Chapter 1

Introduction

Blanca Jiménez[1] and Joan B. Rose[2]

[1]Instituto de Ingeniería, Universidad Nacional Autónoma de México, Mexico
[2]Homer Nowlin Chair in Water Research, Michigan State University, USA

> The twentieth century witnessed the rapid urbanization of the world's population. The global proportion of urban population increased from a mere 13 per cent in 1900 to 29 per cent in 1950 and, according to the 2005 Revision of World Urbanization Prospects, reached 49 per cent in 2005. Since the world is projected to continue to urbanize, 60 per cent of the global population is expected to live in cities by 2030. The rising numbers of urban dwellers give the best indication of the scale of these unprecedented trends: the urban population increased from 220 million in 1900 to 732 million in 1950, and is estimated to have reached 3.2 billion in 2005, thus, more than quadrupling since 1950. According to the latest United Nations population projections, 4.9 billion people are expected to be urban dwellers in 2030.
>
> United Nations, 2007
>
> www.un.org/esa/population/publications/WUP2005/2005wup.htm

The urban water cycle is a new and relatively artificial concept, that is to say, it is of human design, with the 'natural' water cycle artificially engineered to the same extent that cities are also artificial, built environments. Cities have developed by deforming nature, and with it have built around them the urban water cycle. The validity of the urban water cycle concept can be scientifically argued, but more important is the fact that in practice there are several examples of its impacts. To better understand, analyze and address the need to control such impacts, this book has been written without further detailed discussion of the concept. The book is divided into three sections: (a) a brief description of the risks; (b) different approaches to control them; and (c) case studies around the world.

In the first part, Tuula Tuhkanen describes, in Chapter 2, the potential health effects and impacts on drinking water caused by the disposal of treated and untreated urban wastewater. The chapter contains an analysis of how the misconception of disposal has led society to forget that in fact we are widely recycling water. In Chapter 3, Joan Rose and Sam Farrah present the microbial health risks. A quantitative microbial risk assessment framework is used to present the traditional water-borne diseases, such as typhoid and cholera as well as emerging water-borne pathogens, which provides managers with an approach to use science-based monitoring of water supplies, irrigation waters and ambient recreational waters to demonstrate protection of public health. Chapter 4, by Ines Navarro and Francisco Zagmutt, contains a description of the main sources responsible for the increased levels of inorganic and organic compounds in water resources and drinking water. The major exposure routes to chemical compounds in urban populations are also described. Additionally, this chapter includes examples illustrating the different situations in developed and developing countries.

Claudia Sheinbaum presents a review of the climate change impacts on the hydrological cycle and the potential risks to urban areas, in Chapter 5. Using the urban water cycle adaptation concept, she proposes that strategies should integrate climate and water responses to it.

Part II begins with Francisco Cubillo's contribution. In Chapter 6 he describes how the balance between availability and demand needs to be maintained by the urban water service. The solutions offered include methodological and operational aspects. In Chapter 7, considering a non-conventional approach, Blanca Jiménez describes different sources that pollute urban water. To cope with them, she proposes an ample set of measures that go beyond the simplistic approach of treating wastewater in plants. Finally, in this section, because during water and wastewater treatment by-products are formed, Stephen Smith presents, in Chapter 8, the way to treat sludges and eventually to revalorize them. He explores the use of a dual-barrier approach sludge treatment, land use restrictions and a pollutant source control.

In Part III, Ben van der Merwe describes an amazing example of how short the urban water cycle can be under water shortage circumstances, in Chapter 9. The case of Windhoek, Namibia, is a unique example at a world level where water is constantly being used in an almost closed cycle. The reclamation of water over a period of 37 years has demonstrated that water of acceptable quality can consistently be produced from mainly domestic effluent. Pay Drechsel, Liqa Raschid-Sally and Robert Abaidoo present how simple and important it can be to reduce risks arising from the revalorization of untreated wastewater for agricultural irrigation. Using Accra as an example, the authors describe different options (such as irrigation methods, timing and changing types of crops) to reduce health risks caused by wastewater use.

Industries are also an important component in the urban water cycle. To illustrate different ways to better use water and reduce industrial pollutant discharges to the environment, Tuula Tuhkanen presents the water risks associated with urban landfills, in Chapter 11. To avoid the exposure of people living near landfills to chemicals released into the air, water and soil, she suggests different engineering and operational barriers.

Urban water infrastructure is also a source of concern. In Chapter 12, Gordon Garner describes the experience of the sewer explosion of Louisville, Kentucky US, which was due to the industrial use and abuse of the municipal infrastructure, and how to reduce such a risk. He also discusses the importance of industrial pretreatment requirements and monitoring programmes to avoid chemical catastrophes in sewers. Darren Saywell, in Chapter 13, analyzes how to put in place responses to man-made and natural disasters, in order to be efficient and coordinated. Based on experience mainly obtained from the 2004 Asian tsunami disaster response, lessons learned are described to better set up actions in post-disaster scenarios. Finally, in Chapter 14, John Anderson uses Australia as an example to illustrate different approaches to managing climate change risks in the urban sector. The chapter describes how stochastic analysis of system scenarios can help identify drought and climate change risks, and the economic benefits of water conservation and water reuse.

The urban water cycle, its effects, impacts and management strategies, are all concepts that certainly need further development and understanding, to be prepared to face the increasing urbanization that the world is experiencing. We hope this book will contribute to these efforts.

Chapter 2

Drinking water – Potential health effects caused by wastewater disposal

Tuula Tuhkanen

Tampere University of Technology, Institute of Environmental Engineering and Biotechnology, Finland

ABSTRACT: Every year, direct and indirect contamination of drinking water by wastewater causes innumerable water-related diseases. The situation varies dramatically among developing and developed countries. The presence of pathogens of fecal origin causes millions of deaths in developing countries. Preventing the outbreak of water-related diseases also requires constant vigilance of the water supply chain in developed countries. This chapter introduces the most common microbiological risks related to the contamination of drinking water by wastewater of human origin. The occurrence and fate of pharmaceuticals in wastewater, raw water and treated drinking water will be introduced in more detail, since it is a well-studied subject and gives a good example of the transportation and fate of substances in the urban hydrological cycle. The proper disposal and treatment of wastewater combined with the treatment of drinking water would improve the quality of life, particularly in developing countries. The reuse and recycling of wastewater is a necessity in arid and semi-areas. Special risk mitigation measures can reduce microbial and chemical risks to an acceptable level. New water risk management tools, such as multi-barrier water risk control and the Hazard Analysis and Critical Control Point (HACCP) method will be introduced.

2.1 INTRODUCTION

Every year millions of people suffer from water-related diseases. The contamination of drinking water by intestinal pathogens is one of the main oral pathways into the body. The contamination of drinking water can be unintentional or can take place after the intentional discharge of partially treated wastewater into water sources. The microbiological hazards are much more common than the chemical ones. The occurrence of toxic chemicals can, however, constitute a significant risk in long-term water consumption (Toze, 2006).

The systematic screening of the source and composition of wastewater, and the possible treatment of drinking water prior to its use, can provide an estimate of the quality and magnitude of the possible risks. New tools have been introduced, particularly by WHO (2005), to assess and control risks related to drinking water. Water Safety Plans (WSPs) and the Hazard Analysis and Critical Control Point (HACCP) approach can provide powerful tools for risk reduction. Proper treatment, disposal and control measures can also significantly reduce the risks. When the biggest hazards are known, resources for the risk reduction can be allocated in the most cost-effective way.

Even though microbial diseases caused by enteric bacteria, viruses, protozoa and helminths are one of the main killers of people, particularly in the developing countries, the occurrence of chemicals in the urban hydrological cycle should not be overlooked. In this presentation the main results of the US project 'Assessment of technologies for the removal of pharmaceutical and personal care products in sewage and drinking-water facilities to improve indirect potable water reuse', US 5th framework project, EVK1-CT-2000-00047, Poseidon, are presented. Poseidon focused on the assessment and improvement of technologies for the removal of ingredients of Pharmaceuticals and Personal Care Products (PPCPs) from wastewater and drinking-water facilities, to prevent the contamination of receiving waters, groundwater and drinking water by indirect potable water reuse of treated municipal wastewater (Zuger, 2000). The occurrence of pharmaceuticals and personal care products in the urban hydrological cycle is not, perhaps, the major problem globally, but these compounds can be used as an indicator of the fate and removal of chemicals in the environment.

2.2 DIRECT AND INDIRECT WASTEWATER REUSE

The limited quantity of unpolluted water as a resource for drinking-water production is one of the major challenges for the future worldwide. Indirect reuse can increase the water supply in areas in which the growth of the urbanized population exceeds the quantity of natural water sources. Currently, many communities in Europe and worldwide use production water resources that contain a significant percentage of treated wastewater, or which are affected by non-point load from agriculture, for drinking-water purposes (Bixio et al., 2006).

Throughout the US and many parts of the world, treated wastewater effluents comprise small (10% to 25%) to significant (>75%) percentages of the flow in surface waters that are used downstream as potable water supplies. In other settings, subsurface septic systems can affect domestic water supply wells too (Szabo, 2004). During seasonal or prolonged droughts, the contribution of wastewater effluents to drinking-water sources further increases.

The effect of indirect or direct contamination of drinking water by municipal, industrial or agricultural wastewater from point or non-point sources is perhaps the biggest threat for a safe and aesthetically pleasant drinking-water supply. The following drinking-water quality requirements should be considered:

● Protection from the immediate threats to health from pathogens or toxic concentrations of chemicals.
● Protection from short-term exposure to chemicals.
● Lifetime protection from the consumption of contaminants.
● Achieving an aesthetically acceptable supply of water at the tap.
● Protection from the contamination of water by plumbing fittings and water tanks/reservoirs.

In practice, the absolute priority must be protection against pathogens. Most chemicals in drinking water only become a health concern after a certain number of years of exposure. (There are exceptions such as nitrate and cyanotoxins.) For many contaminants, there will be other exposure routes besides drinking water such as food, inhalation and dermal contact.

2.3 MICROBIOLOGICAL RISKS

Wastewater is contaminated with pathogenic microorganisms in proportion to the health of the local community. The most common human microbial pathogens in water are enteric in origin. They enter the environment in the feces of infected hosts and can enter drinking water either directly through defecation into water and contamination from sewage or from runoff from soil and contaminated surfaces. Enteric pathogens in water include viruses, bacteria, protozoa and helminth eggs. The risk of water-borne pathogenic infections depends on the number of pathogen and their dispersion in the water, the infective dose required and the susceptibility of the exposed population. In the case of intentional water recycling, the inadequate risk control methods of wastewater reuse may endanger several contact groups, such as farmers and their families, residents in the vicinity, handlers, and sellers and consumers of the agricultural products. The possible transmission of disease may occur not only through contact or consumption of contaminated drinking water or products, but also through bioaerosols that enter the human body through the mouth and lungs (Carducci et al., 2000).

2.3.1 Viruses

Enteric viruses are the smallest pathogens found in the water. They are obligate intercellular parasites that can replicate only by forcing the host cell to produce multiple copies of the virus. Most human enteric viruses have a narrow host range, which means that only human fecal contamination of water needs to be considered. The viruses are highly infectious and it usually requires only a few viruses to cause infection. Host specificity makes their enumeration difficult and their structural simplicity makes chemical and mechanical inactivation complex and costly. Viruses have a greater potential to cause infections in sensitive population groups such as the elderly, children and the immunocompromised. Enteric viruses are those that can enter the body via the oral route. They include adenoviruses, rotaviruses, reoviruses, caliciviruses and enteroviruses (such as Polioviruses, Coxsackie A and B, and Echoviruses). They are commonly associated with wastewater, since they multiply in the gut and are excreted in large numbers in feces. Several outbreaks of water-borne viral gastroenteritis have also been reported in developed countries, such as Switzerland (Häfliger et al., 2000) and Finland (Lahti, 1991).

2.3.2 Bacteria

Bacteria are the most common and abundant of the microbial pathogens in water. Many of them are enteric in origin. The major bacterial diseases associated with the fecal–oral route are the Gram negative bacteria of the family *enterobacteriae*, which cause typhoid and paratyphoid fevers, salmonellosis, shigellosis and cholera. There are, however, bacterial pathogens which cause non-enteric diseases, such as *Legionella* spp., *Mycobacterium* spp. and *Leptospira*. Besides human beings, enteric bacteria can also infect animals, for example, *Salmonella*, *Campylobacter* and *Yersinia* in animals can form an additional contamination source.

2.3.3 Protozoa

Enteric protozoans are unicellular eukaryotes which are obligate parasites. Outside the infected host, they persist as dormant stages called cysts or oocysts. The most common protozoans are *Entamoeba histolytica*, *Giardia intestinalis* and *Cryptosporidium*.

Infection from protozoan pathogens can occur following the consumption of food or water contaminated by the cysts or oocysts, or through person-to-person contact. Even though the main host is human, several domestic and wild animals can also be potential hosts. Protozoans are significantly more infectious than bacteria and they can cause infection when less than 10 cysts/oocysts are ingested.

2.3.4 Helminths

Helminths are pluricellular organisms measuring 1 mm to several metres. Free-life helminths are not generally pathogenic, but those commonly transmitted through waste-water are. Helminths provoke helminthiasis, which causes different kinds of diseases characterized by undernourishment, anaemia and stunted growth. Helminthiasis is wide-spread around the world with an uneven distribution. In developing countries, it affects 25–33% of the population, whereas in developed ones the affected population is at most 1.5% (Bratton and Nesse, 1993). In poor regions with low sanitary conditions, the incidence of helminthiasis reaches 90%. There are several kinds of helminthiasis; Ascariasis is the most common and is endemic in Africa, Latin America and the Far East. There are 1.3 billion infections globally. And, even though it is a disease with a low mortality rate, most of the people affected are children under 15 years with problems of faltering growth and/or decreased physical fitness. Approximately 1.5 million of these children will prob-ably never bridge the growth deficit, even if treated (Silva et al., 1997). Helminthiasis is transmitted through: (a) the consumption of polluted crops; (b) direct contact with pol-luted feces or polluted wastewater; and (c) the ingestion of polluted meat. The infective agent is the egg, which is microscopic and travels in the wastewater. Helminth eggs are present in turbid water and are associated with particle presence (Jiménez, 2007).

Helminth eggs are not well-known in environmental engineering literature. They measure 20–80 μm in size, are 1.06–1.15 in density and are very sticky (Jiménez, 2007). Helminth eggs contained in wastewater are not infective. To be infective they need to develop larva for which the eggs need to remain at a certain temperature and moisture level for 10 days on soils or polluted crops (irrigated with wastewater, for instance). Helminth eggs can survive in water, soil and crops for several months/years (Jiménez, 2007). Due to the difference in health conditions between developed and developing countries, helminth egg content in wastewater and sludge is very different (Table 2.1).

Table 2.1 Helminth ova content in wastewater and sludge from different countries (Jiménez, 2007)

Country/regions	Helminth ova in wastewater, eggs/L	Helminth ova in sludge eggs/g TS
Developing countries		70–735
Mexico	6–330	73–177
Brazil	166–202	75
Egypt		Mean: 67; Max 735
Ghana		76
Morocco	840	
Jordan	300	
Ukraine	60	
United States	1–8	2–13
France	9	5–7
Germany		<1
Great Britain		<6

Helminth eggs are resistant to conventional disinfectants (chlorine, ozone and UV light). As shown in Table 2.1, not all wastewater contains significant amounts of helminth ova, which is why they are not considered in all countries' norms, as is the case for BOD or fecal coliforms. Consequently it follows that technology to treat wastewater should also differ among countries. When wastewater is used for irrigation, WHO (2006) recommends a content of ≤1 eggs/L to irrigate crops that are eaten uncooked. For fish culture, trematode eggs (*Schistosoma* spp., *Clonorchis sinensis* and *Fasciolopsis buski*) must be zero eggs/L as these worms multiply by the tens of thousands in their first intermediate aquatic host (an aquatic snail).

2.4 RISK REDUCTION OF PATHOGENS IN DRINKING WATER

In low-income countries with insufficient or no drinking-water treatment, the risk of pathogens in drinking water is constant. In developed countries, local outbreaks of bacterial and viral gastroenteritis, Giardiasis and cryptosporidiosis occur occasionally and usually after a sudden change of raw water quality or failure of the treatment process.

The survival of pathogens outside the hosts depends on many factors. In untreated water, sunlight, thermal effects, sedimentation, filtration and predation reduce microbial levels. *E. coli* is used as a microbial indicator of the occurrence of fecal contamination. It is of great importance that it survives in the environment and in treatment processes in the same way as pathogenic organisms. It is known, however, that *E. coli* may die off more quickly than other enteric bacteria, viruses, and microorganism cysts and oocysts.

The advantage of municipal sewage is that its quality is constant but the level of fecal contamination is high. The water normally requires effective treatment before it is considered suitable for reuse or for discharge into the environment. In countries where wastewater is collected and treated, up to 90% of the pathogenic organisms are removed efficiently in a conventional activated sludge treatment plant. In most Northern European countries, there is no further disinfection of wastewater or disinfection occurs only during the swimming season or as needed (Water Supply II, 2004). However, in the US, chlorination is widely practised prior to the discharge of water into receiving water bodies. Tertiary treatment with possible additional filtration and disinfection stages increases the effective removal of pathogens by several orders of magnitude. The formation of disinfection by-products is, however, a significant problem. Alternative disinfection methods, based on the use of UV-radiation or ozonation, are being developed.

The more contaminated the raw water of the drinking-water treatment plant is, the more elaborate the control of bacterial and other microorganisms should be. The coagulation and filtration processes remove up to 99.5%, but given the high content (6–7 log), complete removal requires a further disinfection step. Outbreaks of Giardiasis, amebiasis and cryptosporidiosis can also occur via drinking water. Coagulation followed by an effective filtration stage is vital for the control of these highly resistant pathogens. Higher chlorine and other disinfectant doses and residence times are required than is necessary to meet indicator organism standards, as they are more resistant to traditional disinfectants than the bacterial indictor organism *E. coli*.

The removal of viruses in optimal conditions can be also very high. Chemical coagulation acts to adsorb viruses so they can settle out during sedimentation and/or become entrained in filter media. Viruses can survive for substantial periods of time in the adsorbed state. In the absorption matter, for example, flocs of colloidal humic matter and turbidity can also protect viruses from the disinfection agents. The survival of infective viruses through the drinking-water treatment process, such as disinfection, can be explained by the presence of protective material (Vivier et al., 2004).

Membrane techniques are innovative techniques to control the occurrence of chemical and microbiological hazards in the drinking water in several phases of water treatment. They are one of the techniques also indicated for the treatment of water at point-of-use. The iron and turbidity can be removed almost completely and the drinking-water indicator organisms were removed by at least 2.5–3.5 logs (Hofman et al., 1998).

A removable drinking-water purification system applying membrane filtration system was tested under long-term conditions in the military area of Bundeswehr, Germany. It proved to be effective in the treatment of organic, inorganic and microbiological contaminates from highly contaminated surface water (Heberer et al., 2002).

2.5 CHEMICAL RISKS

The occurrence of various anthropogenic compounds in wastewater, surface water and groundwater used as raw water sources for drinking-water production have caused increasing concern for potential adverse health and environmental effects. Wastewater contains a mixture of harmful compounds including heavy metals, and synthetic organic and inorganic chemicals which originate from industry, households and stormwaters. The composition of wastewater originating from industry depends on the branch of manufacturing. Municipal wastewater, on the contrary, has common features all over the world. Besides urine and feces, for example, it contains household chemicals, consumable and personal care products, and pharmaceuticals. The grey water, which originates from the kitchen, bathroom or laundry, can contain over 900 synthetic organic compounds or xenobiotics (Erikson et al., 2002). The concentrations of phenolic compounds, which can have industrial, natural, municipal or agricultural origin, can be up to a hundred micrograms per litre (Davi and Gnudi, 1999).

The US-Geological Survey studied the occurrence of organic wastewater contaminants (OWCs) from streams susceptible to contamination i.e. downstream of intense urbanization and livestock productions. The OWCs were found in 80% of the samples. The most frequently detected groups of OWCs were steroids, non-prescription drugs, insect repellents, detergent metabolites, disinfectants, plasticizers, fire retardants, antibiotics and insecticides. The percentage of total measured concentration of total OWCs was highest for degradation products of detergents, steroids, plasticizers, non-prescription drugs, disinfectants and antibiotics. The most frequent compounds were coprostanol (fecal steroid), cholesterol (plant and animal steroid), N,N-diethyl toluamide (insect repellent), caffeine (stimulant), triclosan (antimicrobial disinfectant), tri(2-chloroethyl) phosphate (fire retardant) and 4-nonylphenol (non-ionic detergent metabolite). The measured concentrations were generally low and rarely exceeded drinking-water guidelines. Many compounds, however, do not have established guideline levels. Even though the concentrations of single compounds were low, there is still

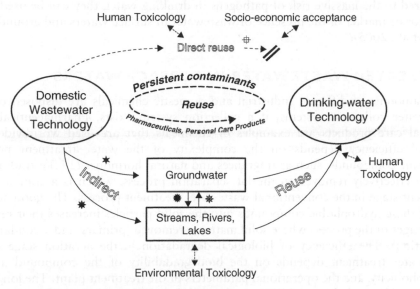

Figure 2.1 Occurrence and fate of pharmaceutical and personal care products in the environment

Source: Poseidon 2006. EVK1-CT-2000-00047

the risk of a possible interactive and cumulative effect from a complex mixture of OWCs (Koplin et al., 2002).

Approximately 3,000 different pharmaceutical ingredients are used in the US today, including painkillers, antibiotics, antidiabetics, beta-blockers, contraceptives, lipid regulators, anti-depressants, antineoplastics, tranquilizers, impotence drugs and cytostatic agents. As these compounds are frequently transformed in the body, a combination of unchanged pharmaceuticals and metabolites are excreted by humans. Human-use pharmaceuticals enter raw sewage via urine and feces and by improper disposal. These pharmaceuticals are discharged from private households and hospitals, and eventually reach municipal wastewater treatment plants. If PPCPs are only partially eliminated, residual quantities enter ambient waters or groundwater. However, direct inputs into natural waters are also possible through stormwater overflow and leaks in the sewer system. The release, occurrence and fate of PPCPs in the urban hydrological cycle is presented in Figure 2.1.

The presence of pharmaceuticals and oestrogenic compounds in natural and drinking water has been reported in recent years. Sewage treatment plants were pointed out as the one of the major sources of the discharge for PPCPs to the environment. Although the compounds have been found in the environment at very low concentrations, damages to man and other living organisms cannot be ruled out, as demonstrated for other xenobiotics. Even though intensive research has been done, more data is still needed to make a solid environmental and health risk assessment of the role of pharmaceuticals as environmental contaminants (Kümmer, 2004). PPCPs often have harmful physico-chemical behaviour, e.g., lipophilicity, in order to pass through membranes, bioaccumulation potential and biological activity in low concentrations. Even though the PPCP compounds might not pose a significant health risk, particularly compared to the massive risk of pathogens in drinking water, they can be used as an indicator or marker of the presence of wastewater in surface waters and groundwater (Fenz et al., 2005).

2.6 TREATED WASTEWATER IN SURFACE WATERS

Large amounts of different industrial and domestic chemicals are disposed of into wastewater collection systems, if a collection system exists. Pharmaceuticals and personal care products are examples of chemicals that are used worldwide. The removal efficiency depends on the complexity of the water treatment process. Lipophilic compounds, such as fragrances and natural hormones (17 β-estradiol), are already effectively removed in the fat separation process, which is a common pre-treatment stage in the conventional wastewater treatment process. The good adsorption of these hydrophobic compounds into the solid particles increases their removal at all stages of the process where solid matter is removed (primary and secondary sedimentation). The efficiency of biological degradation in the aeration stage of the wastewater treatment depends on the biodegradability of the compound and its hydrophobicity, and the operational parameters of the treatment plant. The long solid retention time and the combination of anoxic/aerobic steps generally increase the removal of refractory pharmaceuticals (Carbella et al., 2004).

In developed countries, water treatment consists of mechanical pre-treatment, biological treatment and chemical precipitation of the phosphorous. Even in developed but highly populated countries in Central Europe, the occurrence of pharmaceuticals in raw water sources and in treated drinking water have caused concern. The measured concentrations vary from low nanograms to several micrograms per litre (Ternes, 1998).

The compilation of the concentration of selected pharmaceuticals in untreated and treated wastewater and in surface waters is presented in Table 2.2.

The drug residues of 11 pharmaceuticals lipid regulators (Diclofenac), anti-inflammatory (Ibuprofen, Ketoprofen, Naxophen), and two drug metabolites were detected in raw sewage, treated wastewater and river water in the state of Rio de Janeiro, Brazil. The median concentrations in the untreated water ranged from 0.1 to 1 µg/L. The removal rates in the wastewater treatment process varied from 12 to 90%. The concentrations in the river water affected by treated municipal sewage ranged from 0.02 to 0.04 µg/L. The values were the same as those found in wastewater and surface waters of Central Europe. The compounds followed could be detected in the raw water of the local water treatment plant only sporadically (Stumpf et al., 1999).

The consumption of the most commonly used pharmaceuticals and their occurrence in raw and treated sewage and in receiving surface waters was studied in Finland as part of the EU-project, Poseidon. Most of Finland is sparsely populated and wastewater treatment is efficient. Over 90% of the population is connected to a municipal wastewater treatment system. The treated effluents go to lake systems with high dilution factors and long residence times. There are, however, some hot spots with a dense population and very limited raw water resources. The south-west coastal area in the vicinity of the Turku municipality suffers from a lack of pristine raw water sources. The treated effluents go to rivers and the next city downstream uses the diluted wastewater as a raw water source. This is a very typical situation in Central Europe, in the catchment area of the River Rhine for example.

Table 2.2 Median (maximum) concentrations in Germany (GER), Austria (AUT), Poland (PL), Spain (ES), France (FR), Switzerland (CH) and Finland (FIN). Concentrations in WWTP influents and effluents and in surface waters are given in ng L^{-1} (Poseidon, Final Report 2006)

PPCP	Location	GER	AUT	PL	ES	FR	CH	FIN
Diclofenac	influent	3500 (28000)	3100 (6000)	1750 (2000)	n.d.	n.a.	1400 (1900)	350 (480)
	effluent	810 (2100)	1500 (2000)	n.a.	n.d.	295 (300)	950 (1140)	250 (350)
	river	150 (1200)	20 (64)	n.a.	n.a.	18 (41)	20–150	15 (40)
Ibuprofen	influent	5000 (14000)	1500 (7200)	2250 (2800)	2750 (5700)	n.a.	1980 (3480)	13 000 (19 600)
	effluent	370 (3400)	22 (2400)	n.a.	970 (2100)	92 (110)	<50 (228)	1300 (3900)
	river	70 (530)	n.d.	n.a.	n.a.	23 (120)	n.d.–150	10 (65)
Bezafibrate	influent	4900 (7500)	2565 (8500)	780 (1000)	n.d.	n.a.	n.a.	420 (970)
	effluent	2200 (4600)	103 (611)	n.a.	n.d.	96 (190)	n.a.	205 (840)
	river	350 (3100)	20 (160)	n.a.	n.a.	102 (430)	n.a.	5 (25)
Diazepam	influent	<LOQ	n.d.	n.a.	n.d.	n.a.	n.d.	n.d.
	effluent	<LOQ (40)	n.d.	n.a.	n.d.	n.d.	n.d.	n.d.
	river	n.d.	n.d.	n.a.	n.a.	n.d.	n.d.	n.d.
Carba-mazepine	influent	2200 (3000)	912 (2640)	1150 (1600)	n.a.	n.a.	690 (1900)	750 (2000)
	effluent	2100 (6300)	960 (1970)	n.a.	n.a.	1050 (1400)	480 (1600)	400 (600)
	river	250 (1100)	75 (294)	n.a.	n.a.	78 (800)	30–150	70 (370)
SMX	influent	1370 (1700)	n.d. (470)	1550 (2000)	600	n.a.	425 (570) 1670[a] (1900[a])	n.a.
	effluent	400 (2000)	31 (234)	n.a.	250	n.d.	290 (860) 400[a] (880[a])	n.a.
	river	30 (480)	n.d.	n.a.	n.a.	25 (133)	n.a.	n.a.
Roxithro-mycin	influent	830 (1000)	43 (350)	n.d.	n.d.	n.a.	20 (35)	n.a.
	effluent	100 (1000)	66 (290)	n.a.	n.d.	n.d.	15 (30)	n.a.
	river	<LOQ (560)	n.d.	n.a.	n.a.	9 (37)	n.a.	n.a.
Iopromide	influent	13 000 (22 000)	n.d. (3840)	1330 (2700)	6600	n.a.	810 (7700)	n.a.
	effluent	750 (11 000)	n.d. (5060)	n.d.	9300	n.d.	790 (2000)	n.a.
	river	100 (910)	91 (211)	n.a.	n.a.	7 (17)	n.a.	n.a.
Tonalide (AHTN)	influent	400 (450)	970 (1400)	n.d.	1530 (1690)	n.a.	545 (940)	200 (230)
	effluent	90 (180)	140 (230))	n.a.	160 (200)	n.a.	410 (500)	40 (50)
Galaxolide (HHCB)	influent	1500 (1800)	2800 (5800)	610 (1200)	3180 (3400)	n.a.	1660 (2200)	750 (980)
	effluent	450 (610)	470 (920)	n.a.	500 (600)	n.a.	1150 (1720)	120 (160)

Table 2.3 Concentrations of acidic pharmaceuticals (μg L^{-1}) calculated from the consumption figures reported by the National Agency for Medicines and the measured wastewater concentrations (Lindqvist et al., 2005)

	Ibuprofen	*Naproxen*	*Ketoprofen*	*Diclofenac*	*Benzafibrate*
Calculated	14.7	1.4	0.3	0.3	0.5
Measured	13.1	4.9	2.0	0.35	0.42

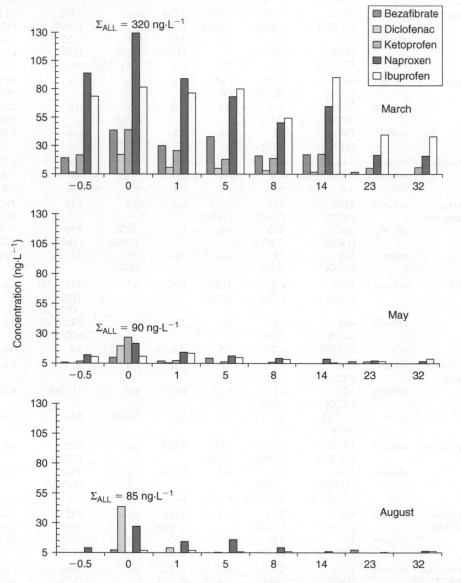

Figure 2.2 Study pharmaceuticals in the influent (A) and the effluent (B) of the sewage treatment plant of Aura. Σ is the sum of all the studied pharmaceuticals. R is the removal of the total load of the studied pharmaceuticals in the treatment process

Source: Vieno et al., 2005

The occurrences of five common medicines in untreated and treated wastewater were studied in south-western Finland. Samples were also taken from seven rivers, which received municipal effluents from the point of discharge and from downstream. It was also possible to theoretically calculate the concentration of individual pharmaceuticals in the wastewater when the amount of pharmaceutical used per day per capita, number of inhabitants and the wastewater flow were known. The calculated and measured concentrations in the wastewater are presented in Table 2.3.

The removal efficiencies of the selected pharmaceuticals in the activated sludge treatment plant varied significantly. The highest removal efficiency present, 92%, concerned the painkiller Ibuprofen and lowest, Diclofenac, was 26%. The removal efficiencies of the Finnish study were slightly better than the literature values.

Low concentrations of the studied pharmaceuticals were found in the samples taken upstream from the effluent discharge point. At the points of effluent discharge, concentrations dramatically increased. The concentration patterns in the river were similar to the patterns in the corresponding wastewater (Vieno et al., 2005).

The concentration of studied pharmaceuticals in the receiving rivers were followed in three seasons: winter, spring and summer up to 32 kilometres from the discharge point of treated waste.

In the wastewater treatment plants, the purification efficiencies of the pharmaceuticals decreased in the winter leading to increased concentrations of the pharmaceuticals in the effluent streams. This can be caused by the decreased nitrification activity of activated sludge during the winter. There is a correlation between the nitrification and effective removal of pharmaceuticals. Accordingly, the highest concentrations in rivers were measured in the winter. Pharmaceuticals were also carried long distances in the winter because of the lack of the main elimination processes, photo degradation, due to the ice cover of the river.

2.7 THE OCCURRENCE OF PHARMACEUTICALS IN DRINKING WATER

The elimination of selected pharmaceuticals by drinking-water treatment processes was studied by the EU-project, Poseidon (Ternes et al., 2002). No significant removal of the hydrophilic pharmaceuticals took place in chemical coagulation. On the contrary, activated carbon filtration and particularly ozonation are very effective methods for eliminating most of the recalcitrant compounds in water (Ternes et al., 2002; Boyd et al., 2003; Huber et al., 2005). The formation of the possible degradation products during the oxidative treatment is a potential draw-back (Andreozzi et al., 2002).

The occurrences and removal of pharmaceuticals were followed in the drinking-water works of the Turku municipality. It is located 32 kilometres downstream from the wastewater treatment plant of a small municipality, Aura. The drinking-water treatment process consisted of two-stage ferric coagulation, granular activated carbon filtration and chlorination. Drinking water contained low concentrations of Ibuprofen and Ketoprofen in the winter samples because of the higher raw water concentrations. Also, the removal efficiency of those compounds was low, between 20–30%, since they are relatively hydrophilic and thus, resistant to coagulation.

The concentrations of the studied painkillers were very low in drinking water. The exposure of pharmaceuticals via drinking-water consumption is insignificant compared to the therapeutic doses of the same pharmaceuticals. However, the pharmaceuticals can be used as indicators of the occurrence and fate of anthropogenic pollutants in the urban hydrological cycle. The Finland results show that the cold season in boreal areas can significantly increase the environmental and health risks of anthropogenic compounds such as pharmaceuticals.

2.8 RISK MANAGEMENT OF MICROBIAL AND CHEMICAL HAZARDS

The prevention of a problem is always better than remediation. Human excreta and urine as such are valuable materials for recycling carbon and nutrients, such as phosphorous and nitrogen, back to food production (Schouw et al., 2002).

Success in the elimination of health risks depends on the means of controlling, containing and mitigating the health hazard agencies generating the risk. This depends largely on the integrated approach, which combines disciplines such as engineering, management, public health, chemistry and agriculture. Health risk reduction is not possible without risk identification, assessment and reduction, which involves regulating, monitoring and enforcement measures.

The health risk depends on the type of contaminant. Microbial pathogens can cause rapid infection if ingested by people in contact with food. In contrast, toxic chemicals tend to have an effect only after constant, long-term exposure over a long period. In drinking-water issues there are various stakeholders, such as consumers and experts (researchers, regulators and suppliers) from different disciplines, such as engineering, chemistry, public health, economics and law. Risk perceptions and uncertainties among the stakeholder groups vary significantly. They are not based solely on numerical and scientific data. Unfamiliar, new and coerced risks are considered more significant than old, familiar and controlled ones. Generally, openness in communication and transparency of decision making and the possibility to participate increase the satisfaction. The scientists are trusted more than the water suppliers and trust among one's own stakeholder group is highest (Baggett et al., 2006).

To avoid any human-health problems caused by wastewater ingredients, a 'multibarrier system' has to be introduced, e.g., placing barriers between the reused water and the members of community (Toze, 2006). This approach implements two basic ideas: if a single natural or technical removal process becomes ineffective, the contamination of drinking water by pathogens or chemicals will be avoided subsequently by other processes. Additionally, the specific removal potencies of the particular steps in the multibarrier system support the efficiency and stability of the entire system, to enhance an efficient and unpolluted water supply, to minimize environmental impacts from wastewater and to avoid potential human-health hazards. In the case of water reuse, barriers can also include restricted irrigation and drying times, preventing access to areas where the water is being used and processing the goods that have been irrigated with reused water.

Within the EU-project Poseidon, many relevant techniques and processes involved in the urban water cycle were assessed regarding their removal efficiency for PPCPs. Besides the treatment of wastewater, natural attenuation processes exist. When lakes are used as an alternative reservoir for treated wastewater within the indirect potable water reuse process, degradation by solar irradiation may play an important role in the elimination of organic pollutants, in addition to microbial degradation and absorption (Toze, 2006).

In the Northern hemisphere, biodegradation and photodegradation processes are significantly hindered during the cold and dark seasons and ice/snow cover.

The degradation of organic matter in waste matter takes place in three phases: first as a result of the decomposition of the easily biodegradable, non-humic organic matter, some of the organic matter can remain in the wastewater. In particular, similar industrial water streams, such as those coming from the paper, pulp and textile industries, can contain very refractory compounds of natural or synthetic origin.

The intentional wastewater infiltration into the soil from septic tanks is also a common practice in the countryside of developed countries, such as Sweden and Finland. If the infiltration system is properly located, constructed and maintained, most of the organic matter is adsorbed and degraded in the soil. The wastewater-induced clogging of the soil can increase the soil biogeochemical activity and can enhance absorption, biotransformation of organic constituents and die-off and inactivation of bacteria and viruses (van Cuyk, 2001). The soil matter also has adsorption capacity with respect to various inorganic compounds, such as phosphorous and harmful metals. When the capacity of soil has been exceeded, the harmful components can enter the groundwater, leading to the contamination of the groundwater (Jacks et al., 2000). Also, unintended wastewater disposal, accidental leaks and leaching of human excreta and animal manure can cause both microbiological and chemical contamination of drinking-water wells (Scandura and Sobsey 1997; Szabo, 2004).

2.9 IMPLEMENTATION OF WATER SAFETY PLANS

The systematic screening of the source and composition of wastewater and possible drinking-water treatment prior to use can provide an idea of the quality and magnitude of the possible risks. New tools are being introduced, particularly by WHO, to assess and control the risks related to drinking water. Water Safety Plans and Hazard Analysis and Critical Control Point approaches provide powerful tools in the risk reduction. When the most important hazards are known, the resources for risk reduction can be allocated in the most cost-effective way.

A Water Safety Plan is a comprehensive risk assessment and risk management approach to help keep the water supply safe and secure from catchment to consumer. It can vary in complexity, depending on the situation. WSPs comprise at least three essential actions:

- System assessment to determine whether the drinking-water supply chain can deliver water that meets health-based targets.
- Identifying control measures that will control identified risks and ensure that the health-based targets are met.
- Management plans describing actions to be taken during normal or incident conditions and documentation of system assessment, monitoring, communication plan and support programme.

More drinking-water risk management data can be found in the WHO documents: Water safety Plans; Managing drinking-water quality from catchment to consumer (WHO, 2005).

2.10 HACCP

A focus of the new WHO (2005) drinking-water guideline is the implementation of a process-controlled, quality management system. This approach is based on the successfully applied HACCP principle. Hazard analysis and critical control point (HACCP) is a concept which was developed for the food industry to enhance food safety, but it has also been used in the water industry, particularly in Austria, Iceland and Australia. It shifts emphasis from resource-intensive end-product inspection and testing to the preventive control of hazards at all stages of the water supply system through systematic identification, evaluation and continuous recording and control as well as the necessary preventive methods which serve the purpose of preventing, eliminating and at least reducing hazards to an acceptable level. The following definitions applied to HACCP:

- Hazard: every biological, chemical or physical characteristic that could pose an unacceptable health risk to the consumer.
- Risk: the assessment of the possible occurrence of a danger or hazard.
- HACCP System: Hazard Analysis and Critical Control Point; recognized international standard for a comprehensive type of quality control. The basis for this control system is a risk analysis of all production and distribution installations.
- CCP: Critical Control point. A critical control point is defined as every point or process in the entire water supply system, the loss of control of which can cause an unacceptable health risk.

For protection from pathogens, numerical standards are not useful since the sampling and analysis process is too slow. For pathogens, the barriers include protection of the catchment, storage ahead of treatment and the treatment stage itself.

To apply HACCP ask three questions:

- What is the hazard?
- How is the hazard fixed?
- How do you know that the hazard is fixed?

HACCP is a systematic method of identification, evaluation and control of, originally food, safety hazards. The five preliminary tasks are:

- Assemble an HACCP team.
- Describe a product.
- Identify the intended use.
- Construct a flow diagram.
- On-site confirmation of flow diagram.

The framework of HACCP consists of seven principles:

- Perform a hazard analysis.
- Determine the critical points.

- Establish one or several critical limits.
- Establish a CCP monitoring system.
- Establish corrective action to be taken if monitoring indicates that a specific CCP is no longer under control.
- Establish a procedure of verification to confirm that the HACCP system works successfully.
- Introduce a documentation system that takes into account all processes and records them in accordance with the principles and their applications.

2.11 HAZARD ANALYSIS

Hazards have to be analyzed and control points have to be identified. Hazards are microbiological, chemical, physical or radioactive parameters, which may be present and affect the consumer's health. The causes of introduction of the hazards can be:

- natural (soil properties, animals, rainfall, etc.)
- accidental, uncontrolled events
- vandalism
- sabotage
- natural disaster
- operational parameters
  - inadequate repair or maintenance
  - inadequate disinfection
  - failure of monitoring and alarm
  - build up of biofilms
  - corrosion.

The hazard analysis and critical control point system is a systematic approach to controlling safety hazards in a process by first identifying the hazards, their severity and likelihood of occurrence. Then, critical control points and their monitoring criteria are identified to establish controls that will reduce, prevent or eliminate the hazards identified.

Ensuring that the use of treated wastewater in an indirect water reuse cycle does not result in adverse health effects, a quality management system, as used in the HACCP approach, is needed. A water safety plan should be implemented with the objective of a well-adjusted evaluation of hazards for each individual indirect potable water reuse system, starting at the source and going to the consumer, to adjust a quality management system, based on monitoring plans, control systems and corrective action plans (Salgot et al., 2006).

The operation of wastewater treatment involves many internal and external risks, the impacts of which have to be analyzed along with the probability of occurrence and the level of damage. If attention is paid to the systematic and complete risk identification, the weak points can be improved and the risks avoided (Wagner and Stube, 2005).

2.12 CONCLUSIONS

When wastewater is used directly or indirectly for the irrigation or recharge of groundwater resources, and particularly for drinking-water production, special attention is needed to reduce the health risk to an acceptable level. The removal of the microbial and chemical impurities in the wastewater usually requires several treatment stages. A multi-barrier approach is particularly useful in when it concerns wastewater reuse and recycle. Besides traditional engineering tools, such as wastewater and drinking-water treatment, policy level and behavioural risk reduction measures are also needed.

REFERENCES

Andreozzi, R., Marotta, R., Pinto, G. and Pollio, A. 2002. Carbamazepine in Water, Persistence in the Environment, Ozonation and a Preliminary Assessment on Algal Toxicity, *Water Research*, Vol. 36, No. 11, pp. 2869–2877.

Baggett, S., Jeffrey, P. and Jefferson, B. 2006. Risk Perception in Participatory Planning for Water Reuse, *Desalination*, Vol. 187, No. 1–3, pp. 149–158.

Bixio, D., Thoeye, C., De Konig, J., Joksimovic, D., Savic, D., Wintgens, T. and Melin, T. 2006. Wastewater Reuse in Europe, *Desalination*, Vol. 187, No. 1–3, pp. 89–101.

Boyd, G.R., Riesman, H., Grimm, D.A. and Mitre, S. 2003. Pharmaceuticals and Personal Care Products (Peps) in Surface and Drinking Water in Louisiana, USA and Ontario, Canada, *The Science of the Total Environment*, Vol. 311, pp. 135–149.

Bratton, R. and Nesse, R. 1993. Ascariasis: An Infection to Watch for in Immigrants, *Postgraduate Medicine*, Vol. 93, pp. 171–178.

Carbella, M., Omil, F., Lema, J.A., Lombart, M., Garcia-Jares, C., Rodrigues, I., Gomez, M. and Thernes, T. 2004. Behavior of Pharmaceuticals, Cosmetics and Hormones in a Sewage Treatment Plant, *Water Research*, Vol. 38, pp. 2918–2926.

Carducci, A., Tozzi, E., Rubelotta, E., Casini, B., Cautini, L. Rovini, E., Muscillo, M. and Pacini, R. 2000. Assessing Airborne Bacterial Hazards from Urban Wastewater Treatment, *Water Research*, Vol. 34, pp. 1173–1178.

Davi, M.L. and Gnudi, F. 1999. Phenol Compounds in Surface Water, *Water Research*, Vol. 33, pp. 3123–3219.

Erikson, E., Auffarth, K., Henze, M. and Ledin, A. 2002. Characteristics of Grey Wastewater, *Urban Water*, Vol. 4, Issue 4, pp. 85–104.

Fenz, R., Blaschke, A.P., Clara, M., Kroiss, H., Masher, D. and Zasner, M. 2005. Quantification of Sewer Exfiltration using the Anti-epileptic Drug Carbamazepine as Marker Species for Wastewater, *Water Science and Technology*, Vol. 52, No. 9, pp. 209–217.

Häfliger, D., Huber, D. and Luthy, J. 2000. Outbreak of Viral Gastroenteritis due to the Sewage-contaminated Drinking Water, *Journal of Food Microbiology*, Vol. 52, pp. 123–126.

Heberer, T., Feldmann, D., Reddsern, K., Altmann, H-J. and Zimmermann, T. 2002. Production of Water from Highly Contaminated Surface Waters: Removal of Organic, Inorganic and Microbial Contaminants Applying Mobile Membrane Filtration Units, *Acta hydrochimica et hydrobiologica*, Vol. 39, No. 1, pp. 24–33.

Hofman, J.A.M.H., Beumer, M.M., Baas, T.E., van der Hoek, J.P. and Koppers, H.M.M. 1998. Enhanced Surface Water Treatment by Ultrafiltration, *Desalination*, Vol. 119, Issue 1–3, pp. 113–125.

Huber, M., Korhonen, S., Ternes, T. and von Gunten, U. 2005. Oxidation of Pharmaceuticals during Water Treatment with Chlorine Dioxide, *Water Research*, Vol. 39, pp. 3607–3617.

Jacks, G., Forsberg, J., Magoub, F. and Palmqvist K. 2000. Sustainability of Local Water Supply and Sewage System – A Case Study in a Vulnerable Environment, *Ecological Engineering*, Vol. 15, pp. 147–153.

Jiménez, B. 2007. Helminth Ova Removal from Wastewater for Agriculture and Aquaculture Reuse, *Water Science and Technology*. Vol. 55, No. 1–2, pp. 485–493.

Koplin, D.W., Furlong, E.T., Meyer, M.T., Thurman, E.M., Zaugg, S.D., Barber, L.B. and Buxton, H.T. 2002. Pharmaceuticals, Hormones, and Other Organic Wastewater Contaminants in U.S. Streams, 1999–2000: A National Reconnaissance, *Environmental Science and Technology*, Vol. 32, No. 18, pp. 1202–1211.

Kümmer, K. (ed.) 2004. *Pharmaceuticals in the Environment – Sources, Fate, Effects and Risks.* 2nd edn. Springer, Berlin, p. 527.

Lahti, K. 1991. The Waterborne Infections in Finland 1989–1990, *Vesitalous.* Vol. 32, pp. 30–32.

Lindqvist, N., Tuhkanen, T. and Kronberg, L. 2005. Occurrence of Acidic Pharmaceuticals in Raw and Treated Sewages and in Receiving Waters, *Water Research*, Vol. 39, pp. 2219–2228.

POSEIDON. Assessment of technologies for the removal of pharmaceuticals and personal care products in sewage and drinking water facilities to improve the indirect potable water reuse. Contract No. EVK1-CT-2000-00047, http://poseidon.bafg.de/servlet/is/2888/.

Salgot, M., Huertas, E., Weber, S., Dott, W. and Hollender, J. 2006. Wastewater Reuse and Risk: Definition of Key Objectives, *Desalination*, Vol. 187, Issue 1–3, pp. 29–40.

Scandura, J.E. and Sobsey, M.D. 1997. Viral and Bacterial Contamination of Groundwater from On-site Sewage Treatment System, *Water Science and Technology*, Vol. 35, No. 11–12, pp. 141–146.

Schouw, N.L., Danteravanich, S., Mosbaek, H. and Thell, J.C. 2002. Composition of Human Excreta – A Case Study from Southern Thailand, *The Science of the Total Environment*, Vol. 286, pp. 155–166.

Silva, N., Chan, M. and Bundy, A. 1997. Morbidity and Mortality due to Ascariasis: Re-estimation and Sensitivity Analysis of Global Numbers at Risk, *Tropical Medicine & International Health*, Vol. 2, No. 6, pp. 519–528.

Stumpf, M., Ternes, T., Vilken, R.D., Rodrigues, S. and Baumann, W. 1999. Polar Drug Residues in Sewage and Natural Waters in the State of Rio de Janeiro, Brazil, *The Science of the Total Environment*, Vol. 225, pp. 135–141.

Szabo, H.M. 2004. *The Effect of Small Scale Wastewater Disposal on the Quality of Well Water.* M.Sc thesis, Tampere University of Technology.

Ternes, T. 1998. Occurrence of Drugs in German Sewage Treatment Plants and Rivers, *Water Research*, Vol. 32, No. 11, pp. 3245–3260.

Ternes, T., Meisenheimer, M.T., McDowel, D., Sacher, F., Brauch, H.J., Haist Gulde, B., Preuss, G., Wilme, U. and Zulei-Seibert, N. 2002. Removal of Pharmaceuticals During Drinking Water Treatment, *Environmental Science and Engineering*, Vol. 36, pp. 3855–3863.

Toze, S. 2006. Water Reuse and Health Risks – Real vs. Perceived, *Desalination*, Vol. 187, No. 1–3, pp. 1–51.

Van-Cuyk, S., Siegrist, R., Logan, A., Masson, S., Fisher, E. and Figueroa, L. 2001. Hydraulic and Purification Behaviors and their Interactions During Wastewater Treatment in Soil Infiltration Systems, *Water Research*, Vol. 35, No. 4, pp. 953–964.

Vieno, N., Tuhkanen, T. and Kronberg, L. 2005. Seasonal Variation of the Pharmaceuticals in Aquatic Environment, *Environmental Science and Technology*, Vol. 39, No. 21, pp. 8220–8226.

Vivier, J.C., Ehlers, M.M. and Grabow, W-O.K. 2004. Detection of Enteroviruses in Treated Drinking Water, *Water Research*, Vol. 38, No. 11, pp. 2699–2705.

Wagner, M. and Stube, I. 2005. Risk Management in Wastewater Treatment, *Water Science and Technology*, Vol. 52, No. 12, pp. 52–61.

Water Supply II, 2004. *Association of Finnish Civil Engineers*, 1st edn. Helsinki, pp. 684.

WHO. 2006. *Guidelines for the Safe Use of Wastewater, Excreta and Greywater. Wastewater Use in Agriculture*, WHO Library Cataloguing-in-Publication Data, Geneva, Vol. 2, p. 213.

WHO. 2005. *Water Safety Plans; Managing Drinking-water Quality from Catchment to Consumer*, 224. Available at: www.who.int/water_sanitation_health/dwq/wsp0506/en.

Zuger, P.S. 2000. Drugs down the Drain, *C& EN*, April 10, pp. 51–53.

Chapter 3

Microbial health risks and water quality

Joan B. Rose[1] and Samuel R. Farrah[2]

[1]Homer Nowlin Chair in Water Research, Michigan State University, 13 Natural Resources
E. Lansing MI 48824, USA
[2]Department of Microbiology and Cell Science, University of Florida, Gainesville, FL, USA

ABSTRACT: The association between water quality and human health has been known for over 150 years. Between 1850 and 1900, studies by Snow, Budd, Escherichia and others led to the development of two important concepts. First, water could transmit microorganisms that caused diseases such as typhoid fever and cholera. Second, coliform bacteria could be used as an indicator of the microbial quality of water.

Since then, several other bacteria, viruses and protozoa have been found to be associated with disease transmission by consumption of contaminated water. Some of these microorganisms that are considered in this chapter include bacteria (*Escherichia coli* O157:H7), viruses (poliovirus and hepatitis virus) and protozoa (*Cryptosporidium* and *Cyclospora* spp.).

Certain diseases, such as typhoid fever and cholera, have been controlled in developed countries but are still problems in some developing countries. Other microbes, such as *Escherichia coli* O157:H7, *Cryptosporidium* and *Cyclospora* spp., continue to be problems for both developed and developing countries. Some of these may cause outbreaks in developed countries but may be endemic in developing countries.

Quantitative models describing the risks associated with pathogens have been developed. These can feed into monitoring programmes to demonstrate risk reduction implemented by the water utility for addressing the potable water supply, irrigation waters and recreational waters, which are also part of the urban cycle. All of the fecal–oral pathogens described herein can be spread through drinking water, irrigation waters (via the food supply) and recreational waters. The water contamination cycle for the urban environment thus goes well beyond the community drinking water. Better assessment of water quality at key locations in the environment and strategic monitoring should be implemented. Findings and recommendations include:

1. Cholera and typhoid remain the classical water-borne diseases and these are readily monitored for. The health target should be eradication of these diseases. It is clear that any area with cholera and typhoid cases does not have adequate separation of drinking water and sewage and this should be immediately investigated with appropriate monitoring within a risk framework to examine vulnerabilities, so that corrective action can be implemented.
2. Indicators are useful, although it is now possible for the industry to examine specific water-borne disease pathogens. Risk assessment and probability of infection models can then be, and should be, used by the water industry to address public health based water quality monitoring with established management plans.
3. There are many stakeholders, in addition to the public, involved in the urban water cycle, including wastewater, agriculture, drinking water and public health. The water industry, in

particular, needs to form better partnerships with public and environmental health professionals so that water-borne disease and risk assessment methods can be appropriately used and communicated.

3.1 INTRODUCTION

Global goals for public health and the benefits of 'safe' water have been understood and studied since before the time of John Snow, who investigated cholera in London, reporting that this disease was transmitted through the drinking-water supply (1849). Robert Koch identified the bacterium *Salmonella* as the cause of typhoid in 1856; he suggested that typhoid fever was spread by feces and presented a scientific method to identify 'contagious agents', known as Koch's postulates (1876), which moved the study of cause-and-effect forward (Beck, 2000). Typhoid fever symptoms were described in 1659, but it was not until 1856 that William Budd first suggested that typhoid was contracted through contaminated water. The idea that typhoid was water-borne did not gain acceptance by the water industry or public health until 1873, and this concept was finally proven in 1884. This was some 30 years after water-borne cholera was established. The indicator concept for 'water quality' associated with fecal pollution came along with the development of the gram stain and identification of the *Escherichia coli* (1884 and 1885, respectively), and by 1910 the 'coliform' was being used to address health issues related to the fecal contamination of water. In that same period (1884), Koch also isolated a pure culture of *Vibrio* and Georg Gaffky isolated a pure culture of the typhoid bacillus (Beck, 2000), and the study of both cholera and typhoid began worldwide (Europe, Africa and India). Now more than 100 years later, with the advances in microbiology and engineering, one wonders why scientific theory and practice have neglected to address water-borne disease and have failed to address safe water at the global level.

Pollution of the water environment is tied directly to human activities, types of water use and the types of pollution prevention strategies employed. Wastewater and sewage discharges are the predominant sources of water-borne disease-causing pathogens. Water pollution and treatment as well as water-borne disease may be seen as a local water issue; however, we would contend that this is truly a global problem. There are 215 international river basins where the waters are shared by more than one country (Gleick, 1993). There are 13 international water basins worldwide shared by five or more nations, including the Danube, Niger, Nile, Zaire, Rhine, Zambezi and Amazon. Thus, while quantity has been addressed at the global scale between countries sharing the same watershed, water quality often remains under the jurisdiction of the local community.

In the latest International Health Regulations (2005), the World Health Assembly has begun to develop a framework for surveillance and reporting of diseases associated with what is defined as 'public health emergencies of international concern' (Baker and Fidler, 2006). This includes cholera and poliovirus, two water-borne agents, but mostly addresses respiratory agents such as influenza, and emerging, rarer and unusual agents including Ebola and SARS. The World Health Organization (WHO) has now estimated that there are around 250 million cases of water-related diseases (e.g., Malaria) annually in addition to 4.37 billion cases of diarrhoea associated with contaminated water (UNDP, 2003; WHO, 2003). Finally, the United

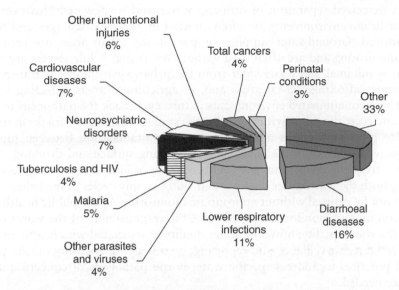

Other unintentional injuries 6%

Total cancers 4%

Cardiovascular diseases 7%

Perinatal conditions 3%

Other 33%

Neuropsychiatric disorders 7%

Tuberculosis and HIV 4%

Malaria 5%

Diarrhoeal diseases 16%

Lower respiratory infections 11%

Other parasites and viruses 4%

Figure 3.1 **Percentage of the disease burden (in DALYs*) attributable to environmental factors**

Note: DALY (disability adjusted life years) is a measurement for diseases which is the sum of the years of life lost
due to premature mortality (YLL) in the population and the years lost due to disability (YLD) for incident
cases of the health condition (WHO, www.who.int/healthinfo/boddaly/en/)

Source: Pruss-Ustun A. and Corvalan C., 2007

Nations has set the Millennium Development Goals (MDG; UN, 2000) for access to
safe water and appropriate sanitation.

Most recently, the World Health Organization (WHO, Pruss-Ustun and Corvalan,
2007) has provided an analysis showing that 24% of all deaths in children under the
age of 15 are due to environmentally related diarrhoea, malaria and respiratory infections. Environmental risk factors associated with water, sanitation and hygiene are
related to diarrhoeal diseases (and mentioned specially are trachoma, schistosomiasis,
ascariasis, trichuriasis and hookworm). It is interesting to note that the five parasites
and worms previously mentioned are associated with exposure to feces in general, and
remain a problem in the developing world, particularly in rural settings with no sewers; however, it is the enteric bacteria, protozoa and viruses that are of greatest concern
within the urban water cycle and reporting of these other enteric pathogens is rare.
Figure 3.1 shows the recent estimates of disease burden associated with environmental risk factors. These global estimates show the magnitude of the problems but do not
address with enough resolution the urban environment where large populations are
affected by poor water quality and where there is little ability to account for and identify public health risks.

Other includes: road traffic accidents, chronic obstructive pulmonary disorder,
other injuries, cataracts, malnutrition, etc.

The urban and peri-urban environments have unique issues with regard to water
pollution and water-borne disease. Plumbing and piping systems are in place. Thus,

there is a perceived separation of drinking water and wastewater. However, source waters for urban environments are often affected by sewage discharges and these are not monitored. Groundwater supplies supplementing urban areas are seen as safe without monitoring and are often used without treatment. While sewers are in place, treatment is minimal and wastewater from the urban environment is transported in large volumes affecting coastal areas and, in agricultural areas, affecting irrigation waters. These contaminated environments in turn cycle back the pathogens to the city via food and recreational activities. Finally, the system leaks due to cracks in the pipes, inadequate pipe connections and specific cross-contamination between pipes, and thus, sewage discharges to the environment affecting surface and drinking waters is probably a frequent event. The true health benefits of investment in engineering and managing both the wastewater disposal and water supply cycles in the urban environment will not be realized without appropriate monitoring. While public health and the medical community monitor the people, it is the responsibility of the water utility to monitor the water quality; however, water quality is associated with health and safety. To achieve better surveillance and reporting, a strategy for focusing on the scientific data and practices to address specific water-borne pathogens of concern and water quality are needed.

Quantitative microbial risk assessment (QMRA) procedures which address hazard identification, pathogen dose–response relationships, exposure assessment and risk characterization have been defined and used for a variety of pathogens (Haas et al., 1999). QMRA can be used to calculate the public health risk associated with exposure to a specified level of pathogens in water. Probability of Infection models, based on levels of contamination in drinking water (exposure to 1 to 1000 organisms, for example), will be presented for each pathogen discussed. There are two mathematical formulas that form the basis of the probability of infection, (an exponential model, which describes dose and response with a single parameter and the beta-poisson, which is described by two parameters), and these models can estimate the likelihood of infection and disease after ingestion of varying concentrations of specific pathogens. Thus, a risk framework, which will connect the specific microbial hazard (pathogen) to the exposure (drinking water or recreating) and to risk management, will be presented for a variety of water-borne pathogens. Through appropriate water quality monitoring and risk assessment, the risk reduction can be demonstrated by the water manager. For the water industry to fully take advantage of the advances made in both microbial monitoring and control, probability of infection models and QMRA methods should be learned and then used to address microbial risks and water safety.

3.2 THE TRADITIONAL ICONS OF WATER-BORNE DISEASE

The traditional water-borne disease agents are the bacteria, and the severe health consequences of cholera or typhoid motivated large urban areas that experienced outbreaks to begin chlorinating water supplies and to examine the protection of the source water or use an improved water source. It was often more difficult to address discharges of sewage and watershed protection; indeed, as populations grew and borders were defined, the location of new sources has been increasingly difficult. Thus, attention must be paid to the wastewater side of the equation. Any community with these diseases does not have adequate separation of sewage and water supply, and

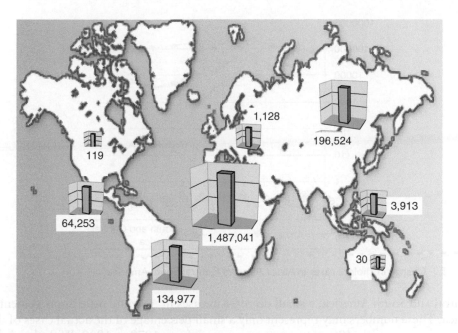

Figure 3.2 Reported Cholera, 1995–2005
Source: WHO, www.who.int

sewage and source waters should be monitored to provide information for risk assessment which can then be used in risk management choices.

3.2.1 Cholera

Cholera is caused by the bacterium *Vibrio cholerae*. An acute diarrhoeal disease that is life threatening. It is one of the few diseases that the International Health Regulations require governments to report. Cholera-like diseases were described early in the history of humans in China and Asia, and were certainly in Greece by 400 BC (Sherman, 2006). Massive outbreaks, killing as many as 60,000, occurred in India in 1768. These spread across Europe and reached all the way to England, and, by 1832, had reached the Americas. This organism spread quickly along major shipping routes and hit port cities, and is the first to be described in terms of a 'pandemic', that is it spread throughout the globe. Once disinfection of drinking water with chlorine began, cholera was readily eliminated in the developed world (as seen in Figure 3.2).

The disease is found throughout the world but endemic disease has essentially been eradicated in North America, Europe, Australia and New Zealand. Figure 3.2 shows the cases reported to WHO from 1995 to 2005. The disease remains problematic in Africa and Asia with average case–fatality ratios of 2.1–4.5, but reported on occasion as high as 23.6 (www.who.int/emc/diseases/cholera). The last global pandemic occurred in the 1990s. The disease spread through every country in South America in one year, resulting in a total of 1,041,422 cases and 9,642 deaths from 1991 to 1994 (MMWR, 1995). Figure 3.3 shows the decrease in cases over the last 15 years in

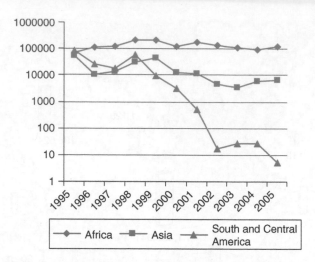

Figure 3.3 Change in cholera cases in Africa, Asia and Central/South America

Central and South America, a small decrease in cases in Asia and little improvement in Africa. These numbers may represent only a small percentage of the actual cases of disease; on average only 40 countries were reporting from 1970 to 1990. With the global epidemic in the 1990s, this jumped up to a peak of 95 countries by 1995 and is currently at 56 countries (WHO, 2004). One of the explanations for the success in the Americas is a renewed focus on drinking water disinfection and sanitation.

The application of scientific methods for the investigation of water-borne cholera using the last 200 years of theory has furthered our current understanding and control of the disease and its spread. Genetic variants of *Vibrio cholera*, including the 01 classical biotype, which was displaced by the El Tor biotype in the 1960s, and finally, in 1992, the emergence of the 0139 Bengal biotype, have developed and been described (Thompson et al., 2006). Monitoring for the *Vibrio* bacterium in water is rarely done. However, scientists who have been monitoring have shown that the bacteria resides in the marine environment and can reproduce in zooplankton that serve as a reservoir of infection in the populations that live at the interface between marine and fresh water systems.

Although cholera is primarily a water-borne disease, it can also be spread via food (particularly rice and shellfish). Outbreaks are documented, however, investigations are rare; thus, information on the source and cause of the contamination is generally not known. Seasonal trends are seen in most areas, with outbreaks being associated with rainfall (Alam et al., 2006). Particularly during the storm season, a significant number of outbreaks have been reported. Disinfection of drinking water provides one barrier, however, during high rain events, distribution system contamination occurs and increased contamination of source water overwhelms the treatment processes. It has also been shown that biotype 0139 may be more resistant to disinfection than previously thought (Clark et al., 1994), requiring longer disinfection contact times. The combination of *Vibrio* excretion by the population, lack of sewage treatment and mixing of sewage and water supplies continues the cycle of disease. The level of contamination can be related to probability of infection, thus, at measurable levels of concern (e.g., 10

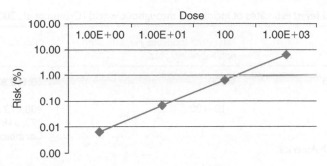

Figure 3.4 Probability of infection (Pi) of *Vibrio cholera* associated with ingestion (*d* = dose) of 1, 10, 100 and 1000 viable bacteria (Beta-poisson model Haas et al., 1999: $\alpha = 0.5487$ and $N_{50} = 2.13 \times 10^4$)

to 1000 bacteria per litre), there is a likelihood of the disease reaching above 1% to 5% levels if the water is not adequately treated (Figure 3.4). This means that water-borne outbreaks are more likely to be documented (reaching levels of between 20 and 40% attack rates; e.g., those ill/those exposed) with multiple days of exposure to the contaminated water. It is recommended that monitoring of sewage and waterways should be initiated for the larger regional water projects, particularly in Africa and Asia where *E. coli* levels are high and cholera is endemic. That monitoring should be enhanced during flooding disasters and major rainy seasons. This will provide the necessary data to promote the development of proactive approaches to disrupt the water-borne disease cycle.

$$Pi = 1 - \left[1 + \frac{d}{N_{50}}(2^{1/\alpha} - 1)\right]^{-\alpha}$$

3.2.2 Typhoid

Typhoid is a disease caused by the bacterium *Salmonella enteric serotype Typhi*. Like cholera, this pathogen was described in the early history of humans. It has been speculated that the Plague of Athens was one of the first well-documented sewage-associated water-borne outbreaks, with typhoid as one of the major diseases described (Rose and Masago, 2007). The pathogen is spread by the fecal–oral route and is associated with severe illness and mortality. There is no animal reservoir and fecally contaminated water is the primary route of exposure.

Typhoid is not reportable at the global scale. The current disease burden of typhoid is estimated at 17 million illnesses and 600,000 deaths per year (Crump et al., 2004) (WHO www.who.int/vaccine_research/diseases/typhoid/en/). However, the estimates from published reports suggest that south central and south-east Asia have the highest rates of new cases, with greater than 100 cases per 100,000 population per year. In contrast, there are only a few hundred cases of typhoid reported per year in Australia, Europe, New Zealand and North America (<1–10). Incidence in the US is now

Table 3.1 Typhoid fever: estimates of incidences throughout world (Crump et al., 2004)

Region	Estimated Incidence Per 100,000 population Per year	Notes on numbers of cases
South central Asia	>100	10,118,879 Asia
South-east Asia		
Northern Asia	10–100	408,873 Africa
South-west Asia		273,518 Latin America;
Africa		Caribbean
Central and South America		
Caribbean		
Oceania		
Australia	<10	19,144 Europe
New Zealand		4656 Oceania
Europe		453 North America
North America		

0.1 (per 100,000), with a total of 322 cases in 2004 and estimates are that 75% of the cases are imported (MMWR, 2006b) (Table 3.1).

Water-borne typhoid continues to be documented throughout Asia, Africa and South America. While there has been a focus on safe water vessels in individual homes, often the cause of the disease was found to be associated with consumption of water outside the home, such as in schools or the work place (Srikantiah et al., 2007; Luby et al., 1998). In addition, multi-drug resistance is an emerging problem (Mermin et al., 1999; Walia et al., 2005), with resistance to ampicillin, chloramphenicol and trimethoprim-sulfamethoxazole being documented. Thus, an emphasis must be placed on safe water at the community level, in order to eradicate the disease.

It is curious that cholera was eliminated long before typhoid in the developed world and is still more difficult to control. This may be due to a number of reasons. For example, survival in fresh water and wastewater is enhanced (*Salmonella* survives better in fresh waters; *Vibrio* prefers saline waters) (Alam et al., 2006). In addition, carrier states for typhoid exist in the population so that infected people will be constantly contributing the typhoid bacteria consistently to the sewage. The potency for *Salmonella* (related to probability of infection) is slightly lower than *Vibrio*, however, at measurable levels of concern (100 to 1,000 bacteria per litre), there is a likelihood of the disease reaching 1% if the water is not adequately treated (Figure 3.5) and this would undoubtedly lead to significant levels of disease with multiple days of exposure.

$$\mathrm{Pi} = 1 - \left[1 + \frac{d}{N_{50}}(2^{1/\alpha} - 1)\right]^{-\alpha}$$

3.2.3 Hepatitis

Hepatitis remains a global problem induced by several different viruses. Water-borne hepatitis is caused by Hepatitis A and E (HAV and HAE) which are usually spread by the fecal–oral route through contaminated food or water (Purcell, 1994). These

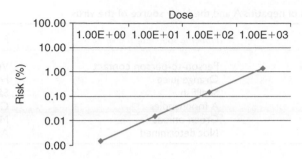

Figure 3.5 Probability of infection (Pi) of *Salmonella typhosa* associated with ingestion of (d = dose) 1, 10, 100 and 1000 viable bacteria (Beta-poisson model Haas et al., 1999: $\alpha = 0.2129$ and $N_{50} = 7.37 \times 10^5$)

viruses act as nano-particles in the environment and may differ slightly in their shape, the nature of their genome, their mechanism of replication and in the nature of the disease they cause.

3.2.3.1 Hepatitis A

Hepatitis A virus is approximately 30 nm in diameter. It contains a single-stranded segment of positive-sense RNA. It is classified in the family *Picornaviridae* and the genus *Hepatovirus*. It is similar to enteroviruses in its size and nucleic acid. It differs from most enteroviruses in that it does not readily cause cytopathic effects in cell culture.[1]

Hepatitis A is usually transmitted by the fecal–oral route (Purcell, 1994). However, it can also be transmitted by sexual activities (Mazick et al., 2005). One of the factors facilitating its transmission is its heat stability. Hepatitis A is generally more resistant to inactivation by heat and other environmental factors than are other similar viruses (Parry and Mortimer, 1984). This ability to survive outside the host is one factor that permits HAV to remain stable in the environment and spread readily through contaminated shellfish, water and other foods.

The course of HAV infection in patients is variable. In general, HAV infects the intestinal tract first. At this stage in the disease, it causes few symptoms, but is excreted in the feces and can be transmitted to others. The incubation period may last from 10 to 50 days, and is an important factor in the transmission of HAV. Later, the virus enters the bloodstream. At this stage, it can infect the liver and is recognized by the body's immune system. Symptoms may develop before liver disease is observed. The early symptoms may be general, and include fatigue, loss of appetite, nausea and vomiting, and fever. This phase may last from a few days to more than one week. Later, damage to the liver is observed. At this stage, the patient may appear jaundiced and blood tests for liver damage (increased bilirubin and liver enzymes) are positive. The

---

[1]Hepatitis B, C and D viruses (HBV, HCV and HCD) are usually transmitted by blood or blood products. Hepatitis B and C may cause acute illness and also long-term, chronic infections. Hepatitis D requires the presence of Hepatitis B virus for its replication. It can co-infect cells along with HBV or it can produce a superinfection by infecting cells already infected with HBV. In either case, it can increase the severity of HBV infections. These viruses will not be considered further.

Table 3.2 Outbreaks of hepatitis A and the likely source of the virus

Location	Probable source of the virus	Reference
Rio de Janeiro, Brazil	Person-to-person contact	Villar et al., 2002
Egypt	Orange juice	Frank et al., 2007
Spain	Shellfish	Sánchez et al., 2002
Italy	A foodhandler	Chironna et al., 2004
Pennsylvania, USA	Green onions	MMWR, 2003
India	Not determined	Arankalle et al., 2006

symptoms coincide with the development of the patient's immune response to the virus. It is thought that the patient's immune response contributes to the production of symptoms and to the clearing of the virus. The appearance of an immune response and symptoms are associated with a decrease in the presence of fecal HAV. Although greatly decreased, there may be sufficient HAV in the feces for the infected individuals to remain infectious.

Overall, mortality is relatively low, occurring in 0.2% of cases where jaundice is observed. However, the patient's age is an important factor in determining the severity of the disease. Many children have asymptomatic infections, and older patients are more likely to have severe or fatal infections (Purcell, 1994).

Hepatitis A is endemic in many parts of the world. The incidence in Africa and the Middle East has been estimated at up to 60 cases per 100,000 compared with 10 per 100,000 in North America (Hadler, 1991). In developing countries, the percentage of individuals exposed to the hepatitis A virus increases throughout life. For example, Barzaga et al. (1996) found that the percentage of individuals who were seropositive to hepatitis A increased from 10% in children less than five years old to 89–96% in adults over 40 years old. The endemic rate of hepatitis can be reduced by socio-economic development. Jacobsen and Koopman (2005) found that increased access to clean water was a factor in reducing the rate of hepatitis infections. The introduction of hepatitis A immunization can dramatically decrease the incidence of hepatitis A infection. For example, Bialek et al. (2004) found a 20-fold decrease in the incidence of hepatitis A in American Indians and Alaska Natives between 1997 and 2001. The decrease was mainly associated with an increase in immunization with the hepatitis A vaccine.

Outbreaks of hepatitis A still occur in both developed and developing countries. Many have been associated with specific foods. Some examples are shown in Table 3.2.

The probability of infection for HAV is high, and thus, low contamination levels have great potential to initiate infection and disease in susceptible individuals. Thus, at measurable levels of concern (1 to 10 viruses per litre), there is a likelihood of the disease reaching 50 to 99% if the water is not adequately treated (Figure 3.6). Widespread jaundice may not be apparent in the population even though HAV levels are high and infection is high, as young children, when infected, rarely exhibit any symptoms (considered asymptomatic) and, after infection in their youth, as adults they acquire immunity to subsequent exposures. Thus, HAV can be spread relatively easily, as children may be excreting and spreading the virus, but symptoms are not apparent.

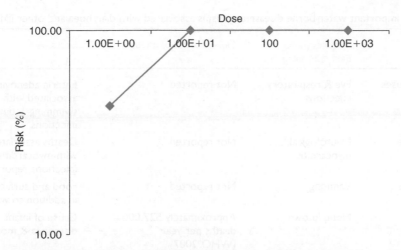

Figure 3.6 Probability of infection (Pi) of Hepatitis A associated with ingestion of (*d* = dose) 1, 10, 100 and 1000 viable viruses (Exponential model Haas et al., 1999: k = 1.8229; Pi = 1 − EXP [− 1/k * *d*])

3.2.3.2 *Hepatitis E*

One of the largest water-borne outbreaks ever documented globally was associated with Hepatitis E virus (HEV). This outbreak occurred in 1955 in New Delhi, India and it was first thought that hepatitis A was the causative agent, but subsequent studies identified hepatitis E as the virus responsible for the outbreak (Viswanathan, 1957). A sandbar had developed in the river used for the city's drinking-water supply and this diverted raw sewage into the drinking-water supply. Approximately 30,000 cases of hepatitis were reported. Hepatitis E is similar to hepatitis A in that it is transmitted through fecally contaminated food and water from human feces. Thus, both water supplies and irrigation waters are at risk.

The hepatitis E virus (HEV) is a non-enveloped virus approximately 32 nm in diameter. It has a positive-sense, single-stranded RNA genome. The time course and nature of HEV disease is similar to that of the HAV disease. Initially, the virus replicates in the intestinal tract and is shed in stool samples. Later, the virus invades the bloodstream and infects the liver. The development of an immune response is associated with a decrease in the virus in stool samples and the appearance of clinical symptoms. As with HAV, HEV infections in children are likely to be asymptomatic. In patients who exhibit symptoms, infections with HEV are generally more severe than infections with HAV. The mortality rate is in the range of 0.5 to 4.0%. An exception is found with pregnant women who are infected in the second or third trimester. The mortality rate for these women is around 50% (Purcell, 1994). As with HAV, HEV is an acute disease. Chronic inflammation of the liver or a chronic carrier state does not develop. Probability of infection models for HEV have not been developed. Often when this is the case one would use the most infectious virus model (rotavirus) to remain conservative.

3.2.4 Generic diarrhoea

Diarrhoea is the most common symptom used to identify water-borne disease. More than 140 types of enteric viruses can be found in fecally contaminated waters and may

Table 3.3 Important water-borne disease organisms associated with diarrhoea and other illnesses

Microbe	Diseases (other than diarrhoea)	Global burden of disease	Issues
Adenoviruses	Eye & respiratory infections	Not reported	Enteric adenoviruses associated with swimming-acquired infections
Coxsackie viruses	Neurological, myocarditis	Not reported	Deaths associated with myocarditis infections reported
Norovirus	Vomiting	Not reported	Food and surfaces in addition to water
Rotavirus	None known	Approximately 527,000 deaths per year (WHO, 2007)	Cause of infant diarrhoea & mortality
Poliovirus	Paralysis	7601 reported cases (2000–05) (WHO statistics, 2007)	Polio remains endemic in Nigeria, India, Pakistan and Afghanistan
Campylobacter	Guillain-Barre paralytic syndrome	No estimates globally, associated mostly with food. Highly related to non-disinfected well water	Can be associated with animal wastes, poultry in particular
Shigella	None known Hospitalization rates high	164.7 million cases per year (95%, 163.2 million in developing countries with 1.1 million deaths) (Kotloff et al., 1999)	High severity Human specific associated with ambient surface waters

cause diarrhoea during the initial infection process. Viruses that infect humans and animals do not occur naturally in the environment and they cannot grow or replicate in environmental waters once they have been excreted by their hosts. The detection of viable enteric viruses in environmental waters is a public health concern due to their low infective dose and their ability to cause other acute and chronic conditions beyond diarrhoea. The most important water-borne viruses include adenoviruses, coxsackie viruses, norovirus, poliovirus and rotavirus (Table 3.3). In addition, there are important bacteria, which also appear to have a low infectious dose, associated with diarrhoea. These include *Campylobacter* and *Shigella*. Other pathogens such as the protozoan parasites and *E. coli* 0157H7 will be described in subsequent sections.

All of these pathogens are restricted to human fecal sources, with the exception of *Campylobacter,* which can originate from animals as well as humans. Rotavirus remains the most common cause of severe diarrhoea in the developing world, particularly for children under five years of age (Wilhelmi et al., 2003). Rotavirus has also been found to have one of the lowest infectious doses of all the viruses studied to date (Figure 3.7, Haas et al., 1999). Studies in Nigeria found enteric Adenoviruses may cause up to 16% of the childhood diarrhoea (Audu et al., 2002), with more frequent

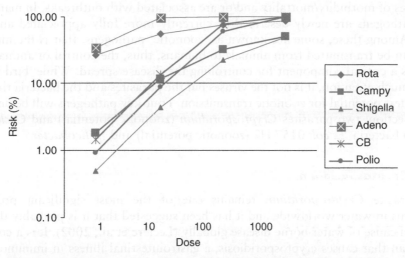

Figure 3.7 Relative probability of infection (Pi) associated with diarrhoeal agents (Rota $\alpha = 0.2531$ and $N_{50} = 6.17$; Campy $\alpha = 0.145$ and $N_{50} = 896$; *Shigella* $\alpha = 0.21$ and $N_{50} = 1120$; Adeno: k = 2.397; CB: k = 69.1; and polio: k = 109.87 see Figures 3.5 and 3.6 for models. Haas et al., 1999; currently no model exists for norovirus)

symptoms associated with poor drinking water. Norovirus causes severe vomiting and gastroenteritis, and new strains appear to be on the increase globally, with Australia, Canada, Japan, New Zealand, US and UK reporting increasing outbreaks in restaurants, nursing homes, schools and cruise ships. Contaminated food is thought to account for 50% of the disease cases. Dysentery caused by *Shigella* remains a serious disease throughout the world. *S. flexneri* remains the primary species of concern, and 61% of the mortality occurs in children under five years of age (Kotloff et al., 1999). Thus, this may be more significant than rotavirus. Despite poliovirus vaccination programmes, 31 countries reported poliovirus infections in 2000; by 2005 this was down to 9. The low potency (probability of infection) means that it is not as easily spread as some viruses, and higher virus levels are needed to continue the spread by environmental contamination. Attention to eradication via vaccination programmes have focused on reducing the spread. Coxsackie viruses have been recently reported as causing a number of deaths in Greece due to myocarditis (Programme for Monitoring Emerging Diseases, PROMED), but global reporting is rare. In the US, *Campylobacter* is the most identifiable cause of groundwater outbreaks (Liang et al., 2006). Campylobacteriosis is generally assumed to be associated with food; global waterborne campylobacteriosis rates are not known.

Thus, even at low levels of contamination (d = dose) (1 to 10 viable organisms per litre), between 1 and 50% probability of infection could occur. In these cases, the use of the *E. coli* indicator system measured in the 100 mL volumes is inadequate to address the risk.

3.3 EMERGING DISEASES AND ZOONOTIC PATHOGENS

The term 'emerging' has been focused on pathogens that are being reported at an increasing rate in the population, are resistant to treatment and vaccinations, have high rates of morbidity/mortality and/or are associated with outbreaks. In many cases these pathogens are newly described or currently more fully appreciated and diagnosed. Among these, some are known as 'zoonotic' pathogens, that is the microbial agent can be transmitted from animals to humans, thus, the control of animal waste becomes a critical component for controlling the disease spread. While 'bird flu' has gained much attention, it is not the viruses but the parasites and the bacteria that have the greatest potential for zoonotic transmission. Four key pathogens will be described in this section: two parasites *Cryptosporidium* (zoonotic potential) and *Cyclospora*; and two bacteria, *E. coli* 0157 H7 (zoonotic potential) and *Helicobacter*.

3.3.1 Cryptosporidium

The parasite *Cryptosporidium* remains one of the most significant protozoan pathogens in water worldwide and it has been suggested that it is probably the most common cause of water-borne disease globally (Leclerc et al., 2002). It is a coccidian protozoan that causes cryptosporidiosis, a gastrointestinal illness in immunocompetent humans that leads to chronic, life-threatening disease in immunocompromised patients and those with AIDS. The main symptom, diarrhoea, may be accompanied by abdominal cramps, loss of appetite, low-grade fever, nausea, vomiting and weight loss. Symptoms may persist from a few days to a few weeks in normally healthy individuals. Currently, no curative therapy exists for individuals with cryptosporidiosis (Carey et al., 2004).

Cryptosporidiosis is a zoonotic disease that also infects domestic (e.g., cattle, sheep, pigs and dogs) and wild (e.g., deer and opossum) animals. There are currently 14 described species included in the genus *Cryptosporidium*, and more than 33 host-adapted genotypes (Xiao et al., 2004). Two species commonly infect humans, *C. hominis*, which naturally infects only humans, and *C. parvum* which infects humans and some other mammals, including ruminants (Sulaiman et al., 1998; Peng et al., 1997). *Cryptosporidium* may be transmitted to humans via several routes because of its wide host range and ubiquitous presence in the environment. *Cryptosporidium* infection is by the fecal–oral route, through consumption of contaminated food or water, or through person-to-person or animal-to-person transmission. Infected cattle serve as an important reservoir of *C. parvum* and are substantial contributors to sporadic human cryptosporidiosis (Casemore et al., 1997). The global inventory of farm animals increased 30% between 1962 and 2002 (FAO, 2003). Cattle, particularly young calves, are recognized as a significant source of *Cryptosporidium parvum* oocysts (Garber et al., 1994).

The oocyst of *Cryptosporidium* is excreted in the feces of humans and animals. It has low infectivity, survives in water for months, is small and able to pass through filters, and is resistant to chlorination. The prevalence is 1 to 4% in Europe and North America, but is much higher in Africa, Asia, and Central and South America (3 to 20%) based on fecal excretion rates (Current and Garcia, 1991). It has been suggested that it is the most commonly identified pathogen associated with diarrhoea, with a 5% prevalence rate in all ages. However, it is estimated that 20% of all childhood diarrhoea is caused by this parasite in the developing world (Mosier and Oberst, 2000). Throughout the world, AIDS patients continue to suffer from *Cryptosporidium* infections (1–3% in Asia, 10–15% in South America and up to 30% in Africa) and

Table 3.4 Occurrence of Cryptosporidiosis in AIDS patients: selected literature

Country	% Infection	Reference
Ethiopia	29	Tadesse and Kassu, 2005
Ghana	28.6	Adjei et al., 2003
Haiti	10	Raccurt et al., 2006
Italy	33	Brandonisio et al., 1993
Korea	1	Lee et al., 2005
Malaysia	3	Lim et al., 2005
Taiwan	0.5–1.2	Hung et al., 2007
Venezuela	15	Certad et al., 2005

BOX 3.1 Recent assessment of water-borne outbreaks of Cryptosporidiosis

Despite widespread infection of cryptosporidiosis, water-borne disease is rarely documented through out the world. Because of the widespread outbreaks in the 1990s, the US and UK have been the most diligent in regard to monitoring and reporting.

- Throughout the world, 325 water-borne parasitic outbreaks have been documented (Karanis et al., 2007).
  - 66% were in the US
  - 30 % were in Europe
  - 24% were in UK
  - 40% and 50% were caused by *Giardia* and *Cryptosporidium,* respectively.
- Recently, swimming-acquired cryptosporidiosis has dominated the water-borne disease statistics in the US (MMWR, 2007).
- In September, 2005, an outbreak was reported in a village in Turkey (near Izmir), probably due to contamination of the public supply by sewage following heavy rainfall. Two parasitic pathogens were detected: *Cryptosporidium* (8%) with a co-infection of *Cyclospora* (5%) (Aksoy et al., 2007).
- In April 2006, a major cryptosporidiosis outbreak was reported in Botswana, and 23,264 cases have been reported across all of Botswana's districts (PUBMED).
- In July 2007 in the village of Bani Hassan, Jordan, a major water-borne outbreak of cryptosporidiosis with 800 hospitalizations was reported.

---

associated chronic diarrhoea (Table 3.4). It is suggested that 60–90% of HIV-infected patients in Africa are suffering from diarrhoea, as defined by WHO classification. A lower occurrence of infections was found in populations in Asia, perhaps due to the practice of consuming boiled water.

Cryptosporidium has a low infectivity and thus, at low levels of contamination (1 to 10 oocysts), anywhere from 0.5 to 5% of the population could become infected with a single exposure (Figure 3.8). Because of its resistance to chlorination, the bacteria indicator system using *E. coli* will not be appropriate for monitoring the drinking-water risks associated with this parasite.

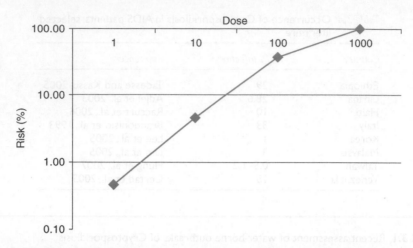

Figure 3.8 Probability of infection of *Cryptosporidium* associated with ingestion of 1, 10, 100 and 1000 viable oocysts (Exponential model Haas et al., 1999: k = 238; Pi = 1 − EXP [−1/k * d])

3.3.2 *Cyclospora*

Cyclospora infections are caused by a protozoan parasite, *Cyclospora cayetanensis*. It is a cyst-forming coccidian protozoan. Oocysts of *C. cayetanensis* are spherical and have a diameter of 7.5 to 10 μm. *C. cayetanensis* is a human parasite. Humans acquire *C. cayetanensis* from water or food that has been contaminated with human feces. Once inside a host, the oocysts excyst and release sporozoites. These sporozoites infect epithelial cells of the gastrointestinal tract. At some stage in its life cycle, *C. cayetanensis* forms zygotes that mature into oocysts that rupture the host cell's membrane. These oocysts are released in stool samples, but are not immediately infectious. In the presence of atmospheric oxygen, the oocysts sporulate and then become infectious. This process can take from one to two weeks at 30°C. Since this sporulation occurs outside the human host, direct transmission of *C. cayetanensis* between people is not likely. Thus, the environment plays a key role in the spread of the infection.

C. cayetanensis causes an intestinal disease. Symptoms may be typical of intestinal diseases, and include watery diarrhoea, abdominal bloating and cramps. Symptoms may also be more generalized. These generalized symptoms may include appetite and weight loss, nausea and low-grade fever. In more severe cases, greater weight loss, vomiting and muscle aches may be present. Without antibiotic treatment, the symptoms may persist for several weeks in normal patients and longer in immunocompromised individuals. Treatment with trimethoprim-sulfamethoxazole reduces the symptoms and the excretion of oocysts.

C. cayetanensis has been found to cause outbreaks in developed countries. These have generally been associated with eating fresh fruits (Table 3.5). Endemic infections with *C. cayetanensis* have been found in several developing countries (Table 3.6). In these countries, many infections are asymptomatic. The infectivity is not known but it may be similar to other parasites. The contamination comes from human feces; there is no known animal reservoir for *Cyclospora*.

Table 3.5 Outbreaks of *Cyclospora cayetanensis*

Location, date	Vehicle	Reference
Ontario, Canada, 1998	Fresh raspberries	MMWR, 1998
Northern Virginia, Washington, DC, USA, 1997	Fresh basil	MMWR, 1997
South Carolina, USA, 1996	Raspberries, strawberries and potato salad	MMWR, 1996
Guatemala	Raspberry juice	Puente et al., 2006

Table 3.6 Areas with endemic infections of *Cyclospora cayetanensis*

Country	Endemic level	Reference
Jordan	6% of patients with gastroenteritis had *C. cayetanensis* oocysts	Nimri, 2003
Venezuela	6.1% of people surveyed were infected with *C. cayetanensis*	Chacin-Bonilla et al., 2003
Guatemala	*C. cayetanensis* was found in 2.3% of people surveyed	Bern et al., 1999
Peru	Children had an average of 0.2 episodes of cyclosporiasis per year	Bern et al., 2002

Because sporulation and development of infectivity occurs in the environment, it is believed that the widespread infections (related to the widespread contamination of the fruits and vegetables) were associated with poor irrigation water quality and not through handling.

3.3.3 E. coli O157:H7

Escherichia coli is found in the intestinal tract of warm-blooded mammals. Because of its association with feces, it has been used as an indicator of fecal pollution. Although most individuals maintain a population of *E. coli* in their intestines without showing any symptoms of disease, *E. coli* can be a serious pathogen in three general circumstances. *E. coli* strains may cause urinary tract infections, they may cause intestinal disease, and some may be invasive and cause septicaemia and meningitis. The ability of different *E. coli* strains to cause these different infections is related to their possession of surface structures that permit adhesion to certain cells and to their ability to produce toxins. *E. coli* strains can be grouped into serotypes, based on their surface antigen (O) and the antigenic characteristics of their flagella (H). The different antigens have been numbered, and certain antigenic types have been found to be associated with different diseases more than are other antigenic types. For example, *E. coli* strains with O6 and H16 antigens are more likely to cause disease by producing toxin than by invading cells.

E. coli strains that cause urinary tract infections are generally better at attaching to cells found in the urinary tract than are other *E. coli* strains. Many of the *E. coli* strains

Table 3.7 Suspected sources of *Escherichia coli* O157:H7 in outbreaks

Location, year	Suspected source of the bacteria	Reference
The Netherlands, 2005	Steak tartare	Doorduyn et al., 2006
Germany, 1995	Sausage	Ammon et al., 1999
France, 2004	Unpasteurized goat's cheese	Espié et al., 2006
Democratic Republic of Congo, 2003	Person-to-person	Koyange et al., 2004
United States, 2002	Ground beef	MMWR, 2002
United States, 1997	Frozen ground beef patties	MMWR, 1997
United States, 1993	Fresh-pressed apple cider	Besser et al., 1993
United States, 1996	Mesclun lettuce	Hilborn et al., 1999
England and Wales, 1995	Person-to-person	Willshaw et al., 1997

Table 3.8 Suspected non-food sources of *Escherichia coli* O157:H7 in outbreaks in the United States

Location, year	Suspected source of the bacteria	Reference
Portland, Oregon, 1991	Lake water used for swimming	Keene et al., 1994
Washington, 1999		Bruce et al., 2003
New York, 1999	Unchlorinated well water	Bopp et al., 2003
Pennsylvania, 2000	Farm animals	MMWR, 2000
North Carolina, Florida and Arizona, 2004–2005	Animals at petting zoos	MMWR, 2006a
Ohio, 2001	Aerosols in a contaminated building	Varma et al., 2003

that cause intestinal disease produce a toxin; one such toxin is the LT toxin. This toxin is similar in structure to that of the toxin produced by *Vibrio cholerae*. *E. coli* strains that produce this toxin cause a watery diarrhoea that can lead to dehydration.

One strain of *E. coli* that has become more important in recent years has O antigen 157 and H antigen 7. This strain, *Escherichia coli* O157:H7, was not observed to be a human pathogen before 1982. Since then it has been implicated is several outbreaks associated with the consumption of contaminated food or water. A major difference between *E. coli* O157:H7 and other *E. coli* strains is that the O157:H7 strain produces a shiga-like toxin (also known as verocytotoxin). This toxin increases the severity of infections with strains that produce it. One of the consequences of infection with strain O157:H7 is hemolytic uremic syndrome. This complication can lead to a fatal infection, especially in young children (1–4 years) and in older patients (61–91 years).

From 1982 to 2002, food was implicated in 52% of the outbreaks of disease by *E. coli* O157:H7 in the United States. Over this period, 31 water-borne outbreaks were reported. Of these, 21 were associated with recreational water and 10 with drinking water. For the outbreaks associated with recreational water, 14 were associated with fresh water lakes or ponds and 7 with swimming pools. A major reservoir for *E. coli* O157:H7 is cattle. It has been found associated with both dairy and beef cattle. Several outbreaks have been associated with food of bovine origin (Table 3.7). Other types of food and person-to-person contact have been also associated with *E. coli* O157:H7 infections in different countries (Table 3.7). Some non-food sources of *E. coli* O157:H7 infections are listed in Table 3.8.

3.3.4 Helicobacter

Helicobacter pylori is an emerging bacterium, and is now recognized as a cause of peptic ulcers and chronic gastritis and may be related to stomach cancers associated with MALT lymphomas (Aruin, 1997; Blaser, 1990). Globally, up to 50% of the population may be infected and not realize it, as no symptoms are apparent (Dunn et al., 1997; Feldman et al., 1997). There is still some speculation about its transmission (Munnangi and Sonnenberg, 1997), yet fecal–oral and oral–oral are the most likely routes (Hegarty et al., 1999). There is growing evidence that water-borne disease transmission occurs for *H. pylori* (Hulten et al., 1996). One of the best studies, conducted in Germany, demonstrated the relationship between contamination of well water and colonization in the population (Rolle-Kampczyk et al., 2004).

There are a number of bacteria, including *H. pylori* that probably enter into a viable but nonculturable (VBNC) state in the environment, and thus, routine culture methods cannot be used to evaluated water contamination (Oliver, 2002). Thus, molecular techniques, particularly the method which replicates DNA known as Polymerase chain reaction (PCR), are used for monitoring polluted water. Nayak and Rose (2007) examined conventional culture and PCR in well water and reported that the bacterium. *H. pylori* remained culturable for five days at 4°C as opposed to only one day at 15°C, but was detectable by PCR without any loss of signal of concentration. *H. pylori* was found 84% of the time in sewage, at levels of 2 to 28 cells ML^{-1}.

Preliminary studies have reported that the prevalence of *H. pylori* increases with age (Al-Moagel et al., 1990; Graham et al., 1988), but detailed information on the prevalence of the bacterial infection in any defined population and on the factors that may influence these distribution patterns remains scanty. In developing countries, the organism is acquired early in childhood and up to 90% of children are infected by age five (Thomas et al., 1992). In the United States, few infections appear to occur during childhood, and the organism has a projected incidence of about 0.5 to 1.0% per year with about 50% incidence by age 60 (Blaser et al., 1991). The high incidence in both developing and developed countries is in poorer, lower socio-economic groups, for which crowding and poor sanitation appear to be risk factors. Globally, levels of gastric cancers are decreasing, probably due to improved hygiene. Thus, in countries where untreated sewage contamination of water supplies is common, *H. pylori* may remain an important cause of gastric cancers.

Dose-infectivity models have not been developed yet for *H. pylori*, as this bacterium is related to *Campylobacter* one could suggest that sewage pollution of water supplies is leading to widespread infection in the community.

3.4 RISK ASSESSMENT AND CONTROL OF WATER-BORNE PATHOGENS

An improved discourse with better resolution of information is needed to move the agenda on water and health to an action-oriented level for the utility manager, justifying investment in the urban wastewater and drinking-water cycles. This can only be achieved through improved monitoring and reporting with the goal of eradication of specific water-borne diseases. This can be done in combination with renewed efforts toward the assessment of source water quality, watershed impacts and beachshed impacts with the implementation of Water Safety Plans (WHO, 2005). Thus, the connection between

Table 3.9 Development of a community report card for adequate attainment of key parameters influencing sustainable access to 'Safe Water'

Measurable Criteria

ACCESSABLE	At home	At health care facilities/clinics	At schools	At other community venues	At work venues

Duration (hours/ daily); Tap or storage container; Source and quality

SUSTAINABLE	Year to year fluctuations & Impact by climate	Can grow to meet new community objectives	Can be repaired and maintained	Protected from waste discharges	Adaptable to new technology

Monitoring data needed for quality and quantity; Watershed; Development of Water Safety Plans Technology platforms;

SAFE	Water-borne outbreaks investigated	Water-borne typhoid & cholera eliminated	E. coli, arsenic, nitrate, fluorides levels evaluated	Treated & disinfected to meet water quality goals	Assessed for impacts via storms and flooding

Monitored and Reported

health status in communities and water quality need to be strengthened and reported. All water projects should support water quality monitoring as well as reporting of and investigation of water-borne diseases and outbreaks. Accountable measures of improvement in health status should be a part of any 'sustainable access to safe water' programme and thus it is proposed that a simple report card (Table 3.9) could be used, in which one is graded in a relative fashion with specific criteria to address progress toward safe water with a partnership between the health department and the water utility. This will involve assessment, monitoring, investigation and reporting.

In addition to the potable water supply report card, a programme for assessing risks to irrigation waters and recreational waters is needed. This may best be done through prevention and protection approaches with a focus on sewage discharge and treatment. All of the fecal–oral pathogens described herein can be spread through drinking water, irrigation waters that affect the agricultural produce and recreational waters. The water contamination cycle for the urban environment goes well beyond the community drinking water (Figure 3.9). There is no adequate programme for monitoring irrigation waters, thus one would need to be established. For recreational waters, WHO has proposed an enterococci indicator criterion with levels from $\leqslant$40; 41–200; 201–500 to >500 CFU/100 mL for 'Very Good' water quality, with risks of any gastrointestinal infection of <1%; 'Good' water quality, with risks 1–5%; 'Fair' water quality, with risks 5–10% and 'Poor' water quality, with risks >10%, respectively for swimming and (small ingestion exposures per swimming event ~50 to 100 mL) (WHO, 1999 and 2003).

3.4.1 Use of quantitative microbial risk assessment

WHO has supported the QMRA framework, particularly for resistant organisms such as *Cryptosporidium* (Medema et al., 2006) and viruses, for which the bacterial indicator system may not provide the appropriate measure of exposure and risk. The Probability of Infection models, based on levels of contamination in drinking water

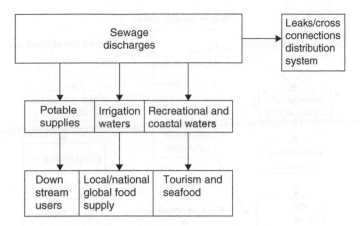

Figure 3.9 The urban water–sewage cycle: pathways for disease spread

(exposure e.g., 1 to 1,000 organisms) have been presented in each pathogen section. These models can be used to examine even low levels of contamination (1 pathogen per litre or 10 L), which in most communities may result in endemic disease and not outbreaks, assisting health surveillance strategies and investment in pollution prevention approaches. Risk reduction can be demonstrated using systems that reduce fecal loading to ambient waters and drinking waters. Thus, it is recommended that monitoring and control measures be directed toward reduction of wastewater and human fecal loading to water supplies.

Initially bacterial indicators would be used to identify and rank polluted sites. This would lead to strategic and specific pathogen monitoring to identify hazards. Loading and transport would need to be examined to fully characterize exposure, under the Water Safety Plans; this would be part of sanitary surveys and potentially identify critical control points. Rainfall and flood event monitoring would be required and this could be used to address surface and groundwater vulnerability (Figure 3.10). Thus, appropriate risk management strategies could be implemented.

3.4.2 Interventions to reduce enteric diseases

The characteristics of pathogens associated with risks to drinking water, irrigation water and recreational water are:

- incidence in population and concentrations in sewage
- probability of infection
- survival in the environment
- maintenance of levels in reservoirs (animals, sediments and biofilms)
- resistance to treatment and in particular disinfection.

Intervention procedures have been applied in developing countries to reduce enteric disease. Some of the methods that have been applied are presented in Table 3.10.

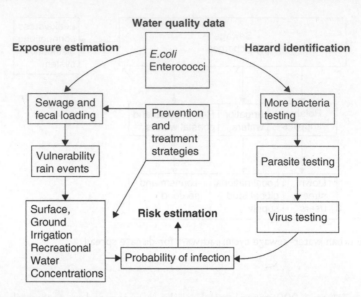

Water quality data

Figure 3.10 Risk estimation and management using strategic monitoring and probability of infection
       models

Table 3.10 Possible intervention practices to reduce enteric disease

Hand washing
Public latrines
Hygiene education
Installation of wells with household connection
Connection to a municipal water supply
Point-of-use disinfection
Central collection and treatment of wastewater

Clasen et al. (2007a) considered the cost-effectiveness of different intervention procedures for reducing diarrhoeal disease in developing countries. These authors concluded that several methods were cost-effective on the basis that they reduced
healthcare costs. In areas where resources were limited, household chlorination was
the most cost-effective intervention. The literature on intervention studies in developing countries was reviewed and analyzed by Fewtrell et al. (2005). A major conclusion
of this study was that multiple interventions were generally better than a single intervention in reducing diarrhoea. These multiple interventions consisted of improving the
water supply, increased sanitation and improved hygiene procedures. Similar results
on the importance of multiple interventions were found by Eisenberg et al. (2007).
These authors found that the sanitation and hygiene conditions could influence the
benefits obtained by improving the water supply. Improving the water quality was
most beneficial when sanitation and hygiene conditions were also improved. In an
analysis of 42 controlled trials, Clasen et al. (2007b) concluded that diarrhoea in children and adults could be reduced by interventions that improved the microbial quality of water.

Table 3.11 Examples of relative resistance to various treatment processes (Estimated ranges of reductions applied to drinking water)

Microbe	Sand Filtration	Chlorination (free chlorine 1–2 mg/L)	Chloramination* Need 60 to 120 min contact times	UV* Doses 20 to 119 mW/cm²	Ozone 0.35 mg/L
Bacteria Including E. coli Helicobacter, Salmonella, Vibrio	90–95%	99.999%	90–99.9%	99.9%	99.999%
Parasites Cryptosporidium	99%	0%	0%	99.9%	90%
Viruses (including Poliovirus and HAV)	30%	99.9%	99%	99%	99%
Adenovirus	30%	99.99%	50%–90%	50%–90% >30 mW s/cm²	99%

* May be less when applied to wastewater with more suspended solids

In the urban environment, the goals are to address sewer, wastewater treatment, water reclamation and reuse, source water protection, drinking-water treatment, and distribution system integrity in a holistic manner. Monitoring programmes are established for examining first, occurrence and impacts and second, to address management strategies. These pathogens can be controlled and reduced by physical removal (e.g., sedimentation and filtration) or killed via disinfection (e.g., chlorination, UV and ozone). Many of these microorganisms are resistant to removal because of their size; sedimentation, such as in wastewater lagoons or sedimentation basins, removes only a small percentage. In particular, viruses, due to their size, have always been seen as the greatest risk to physical barriers, as they readily pass through sand filtration barriers, pass through soil systems to groundwater and are not removed by sedimentation. Thus, membrane filtration processes, such as ultra-filtration or reverse osmosis, are needed to remove viruses. However, many viruses are highly susceptible to inactivation associated with disinfection. Disinfection has historically been used as one of the major barriers to kill and destroy the viability of microbes (Baker et al., 2002; Meunier et al, 2006; Thurston-Enriquez et al., 2003; Yates et al., 2006). Yet, depending on the disinfectant used, different groups of pathogens will be more or less resistant. For example, Cryptosporidium is completely recalcitrant to any type of water chlorination but it is highly susceptible to UV disinfection, whereas adenoviruses are very susceptible to free chlorine but very resistant to UV disinfection. Ozone is highly effective, however, it is dependent on the hydraulics of the contactor and maintaining a residual for some contact time. Chloramines are less effective than free chlorine and require more contact time. UV intensity can be adjusted to deliver a variety of doses and, while low doses (5–10 mW s/cm²) are capable of reducing the viability of the parasites by 99% to 99.9%, up to 30 mW s/cm² are needed for even 90% reduction of adenoviruses. Therefore, the design, intensity and dose of UV reactors are extremely important.

BOX 3.2 Eradication of typhoid in the Great Lakes, the role of separation of wastewater and drinking water

The understanding and reporting of water-borne outbreaks associated with typhoid, environmental monitoring and finally implementation of water disinfection has eradicated typhoid in the developed world. In 1909, the United States and Great Britain (on behalf of Canada, which became independent in 1931) signed the Boundary Waters Treaty aimed at preventing and resolving disputes between US and Canada over waters forming the boundary between the two countries, and which also established a formal binational body, the International Joint Commission (IJC). The treaty was originally intended to protect lake levels and navigability however, this created the capacity for the IJC to become involved in pollution problems in boundary waters (Binder, 1972). There was a huge interest in water quality and the impacts on water-borne disease, primarily typhoid, in the Great Lakes region. Major outbreaks had occurred in Chicago, Detroit and Milwaukee major urban areas. This instigated one of the most comprehensive bacteriological studies ever conducted. There are several lessons to be learned from the 1914 IJC study. They promoted the use of the most technologically advanced methods; at the time, bacteria samples were often grown in gel rather than agar (Durfee and Bagley, 1997) and funded and set up laboratories for testing. Recommendations from the study which focused on coliform bacteria included the protection of the Lakes from wastewater discharges. Studies on water-borne typhoid in the US and Canada between 1920 and 1930 reported 242 outbreaks, with 9,367 cases; even more surprising, 84,345 cases of dysentery (of an unknown etiology) were associated with these outbreaks (Wolman and Gorman, 1931); in Canada, there were 40 outbreaks and 2,836 cases of typhoid. The high mortality finally precipitated drinking-water treatment (both filtration and disinfection). But more importantly, this changed the way wastewater was handled. In 1833, about 150 people lived in the Chicago area; by 1880 more than half a million people lived and disposed of their waste into the water supply, Lake Michigan. Begun in 1900, it was not until 1922 that the Chicago Drainage canal was completed, taking all wastewater downstream into the Chicago River and to the Mississippi (Capano, 2003). During this time, typhoid dramatically decreased and the disease was finally eliminated.

As disinfectant residuals are typically not produced with UV and ozone, in contrast to the situation with chlorine and chloramines, monitoring water quality is critical.

3.4.3 Vaccinations

3.4.3.1 Immunization

Immunization can be effective in controlling the transmission of some enteric pathogens. In general, immunization is easier to accomplish and more effective in controlling pathogens that enter the bloodstream than in controlling those that are mainly confined to the intestinal tract. In cases where a pathogen must enter the bloodstream before reaching the target organ, the presence of circulating antibodies can neutralize the infectivity of the pathogen and therefore protect the individual. Examples are the protection of neurons from infection by poliovirus, the protection of the liver from hepatitis A and protection against the more severe aspects of systemic infection with *Salmonella enterica* serotype Typhi.

Some examples of diseases that may be controlled with vaccines are shown in Table 3.12. Immunization against poliovirus has greatly reduced the incidence of this

Table 3.12 Water-borne diseases that may be controlled by immunization

Disease	Microbe
Paralysis, neurological problems	Poliovirus
Hepatitis	Hepatitis A, Hepatitis E
Fever, generalized infection	Salmonella enteric serotype Typhi
Gastroenteritis	Vibrio cholerae
Gastroenteritis	Human rotavirus

Table 3.13 Water-borne diseases not controlled by immunization

Disease	Microbe
Gastroenteritis	Noroviruses
Gastroenteritis	Certain adenovirus serotypes
Gastroenteritis	Cryptosporidium parvum
Gastroenteritis	Cyclospora cayetanensis

disease, and there is hope for its eventual elimination (Kimman and Boot, 2006). Likewise, where immunization against hepatitis A has been conducted, the number of cases has declined (Zhou et al., 2007). It has been more difficult to develop a vaccine for hepatitis E. However, a vaccine against hepatitis E has been produced and found to be safe and effective in initial trials (Shrestha et al., 2007). An earlier rotavirus vaccine (RotaShield) was found to cause increased rates of a bowel problem, intusecception, in children and was withdrawn from the market (Kang, 2006). Now, a newer vaccine (RotaTeq) has been developed, and found to be safe for use with children (Kang, 2006). A vaccine against *Salmonella enteric* serotype Typhi has been found to offer protection against typhoid fever and to help control outbreaks (Bodhidatta et al., 1987). A vaccine against *Vibrio cholerae* has also been found to reduce the level of cholera in a community (Emch et al., 2006).

Vaccines are not available for many important enteric pathogens (Table 3.13). Some problems with development of useful vaccines include difficulties in growing the pathogen under laboratory conditions or with cloning genes that code for molecules that can be used for immunization. Once produced, vaccines must also be shown to be both safe and effective before they can be used.

3.5 CONCLUSIONS AND RECOMMENDATIONS

The urban environment generally serves larger numbers of people, and thus, the integrity and type of community wastewater and drinking-water systems are of great importance and have a significant influence on the health of the population. Historically, water quality, and the prevention of water-borne cholera and typhoid drove the investment in both drinking-water and wastewater infrastructure. In most areas throughout the world, wastewater is reused (albeit inadvertently), influencing recreational areas and irrigation water, and thus, the safety of the food supply. Thus, the cycle continues. Currently, however, the connection to health and water quality has been reduced as resources have been spent on compliance monitoring rather than strategic monitoring plans, which provide critical information on sources, transport, exposure and risk. Thus, it is recommended that the water industry consider the following:

- Better assessment of water quality at key locations and strategic monitoring should be implemented. This means improvement in water quality laboratories and initially better sampling for *E. coli* and enterococci. In time, sampling for pathogens can be implemented.
- Risk assessment and probability of infection models can be and should be used by the water industry to address water quality monitoring and establish management plans.
- The water industry needs to form better partnerships with public and environmental health agencies.
- Any area with cholera and typhoid does not have adequate separation of drinking water and sewage, and this should be immediately investigated with appropriate monitoring within a risk framework to examine vulnerabilities.

REFERENCES

Adjei, A., Lartey, M., Adiku, T.K., Rodrigues, O., Renner, L., Sifah, E., Mensah, J.D., Akanmori, B., Otchere, J., Bentum, B.K. and Bosompem, K.M. 2003. *Cryptosporidium* Oocysts in Ghanaian AIDS Patients with Diarrhoea, *East African Medical Journal*, Vol. 80, No. 7, pp. 369–372.

Aksoy, U., Akisu, C., Sahin, S., Usluca, S., Yalcin, G., Kuralay, F. and Oral, A.M. 2007. First Reported Waterborne Outbreak of Cryptosporidiosis with *Cyclospora* Co-infection in Turkey, *Euro Surveill*, Vol. 12, No. 2, p. E070215.4.

Alam, M., Hasan, N.A., Sadique, A., Bhuiyan, N.A., Ahmed, K.U., Nusrin, S., Nain, G.B., Siddique, A.K., Sack, R.B., Sack, D.A., Huq, A. and Colwell, R.R. 2006. Seasonal Cholera Caused by *Vibrio Cholerae* Serogroups 01 and 0139 in the Coastal Aquatic Environment of Bangladesh, *Applied Environmental Microbiology*, Vol. 72, pp. 4096–4104.

Al-Moagel, M.A., Evans, D.G., Abdulghani, M.E., Adam, E., Evans, D.J. Jr., Malaty, H.M. and Graham, D.Y. 1990. Prevalence of *Helicobacter* (Formerly *Campylobacter) Pylori* Infection in Saudi Arabia, and Comparison of those with and Without Upper Gastrointestinal Symptoms, *American Journal of Gastroenterology*, Vol. 85, pp. 944–948.

Ammon, A., Petersen, L.R. and Karch, H. 1999. A Large Outbreak of Hemolytic Uremic Syndrome Caused by an Unusual Sorbitol-fermenting Strain of *Escherichia coli* O157:H-, *Journal of Infectious Diseases*, Vol. 179, pp. 1274–1277.

Arankalle, V.A., Devi, K.L.S., Lole, K.S., Shenoy, K.T., Verma, V. and Haneephabi, M. 2006. Molecular Characterization of Hepatitis A Virus from a Large Outbreak from Kerala, India, *Indian Journal of Medical Research*, Vol. 123, pp. 760–769.

Aruin, L.I. 1997. *Helicobacter Pylori* Infection is Carcinogenic for Humans, *Arkhiv Patologii*, Vol. 59, pp. 74–78.

Audu, R., Omilabu, A.S., Peenze, I. and Steele, D.A. 2002. Isolation and Identification of Adenovirus Recovered from the Stool of Children with Diarrhoea in Lagos, Nigeria, *African Journal of Health Sciences*, Vol. 9, No. 1–2, pp. 105–11.

Baker, K.H., Hegarty, J.P., Redmond, B., Reed, N.A. and Herson, D.S. 2002. Effect of Oxidizing Disinfectants (Chlorine, Monochloramine, and Ozone) on *Helicobacter Pylori*, *Applied Environmental Microbiology*, Vol. 68, No. 2, pp. 981–984.

Baker, M.G. and Fidler, D.P. 2006. Surveillance Under International Health Regulations, *Emerging Infectious Diseases*, Vol. 12, No. 7, pp. 1058–1063.

Barzaga, N.G., Florese, R.H., Roxas, J.R. and Francisco, Z.T. 1996. Seroepidemiology of Hepatitis A Virus Among Filipinos Living in Selected Communities in and Around Metro Manila, *Philippine Society for Microbiology and Infectious Diseases*, Vol. 25, Topic I.

Beck, R. W. 2000. *A Chronology of Microbiology*. ASM Press, Washington DC.

Bern, C., Hernandez, B., Lopez, M.B., Arrowood, M.J., de Mejia, M.A., de Merida, A.M., Hightower, A.W., Venczel, L., Herwaldt, B.L. and Klein, R.E. 1999. Epidemiologic Studies of *Cyclospora Cayetanensis* in Guatemala, *Emerging Infectious Diseases*, Vol. 5, pp. 766–774.

Bern, C., Ortega, Y., Checkley, W., Roberts, J.M., Lescano, A.G., Cabrera, L., Verastegui, M., Black, R.E., Sterling, C. and Gilman, R.H. 2002. Epidemiological Differences Between Cyclosporiasis and Cryptosporidiosis in Peruvian Children, *Emerging Infectious Diseases*, Vol. 8, pp. 581–585.

Besser, R.E., Lett, S.M., Weber, J.T., Doyle, M.P., Barrett, J.G., Wells, G. and Griffin, P.M. 1993. An Outbreak of Diarrhea and Hemolytic Uremic Syndrome from *Escherichia coli* O157:H7 in fresh-pressed apple cider, *Journal of the American Medical Association*, Vol. 269, pp. 2217–2220.

Bialek, S. R., Douglas, A., Thoroughman D.A., Hu, D., Simard, E.P., Chattin, J., Cheek, J., Bell, B.P. 2004. Hepatitis A Incidence and Hepatitis A Vaccination Among American Indians and Alaska Natives, 1990–2001, *American Journal of Public Health*, Vol 94, No. 6, pp. 996-1001

Binder, R.B. 1972. Controlling Great Lakes Pollution: A Study in the United States-Canadian Environmental Cooperation, *Michigan Law Review*, Vol. 70, pp. 469–556.

Blaser, M.J. 1990. *Helicobacter Pylori* and the Pathogenesis of Gastro/-duodenal Inflammation, *Journal of Infectious diseases*, Vol. 161, pp. 626–633.

Blaser, M.J., Perez-Perez, G.I., Lindenbaum, J., Schneidman, D., Van Deventer, G., Marin-Sorensen, M. and Weinstein, W.M. 1991. Association of Infection Due to *Helicobacter Pylori* with Specific Upper Gastrointestinal Pathology, *Review of Infectious Diseases, Supplement*, Vol. 8, pp. S704–708.

Bodhidatta, L., Taylor, D.N., Thisyakorn, U. and Echeverria, P. 1987. Control of Typhoid Fever in Bangkok, Thailand, by Annual Immunization of Schoolchildren with Parenteral Typhoid Vaccine, *Review of Infectious Diseases*, Vol. 9, pp. 841–845.

Bopp, D.J., Sauders, B.D., Waring, A.L., Akelsberg, J., Dumas, N., Braun-Howland, E., Dziewulski, D., Wallace, B.J., Kelly, M., Halse, T., Musser, K.A., Smith, P.F., Morse, D.L. and Limberger, R.J. 2003. Detection, Isolation, and Molecular Subtyping of *Escherichia coli* O157:H7 and *Campylobacter jejuni* Associated with a Large Water Borne Outbreak, *Journal of Clinical Microbiology*, Vol. 41, pp. 174–180.

Brandonisio, O., Maggi, P., Panaro, M.A., Bramante, L.A., Di Coste, A. and Angarano, G. 1993. Prevalence of Cryptosporidiosis in HIV-infected Patients with Diarrhoeal Illness, *European Journal of Epidemiology*, Vol. 9, No. 2, pp. 190–194.

Bruce, M.G., Curtis, M.B., Payne, M.M., Gautom, R.K., Thompson, E.C., Bennett, B.A. and Kobayashi, J.M. 2003. Lake-associated Outbreak of *Escherichia coli* O157:H7 in Clark County, Washington, August 1999, *Archives of Pediatrics & Adolescent Medicine*, Vol. 157, pp. 1016–1021.

Capano, D.E. 2003. Chicago's War with Water, on its Way to Pioneering Our Modern Sewer System Chicago Survived Epidemics, Floods, and Countless Bad Plans, *Invention & Technology, Spring*.

Carey, C.M., Lee, H. and Trevors, J.T. 2004. Biology, Persistence and Detection of *Cryptosporidium Parvum* and *Cryptosporidium Hominis* Oocyst, *Water Research*, Vol. 38, pp. 818–862.

Casemore, D.P., Wright, S.E. and Coop, R.L. 1997. Cryptosporidiosis – Human and Animal Epidemiology. Fayer R. (ed.) *Cryptosporidium and Cryptosporidiosis*. Boca Raton, Fla: CRC Press.

Centers for Disease Control and Prevention (USA). 1995. Mortality Morbidity Weekly Report: Update: *Vibrio* cholerae O1 – Western Hemisphere, 1991–1994, and V. Cholerae O139 – Asia, 1994, March 24, 1995. Vol. 44, No. 11, pp. 215–219.

———— 1996. Mortality Morbidity Weekly Report: Outbreaks of *Cyclospora Cayetanensis* Infection – United States, June 28, 1996, Vol. 45, No. 25, pp. 549–551.

—— 1997. Mortality Morbidity Weekly Report: Outbreak of Cyclosporiasis – Northern Virginia-Washington, D.C. – Baltimore, Maryland, Metropolitan area, August 01, 1997, Vol. 46, No. 30, pp. 689–691.

—— 1998. Mortality Morbidity Weekly Report: Outbreak of Cyclosporiasis – Ontario, Canada, May, October 02, 1998, Vol. 47, No. 38, pp. 806–809.

—— 2000. Mortality Morbidity Weekly Report: Outbreaks of *Escherichia coli* O157:H7 Infections Among Children Associated with Farm Visits – Pennsylvania and Washington, April 20, 2000, Vol. 50, No. 15, pp. 293–297.

—— 2002. Mortality Morbidity Weekly Report: Multistate Outbreak of *Escherichia coli* O157:H7 Infections Associated with Eating Ground Beef – United States, July 26, 2002, Vol. 51/29, pp. 637–639.

—— 2003. Mortality Morbidity Weekly Report: Hepatitis A Outbreak Associated with Green Onions at a Restaurant – Monaca, Pennsylvania, November 28, 2003, Vol. 52, No. 47, pp. 1155–1157.

—— 2006a. Mortality Morbidity Weekly Report: Outbreaks of *Escherichia coli* O157:H7 Associated with Petting Zoos – North Carolina, Florida, and Arizona, 2004 and 2005, January 25, 2006, Vol. 54, No. 51–52, pp. 1277–1280.

—— 2006b. Mortality Morbidity Weekly Report: Summary of Notifiable Diseases – United States, 2006. (prepared by Jajosky, P.A., Hall, P.A., Adams, D.A., Dawkins, F.J., Sharp, P., Anderson, W.J., Aponte, J.J., Jones, G.F., Nitschke, D.A., Worsham, C.A., Adekoya, N. and Doyle, T.) June 16, 2006, Vol. 53, No. 53, pp. 1–79.

—— 2007. Mortality Morbidity Weekly Report: Postmarketing Monitoring of Intussusceptions after RotaTeq™ Vaccination – United States, February 1, 2006 – February 15, 2007. Vol. 56, No. 10, pp. 218–222.

Certad, G., Arenas-Pinto, A., Pocaterra, L., Ferrara, G., Castro, J., Bello and A., Núñez, L. 2005. Cryptosporidiosis in HIV-infected Venezuelan Adults is Strongly Associated with Acute or Chronic Diarrhea, *American Journal of Tropical Medicine and Hygiene*, Vol. 73, No. 1, pp. 54–57.

Chacin-Bonilla, L., de Young, M.N. and Estevez, J. 2003. Prevalence and Pathogenic Role of *Cyclospora Cayetanensis* in a Venezuelan Community, *American Journal of Tropical Medicine and Hygiene*, Vol. 68, pp. 304–306.

Chironna, M., Lopalco, P., Prato, R., Germinario, C., Barbuti, S. and Quarto, M. 2004. Outbreak of Infection Hepatitis A Virus (HAV) Associated with a Foodhandler and Confirmed by Sequence Analysis Reveals a New HAV Genotype IB Variant, *Journal of Clinical Microbiology*, Vol. 42, pp. 2825–2828.

Clark, R.M, Rice, E.W., Pierce, B.K., Johnson, C.H. and Fox, K.R. 1994. Effect of Aggregation on *Vibrio* Cholerae Inactivation, *Journal of Environmental Engineering*, Vol. 120, pp. 875–887.

Clasen, T., Cairncross, S., Hiller, L., Bartram, J. and Walker, D. 2007a. Cost Effectiveness of Water Quality Interventions for Preventing Diarrhoeal Disease in Developing Countries, *Journal of Water and Health*, Vol. 4, pp. 599–608.

Clasen, T., Schmidt, W-P., Rabie, T., Roberts, I. and Cairncross, S. 2007b. Interventions to Improve Water Quality for Preventing Diarrhoea: Systematic Review and Meta-analysis, *British Medical Journal*, Vol. 334, pp. 782–794.

Crump, J.A., Luby, S.P. and Mintz, E.D. 2004. The Global Burden of Typhoid Fever, *Bulletin of the World Health Organization*, Vol. 82, No. 5, pp. 346–353.

Current, W.L. and Garcia, L.S. 1991. Cryptosporidiosis, *Clinical Microbiology Reviews*, Vol. 4, No. 3, pp. 225–258.

Doorduyn, Y., de Jager, C.M., van der Zwaluw, W.K., Friesema, I.H.M., Heuvelink, A.E., de Boer, E., Wannet, W.J.B. and van Duynhoven, Y.T.H.P. 2006. Shiga Toxin-producing *Escherichia Coli* (STEC) O157 Outbreak, *Euro Surveill*, Vol. 11, No. 7, pp. 182–185.

Dunn, B.E., Vakil, N.B., Schneider, B.G., Miller, M.M., Zitzer, J.B., Peutz, T. and Phadnisi, S.H. 1997. Localization of *Helicobacter Pylori* Urease and Heat Shock Protein in Human Gastric Biopsies, *Infection and Immunity*, Vol. 65, pp. 1181–1188.

Durfee, M. and Bagley, S.T. 1997. *Bacteriology and diplomacy in the Great Lakes 1912–1920.* Paper prepared for the 1997 Meeting of the American Society for Environmental History, Baltimore, MD March 6–9, 1997.

Eisenberg, J.N.S., Scott, J.C. and Porco, T. 2007. Integrating Disease Control Sstrategies: Balancing Water Sanitation and Hygiene Interventions to Reduce Diarrheal Disease Burden, *American Journal of Public Health*, Vol. 97, pp. 846–852.

Emch, M., Ali, M., Park, J.K., Sack, D.A. and Clemens, J.D. 2006. Relationship Between Neighbourhood-level Killed Oral Cholera Vaccine Coverage and Protective Immunity, Evidence for Herd Immunity, *International Journal of Epidemiology*, Vol. 35, pp. 1044–1050.

Espié, E., Vaillant, V., Mariani-Kurkdjian, P., Grimont, F., Martin-Schaller, R., De Valk, H. and Vernozy-Rozand, C. 2006. *Escherichia Coli O157 Outbreak Associated with Fresh Unpasteurized Goats' Cheese, Epidemiology and Infection*, Vol. 134, pp. 143–146.

FAO (Food and Agriculture Organization of the United Nations). 2003. Available at: http://faostat.fao.org/faostat/apps.fao.org/page.collections?subset=agriculture.

Feldman, R.A., Eccersley, A.J.P. and Hardie, J.M. 1997. Transmission of *Helicobacter Pylori*, *Current Opinion in Gastroenterology*, Vol. 13, pp. 8–12.

Fewtrell, L., Kaufmann, R.B., Kay, D., Enanoria, W., Haller, L. and Colford, J.M. 2005. Water, Sanitation, and Hygiene Interventions to Reduce Diarrhea in Less Developed Countries: A Systematic Review and Meta-analysis, *The Lancet Infectious Diseases*, Vol. 5, pp. 42–52.

Frank, C., Walter, J., Muehlen, M., Jansen, A., Van Treeck, U., Hauri, A.M., Zoellner, I., Rakha, M., Hoehne, M., Hamouda, O., Schreier, E. and Stark, K. 2007. Major Outbreak of Hepatitis A Associated with Orange Juice Among Tourists, Egypt, 2004, *Emerging Infectious Diseases*, Vol. 13, pp. 156–158.

Garber, L.P., Salmen, M.D., Hurd, H.S., Keefe, T. and Schlater, J. L. 1994. Potential Risk Factors for *Cryptosporidium* Infection in Dairy Calves, *Journal of the American Veterinary Medical Association*, Vol. 295, pp. 86–91.

Gleick, P.H. 1993. *Water in Crisis a Guide to the World's Fresh Water Resources*, Oxford University Press, New York, NY.

Graham, D.Y., Alpert, L.C., Smith, J.L. and Yoshimura, H.H. 1988. Iatrogenic *Campylobacter pylori* Infection is a Cause of Epidemic Achlorhydria, *American Journal of Gastroenterology*, Vol. 83, pp. 974–980.

Haas, C.H., Rose, J.B. and Gerba, C.P. (eds) 1999. *Quantitiative Microbial Risk Assessment.* John Wiley and Sons, New York, NY.

Hadler, S.C. 1991. Global impact of Hepatitis A Virus Infection Changing Patterns. Hollinger, F.B., Lemon, S.M. and Margolis, H.S. (eds) *Viral Hepatitis and Liver Disease.* Williams and Wilkins, Baltimore, pp. 14–20.

Hegarty, J.P., Dowd, M.T. and Baker, K.H. 1999. Occurrence of *Helicobacter Pylori* in Surface Water in the United States, *Journal of Applied Microbiology*, Vol. 87, pp. 697–701.

Hilborn, E.D., Mermin, J.H., Mshar, P.A., Hadler, J.L., Voetsch, A., Wojtkunski, C., Swartz, M., Mshar, R., Lambert-Fair, M.A., Farrar, J.A., Glynn, M.K. and Slutsker, L. 1999. A Multistate Outbreak of *Escherichia Coli* O157:H7 Infections Issociated with Consumption of Mesclun Lettuce, *Archives of Internal Medicine*, Vol. 159, pp. 1758–1764.

Hulten, K., Enroth, H., Klein, P.D., Opekun, A.R., Gilman, R.H., Evans, D.G., Engstrand, L., Graham, D.Y. and El-Zaatari, F.A. 1996. *Helicobacter Pylori* in Drinking Water in Peru, *Gastroenterology*, Vol. 110, pp. 1031–1035.

Hung, C.C., Tsaihong, J.C., Lee, Y.T., Deng, H.Y., Hsiao, W.H., Chang, S.Y., Chang, S.C. and Su, K.E. 2007. Prevalence of Intestinal Infection Due to *Cryptosporidium* Species Among Taiwanese Patients with Human Immunodeficiency Virus Infection, *Journal of Formos Medical Association*, Vol. 106, No. 1, pp. 31–35.

Jacobsen, K.H. and Koopman, J.S. 2005. The Effects of Socioeconomic Development on Worldwide Hepatitis A Virus Seroprevalence Patterns, *International Journal of Epidemiology*, Vol. 34, pp. 600–609.

Kang, G. 2006. Rotavirus Vaccines, *Indian Journal of Medical Microbiology*, Vol. 24, pp. 252–257.

Karanis, P., Kourenti, C. and Smith, H. 2007. Waterborne Transmission of Protozoan Parasites: a Worldwide Review of Outbreaks and Lessons Learnt, *Journal of Water Health*, Vol. 5, No. 1, pp. 1–38.

Keene, W.E., McAnulty, J.M., Hoesly, F.C., Williams, L.P., Hedberg, K., Oxman, G.L., Barrett, T.J., Pfaller, M.A. and Fleming, D.W. 1994. A Swimming-associated Outbreak of Hemorrhagic Colitis Caused by *Escherichia Coli* O157:H7 and *Shigella Sonnei*, *The New England Journal of Medicine*, Vol. 331, pp. 579–584.

Kimman, T.G. and Boot, K. 2006. The Polio Eradication Effort has been a Great Success – Let's Finish it and Replace it with Something Even Better, *Lancet Infectious Diseases*, Vol. 6, pp. 675–678.

Kotloff, K.L., Winickoff, J.P., Ivanoff, B., Clemens, J.D., Swerdlow, D.L., Sansonetti, P.J., Adak, G.K. and Levine, M.M. 1999. Global Burden of Shigella Infections: Implications for Vaccine Development and Implementation of Control Strategies, *Bulletin of the World Health Organization*, Vol. 77, No. 8, pp. 651–666.

Koyange, L., Olliver, G., Muyembe, J-J., Kebela, B., Gouali, M. and Germani, Y. 2004. Enterohemorrhagic *Escherichia coli* O157, Kinshasa, *Emerging Infectious Diseases*, Vol. 10, pp. 968–969.

Leclerc, H., Schwartzbrod, L. and Dei-Cas, E. 2002. Microbial Agents Associated with Waterborne Diseases, *Critical Reviews in Microbiology*, Vol. 28, No. 4, pp. 371–409.

Lee, J.K., Song, H.J. and Yu, J.R. 2005. Prevalence of Diarrhea Caused by *Cryptosporidium Parvum* in Non-HIV Patients in Jeollanam-do, Korea, *The Korean Journal of Parasitology*, Vol. 43, No. 3, pp. 111–114.

Liang, J.L, Dziuban, E.J., Craun, G. F., Hill, V., Moore, M.R., Gelting, R.J., Calderon, R.L., Beach, M.J. and Roy, S.L. 2006. Surveillance for Waterborne Disease and Outbreaks Associated with Drinking Water and Water not Intended for Drinking – United States, 2003–2004, MMWR Vol 55/SS-12.

Lim, Y.A., Rohela, M., Sim, B.L., Jamaiah, I. and Nurbayah, M. 2005. Prevalence of Cryptosporidiosis in HIV-infected Patients in Kajang Hospital, Selangor, *Southeast Asian Journal of Tropical Medicine & Public Health*, Vol. 36, Suppl. 4, pp. 30–33.

Luby, S.P., Faizan, M.K., Fisher-Hoch, S.P., Syed, A., Mintz, E.D., Bhutta, Z.A. and McCormick, J.B. 1998. Risk Factors for Typhoid Fever in an Endemic Setting, Karachi, Pakistan, *Epidemiology and Infection*, Vol. 120, No. 2, pp. 129–138.

Mazick, A., Howitz, H., Rex, S., Jensen, I.P., Weis, N., Katzenstein, T.L., Haff, J. and Molbak, K. 2005. Hepatitis A Outbreak Among MSM Linked to Casual Sex and Gay Saunas in Copenhagen, Denmark, *Euro Surveillance*, Vol. 10, pp. 111–114.

Medema, G., Teunis, P., Blokker, M., Deere, D., Davison, A., Charlis, P. and Loret, J. 2006. *WHO Guidelines for Drinking Water Quality Cryptosporidium*, EHC *Cryptosporidium* draft.

Mermin, J.H., Villar, R., Carpenter, J., Roberts, L., Samaridden, A., Gasanova, L., Lomakina, S., Bopp, C., Hutwagner, L., Mead, P., Ross, B. and Mintz, E.D. 1999. A Massive Epidemic of Multidrug-resistant Typhoid Fever in Tajikistan Associated with Consumption of Municipal Water, *Journal of Infectious Diseases*, Vol. 179, No. 6, pp. 1416–1422. Comment in: *Journal of Infectious Diseases*, 1999, Vol. 180, No. 6, pp 2089–2090.

Meunier, L., Canonica, S. and von Gunten, U. 2006. Implications of Sequential Use of UV and Ozone for Drinking Water Quality, *Water Research*, Vol. 40, No. 9, pp. 1864–1876.

Mosier, D.A. and Oberst, R.D. 2000. *Ann. N Y. Acad. Sci.* 916:102-111

Munnangi, S. and Sonnenberg, A. 1997. Time Trends of Physician Visits and Treatment Patterns of Peptic Ulcer Diseases in the Unites States, *Archives of Internal Medicine*, Vol. 175, pp. 1489–1494.

Nayak, A. and Rose, J.B. 2007. Detection of *Helicobacter Pylori* in Sewage and Water Using a New Quantitative PCR Method with SYBR Green, *Journal of Applied Microbiology*, IN PRESS JAM-2007-0053.R1.

Nimri, L.F. 2003. *Cyclospora Cayetanensis* and Other Intestinal Parasites Associated with Diarrhea in a Rural Area of Jordan, *International Microbiology*, Vol. 6, pp. 131–135.

Oliver, J. D. 2002. Public Health Significance of Viable but Nonculturable Bacteria. Colwell, R.R. and Grimes, D.J. (eds) *Nonculturable microorganisms in the environment*, American Society for Microbiology, Washington, D.C., pp. 277–300.

Parry, J.V. and Mortimer, P.P. 1984. The Heat Sensitivity of Hepatitis A Virus Determined by a Simple Tissue Culture Method, *Journal of Medical Virology*, Vol. 14, pp. 277–283.

Peng, M., Xiao, L., Freeman, A.R., Arrowood, M.J., Escalante, A.A., Weltman, A.C., Ong, C.S., Mac Kenzie, W.R., Lal, A.A. and Beard, C.B. 1997. Genetic Polymorphism Among *Cryptosporidium Parvum* Isolates: Evidence of Two Distinct Human Transmission Cycles, *Emerging Infectious Diseases*, Vol. 3, No. 4, pp. 567–573.

Promed. Available at: www.promedmail.org

Pruss-Ustun, A. and Corvalan, C. 2007. Preventing Disease Through Healthy Environments. Towards an Estimate of the Environmental Burden of Disease, WHO, Geneva. Available at: www.who.int

Puente, S., Morente, A., Garćia-Benayas, T., Subritas, M., Gasćon, J. and González-Lahoz, J.M. 2006. Cyclosporiasis a Point Source Outbreak Acquired in Guatemala, *International Society of Travel Medicine*, Vol. 13, pp. 334–337.

Purcell, R.H. 1994. Hepatitis Viruses: Changing Patterns of Human Disease, *Proceedings of the National Academy of Sciences, US*, Vol. 91, pp. 2401–2406.

Raccurt, C.P., Brasseur, P., Verdier, R.I., Li, X., Eyma, E., Stockman, C.P., Agnamey, P., Guyot, K., Totet, A., Liautaud, B., Nevez, G., Dei-Cas, E. and Pape, J.W. 2006 Human Cryptosporidiosis and *Cryptosporidium* spp. in Haiti, *Tropical Medicine and International Health*, Vol. 11, No. 6, pp. 929–934.

Rolle-Kampczyk, U., Fritz, J.G., Diez, U., Lehmann, I., Richter, M. and Herbarth, O. 2004. Well Water – one Source of *Helicobacter Pylori* Colonization, *International Journal of Hygiene and Environmental Health*, Vol. 207, pp. 363–368.

Rose, J.B. and Masago, Y. 2007. A Toast to Our Health: Our Journey Toward Safe Water, *Water Science & Technology: Water Supply*, Vol. 7, No. 1, pp. 41–48. © IWA Publishing 2007 doi:10.2166/ws.2007.005

Sánchez, G., Pintó, Vanaclocha, H. and Bosch, A. 2002. Molecular Characterization of Hepatitis A Virus Isolates from a Transcontinental Shellfish-borne Outbreak, *Journal of Clinical Microbiology*, Vol. 40, pp. 4148–4155.

Sherman, I. W. (ed.) 2006. *The Power of Plagues*. ASM Press, Washington DC.

Shrestha, M.P., Scott, R.M., Joshi, D.M., Mammen, M.P., Thapa, G.B., Thapa, N., Myint, K.S.A., Forneau, M., Kuschner, R.A., Shrestha, S.K., David, M.P., Seriwatana, J., Vaughn, D.W., Safary, S.K., Endy, T.E. and Innis, B.L. 2007. Safety and Efficacy of a Recombinant Hepatitis E vaccine, *The New England Journal of Medicine*, Vol. 356, pp. 895–903.

Srikantiah, P., Vafokulov, S., Luby, S.P., Ishmail, T., Earhart, K., Khodjaev, N., Jennings, G., Crump, J.A. and Mahoney, F.J. 2007. Epidemiology and Risk Factors for Endemic Typhoid Fever in Uzbekistan, *Tropical Medicine and International Health*, Vol. 12, No. 7, pp. 838–847.

Sulaiman, I., Xiao, L., Yang, C., Escalante, L., Moore, A., Beard, C.B., Arrowood, M.J. and Lal, A.A. 1998. Differentiating Human from Animal Isolates of *Cryptosporidium Parvum*, *Emerging Infectious Diseases*, Vol. 4, No. 4, pp. 681–685.

Tadesse, A. and Kassu, A. 2005. Intestinal Parasite Isolates in AIDS Patients with Chronic Diarrhea in Gondar Teaching Hospital, North West Ethiopia, *Ethiopian Medical Journal*, Vol. 43, No. 2, pp. 93–96.

Thomas, J., Gibson, G., Darboe, M., Dale, A. and Waever, L. 1992. Isolation of *Helicobacter Pylori* from Human Feces, *Lancet*, Vol. 340, pp. 1194–1195.

Thompson, F.L., Austin, B. and Swings, J. (eds) 2006. *The Biology of Vibrios*, ASM Press, Washington DC.

Thurston-Enriquez, J.A., Haas, C.N., Jacangelo, J., Riley, K. and Gerba, C.P. 2003. Inactivation of Feline Calicivirus and Adenovirus Type 40 by UV Radiation, *Applied Environmental Microbiology*, Vol. 69, No. 1, pp. 577–582.

United Nations Development Programme (UNDP). 2003. *Human Development Report 2003*. United Nations, New York.

United Nations. 2000. www.unmillenniumproject.org.

Varma, J.K., Greene, K.D., Reller, M.E., DeLong, S.M., Tottier, J., Nowicki, S.F., DiOrio, M., Koch, E.M., Bannerman, T.L., York, S.T., Lambert-Fair, M.A., Wells, J.G. and Mead, P.S. 2003. An Outbreak of *Escherichia coli* O157 Infection Following Exposure to a Contaminated Building, *The Journal of the American Medical Association*, Vol. 290, pp. 2709–2712.

Villar, L.M., da Costa, M., se Paula, V.S. and Gaspar, A.M.C. 2002. Hepatitis A Outbreak in a Public School in Rio de Janeiro, Brazil, *Memorias do Instituto Osealdo Cruz*. Vol. 97, pp. 301–305.

Viswanathan, R. 1957. Infectious hepatitis in Delhi (1955–56): A Critical Study: Epidemiology, *Indian Journal of Medical Research*, Vol. 45, pp. 1–30.

Walia, M., Gaind, R., Mehta, R., Paul, P., Aggarwal, P. and Kalaivani, M. 2005. Current Perspectives of Enteric Fever: a Hospital-based Study from India, *Ann Trop Paediatr*. Sep. Vol. 25, No. 3, pp. 161–74.

Wilhelmi, I., Roman, E. and Sánchez-Fauquier, A. 2003. Viruses causing Gastroenteritis, *Clinical Microbiology and Infection*, Vol. 9, No. 4, pp. 247–262.

Willshaw, G.A., Smith, H.R., Cheasty, T., Wall, P.G. and Rowe, B. 1997. Vero Cytotoxin-Producing *Escherichia coli* O157 Outbreaks in England and Wales, 1995: phenotypic methods and genotypic subtyping, *Emerging Infectious Diseases*, Vol. 3, pp. 561–565.

Wolman, A. and Gorman, A. E. 1931. *The Significance of Waterborne Typhoid Fever Outbreaks, 1920–1930*. William and Wilkins Company, Baltimore.

WHO. 2007. www.who.int/vaccine_research/diseases/typhoid/en/.

World Health Organization Statistics. 2007. http://globalatlas.who.int/globalatlas/DataQuery/default.asp.

World Health Organization. 1999. *Health-based Monitoring of Recreational Water: The Feasibility of a New Approach (The'Annapolis Protocol')*. (WHO/SDE/WSH/99.1). World Health Organization – Sustainable Development and Healthy Environments.

World Health Organization. 2003. *Guidelines for Safe Recreational Water Environment Volume 1 – Coastal and Fresh Waters*. Accessed: July 1, 2007. Available at: http://whqlibdoc.who.int/publications/2003/9241545801_contents.pdf www.who.int/water_sanitation_health/bathing/srwe1/en/.

World Health Organization. 2004. Data Table 14, Global Cholera Cases and Deaths Reported to the World Health Organization, 1970 to 2004. Available at: www.who.int/emc/diseases/cholera/choltbl1999.htm and www.who.int/emc/topics/cholera/wer7631.pdf.

World Health Organization. 2005. *Water Safety Plans*. Available at: www.who.int/water_sanitation_health/ dwq/safetyplans/en/.

World Health Organization. 2007. *Weekly Epidemiological Record*. 82(32): 285–296. Available at: www.who.int/wer.

Xiao, L., Fayer, R., Ryan, U. and Upton, S.J. 2004. *Cryptosporidium* Taxonomy: Recent Advances and Implications for Public Health, *Clinical Microbiology Reviews*, Vol. 17, pp. 72–97.

Yates, M.V., Rochelle, P. and Hoffman, R. 2006. Impact of Adenovirus Resistance on UV Disinfection Requirements: A Report on the State of Adenovirus Science, *Journal of American Water Works Association*, Vol. 98, pp. 93–106.

Zhou, F., Shefer, A., Weinbaum, C., McCauley, M. and Kong, Y. 2007. Impact of hepatitis A vaccination on health care utilization in the United States, 1996–2004, *Vaccine*, Vol. 25, pp. 3581–3587.

Chapter 4

Chemical health risks

Inés Navarro[1] and Francisco J. Zagmutt[2]

[1]Instituto de Ingeniería, Universidad Nacional Autónoma de México, Mexico
[2]DVM, MPVM, Vose Consulting, 2891 20th Street, Boulder, CO, 80304, USA

ABSTRACT: This chapter describes the main sources responsible for the increased levels of inorganic and organic compounds in water resources and drinking water. Six inorganic compounds (arsenic, fluoride, lead, mercury, and nitrates and nitrites) and eight groups of organic compounds (hydrocarbons, chlorinated compounds, volatile organic compounds, solvents, trihalomethanes, pesticides, persistent organic pollutants and emerging chemicals) were chosen to demonstrate the diversity of risk sources (natural releases to aquatic environment, point sources and non-point discharges to surface waters, and releases from hazardous waste disposal sites and chemical spills). The role of the major routes of exposure to chemical compounds in urban populations and the identification of the main adverse health effects that each compound can cause are also described.

Chapter 4 also includes examples that illustrate the compounds most frequently reported in research published in scientific journals, to which population in cities from developed and developing countries are exposed. Some examples reflect national water quality status and the concern with monitoring, analysis and identification of compounds at risk level. Others focus on the hundreds and even thousands of people with some type of disease caused by the exposure to inorganic compounds naturally found in water resources. They also address the diversity of diseases a chemical agent may cause, the severity and endemic characteristics that some diseases have, and the poverty, the low sanitation level achieved, and the poor regulation conditions under which such exposures occur in some developing countries.

The last section includes some issues that should be taken into account in the implementation of a successful risk management strategy to prevent and reduce chemical risks in the urban water cycle. Factors such as low income, poor housing and public services, unplanned urbanization and environmental mismanagement increase the risk. Vulnerability to chemical risks is dynamic and influenced by a multitude of factors, including population growth and regional shifts in population, technology, urban government policies, land and water use trends, and increasing environmental awareness. Consequently, it is important that cities identify plans at all levels, including prevention, mitigation, response and public health advisories, through multidisciplinary and inter-sector interactions, according to local socio-economic and cultural circumstances.

4.1 INTRODUCTION

Nearly half of the world's population now lives in urban areas. A rapid and often unplanned urban growth is frequently associated with poverty, environmental degradation and population demands that outstrip service capacity. These conditions place human health at risk. This chapter addresses the main diseases of concern in public health policy related to the toxic chemical compounds most commonly found in water sources and drinking water in cities around the world.

Chemical health risk is addressed in this chapter as the identification of potential adverse human-health effects resulting from exposure to hazardous chemical compounds in the urban water cycle. A chemical risk is considered as the probability of an adverse health outcome, caused by one or more hazardous chemicals. The characterization of chemical risks is based on the strength of evidence and the nature of the outcomes, as well as on the human exposure situations, the population susceptibility factors and the potential magnitude of the risks. Consequently, a major challenge for risk assessment in general, and for chemical health risks in particular, is to estimate quantitatively the health risk through a procedure, based on four primary components as listed by the National Research Council (NRC, 1998): 1) hazard identification, 2) dose–response assessment, 3) exposure assessment, and 4) risk characterization.

In the case of chemical risk related to water, the hazard identification consists of finding the components in the water that are hazardous and are present in sufficiently high concentrations to adversely affect health. Dose–response assessment for chemical contaminants includes data on the detrimental effect of chemicals and toxicological data, mainly obtained through research on test organisms, which are used to determine the acceptable dose for humans (Aertgeerts and Angelakis, 2003). Exposure assessment is the process of estimating or measuring the magnitude, frequency and duration of exposure to a chemical agent, along with the number and characteristics of the population exposed. Ideally, the sources, pathways, routes and uncertainties are described during the assessment (IPCS, 2004). Exposure assessment is based on exposure scenarios, which are defined as a combination of facts, assumptions and inferences that define a discrete situation where potential exposures may occur. Exposure factors summarize data on human behaviour and characteristics that affect exposure to environmental contaminants (US EPA, 1997a). In addition, risk characterization is the phase of a risk assessment that combines the results of the characterizations of exposure and effects to estimate the risks to each endpoint, and estimates the uncertainties associated with the risks.

When considering the content of this chapter, it must be kept in mind that, on a day-to-day basis, human beings are rarely exposed to a single chemical as occurs during toxicological studies used to determine permitted levels, except perhaps for cases of occupational exposure. In fact, it is imperative to consider all possible sources of exposure (ingestion of drinking water and food, use of personal and home products, and the air, especially in polluted cities). Therefore, this addition of compounds, as well as the varied effects caused by synergetic, neutralizing, potentiating and antagonistic effects between substances that enter the body, makes it very difficult to predict the effects on human beings. Hence, the content of this chapter, as a general approach to characterize chemical health risks, will focus on descriptions of the risk sources, the identification of exposure factors and the human-health consequences for a selected group of chemical compounds of concern, in water sources and drinking water, in most cities around the world.

4.2 HUMAN-HEALTH RISKS

Exposure to chemicals may result in different types of human disease, depending on the chemical compound, the dose (that is, the amount that gets into the body), the exposure route (the pathway through which exposure occurred), the frequency of exposition and the health conditions of individuals. Other factors such as gender, age, ethnicity, daily diet, body weight and height, as well as any kind of abnormalities in health status, will affect the clinical presentation of the disease and its severity. Individual behaviour and social factors may be very important to increase (risk factors) or reduce (mitigation factors) individual level disease risk.

A chemical agent does not produce toxic effects in the human system unless that agent or its metabolic breakdown (biotransformation) products reach an appropriate organ in the body at a concentration and for a length of time sufficient to produce a toxic manifestation. Whether a toxic response occurs depends on the chemical and physical properties of the agent, the exposure scenario, how the agent is metabolized by the system and the overall susceptibility of the individual.

4.2.1 An overview on exposure factors

Diseases cannot occur in the absence of exposure to the toxic chemical. Thus, the primary objectives of exposure assessments are to determine the source, type, magnitude and duration of contact with the chemical of interest.

Toxic agents enter the human body through three major routes or pathways: gastrointestinal tract (ingestion), lungs (inhalation) and skin (dermal) (Covello and Merkhofer, 1993). Toxic agents generally produce the greatest effect and the most rapid response when they are introduced directly into the bloodstream. Therefore, a key step in exposure assessment is determining which exposure pathways are relevant for the scenario being evaluated. For example, if a chemical is ingested through food or drinking water, the primary route of entry is the gastrointestinal tract. If the chemical is present in the air (as a gas, aerosol or particle) the primary route of entry is the lungs.

In addition to the amount of the risk agent to which the individual is exposed, the potency or toxicity of the agent is important. For example, there is sufficient evidence of carcinogenicity in humans for arsenic; another example is the association between levels of lead in blood and clinical anaemia observed at concentrations of lead as low as $1 \mu g/mL$.

Another important factor is the frequency and/or duration of exposure. In humans, the frequency and duration of exposure may be described as acute (resulting from a single incident or episode), subchronic (occurring repeatedly over several weeks or months) or chronic (occurring repeatedly for many months or years) (Klaassen and Watkins III, 2003). In general, adverse health effects may be acute (short term) or chronic (long term). Acute adverse effects occur within seconds or days; for example, skin burns and poisonings. Chronic adverse health effects, by comparison, last longer and develop over a longer period. Examples of chronic effects include cancer, birth defects, genetic damage and degenerative illnesses (Covello and Merkhofer, 1993). Most chemicals undergo metabolic changes in the human body. Metabolic changes can convert chemicals with minor toxicity to more toxic forms; for example, the primary concern regarding nitrate and nitrite in drinking water is the formation of methemoglobin in infants. Metabolic processes can also detoxify chemicals and expedite excretion of potentially harmful substances from the body. In such cases, it is important to consider whether the interval between doses is sufficient to allow for complete repair of the damaged organ or not. Hence, chronic effects may occur if the chemical accumulates in the body (the rate of absorption exceeds the rate of biotrans-formation and/or excretion), if it produces irreversible toxic effects, or if there is insuf-ficient time for the system to recover from the toxic damage within the exposure frequency interval.

The environmental conditions of exposure are also important, including the pres-ence or absence of other risk agents. Additional factors that influence susceptibility to environmental exposures include genetic traits, sex and age, pre-existing diseases, behavioural traits (most importantly smoking and chronic ethanol consumption), coexisting exposures, medications, vitamins and protective measures (Klaassen and Watkins III, 2003). Nutritional factors play an important role in both exposure and susceptibility to various chemical agents. Hence, individual personal habits can have a strong influence on human exposure and are an important source of complexity of exposure estimation.

Another important aspect of chemical health risks is to determine which groups in the population may be exposed to a risk agent. It is well known that pregnant women, very young and very old people as well as people with impaired health may all be par-ticularly sensitive to toxic chemicals (Covello and Merkhofer, 1993). Population under poverty conditions is a target concern in big cities, particularly in developing coun-tries.

When assessing drinking water and food contaminated with one particular chemical agent, food and water storage practices, food preparation and dietary habits have a major influence on the amount of the agent actually consumed. For instance, the estima-tion of consumer exposure to a pesticide in water requires quantitative data on pesticide residues, the amount of drinking water consumed by an average person and the number of people consuming from tap water. This approach has been traditionally used but has important limitations, since using exposure and frequency data from an 'average' person ignores the variability in the exposure patterns among individuals. Furthermore, using single-point estimates (i.e. the mean pesticide exposure per day per person) also ignores the uncertainty the analyst has about the true parameter values. Variability and uncer-tainty analysis can be incorporated in chemical health risk assessment using Monte Carlo simulation methods. The reader is referred to Vose (2000) to read about applied method-ologies in quantitative risk analysis that can be used in health risk assessments.

All the above exposure factors relate the source of risk with the magnitude of the hazard, the exposure conditions and with the characteristics of the population of con-cern. Exposure evaluation often depends on factors that are hard to estimate and for which there are few data. It is common to find detailed information on relatively few people and very limited information on large numbers of people. In general, critical information on the conditions of exposure is often lacking; but exposures to the gen-eral population in urban cities are even less well documented due to the limited avail-ability of systems capable of measuring the exposures to specific risk agents that people are actually exposed to. For that reason, exposure data is frequently identified as the key area of uncertainty in overall risk determination.

4.2.2 Human exposure in urban water cycle

Exposure to chemical compounds in water resources and in drinking water may occur in a variety of ways. The main direct exposure routes for urban people are: exposure to chemicals in drinking water via direct ingestion; ingestion of foods prepared with drinking water; and inhalation and dermal contact during activities such as showering. Indirect exposure to chemicals may occur via fish ingestion or with consumption of produce that is irrigated by untreated wastewater (WHO, 2006a). Therefore, the concentration of chemical compounds in the raw surface water or groundwater is an important factor in estimating concentrations for risk assessment.

Raw surface water or groundwater may be consumed directly or after treatment to remove various types of contaminants. The type of water treatment and the conditions of the water supply system are additional factors to be considered when assessing drinking-water exposure pathways. For example, some compounds, such as disinfection by-products formed during the chlorination treatment process, may be present in drinking water supplied. In some cases, contaminants such as metals may leach into the public water supply from corroded pipes.

To protect human health, the World Health Organization (WHO) has developed risk-based drinking-water standards. The WHO Guidelines for Drinking-Water Quality (WHO, 2004) are used worldwide as a reference in the development of local national standards and as a basis for improved water treatment practices. It contains guideline values established for a number of chemical substances that may be found in drinking water, including inorganic substances, organic substances, pesticides and disinfection by-products. Therefore, a guideline value represents the concentration of a constituent that does not exceed tolerable risk to the health of the consumer over a lifetime of consumption.

Some countries have adopted the WHO Guidelines as the official and legal standards of water quality, while other countries have developed national standards, based totally or partially on them. In spite of this, in some countries a few chemicals have caused widespread health effects in humans as a consequence of excessive exposure through drinking water. These chemicals include fluoride, arsenic and nitrate (WHO, 2004). In some areas, human-health effects have also been associated with lead from domestic plumbing (WHO, 2004).

In the case of some other organic compounds, the exposure to risk agents through ingestion of drinking water for human population in certain urban areas could be stronger, more frequent and for longer periods of time. In fact, the amount and quality of drinking-water intake varies from one city to another. In most of the developing countries studied by Morgenstern et al. (2000), problems involving drinking water are ranked as a high or medium health risk due to contamination, limited coverage and erratic service by water supply systems.

If exposure occurs through surface water resources, then exposure estimation must consider how the risk agent moves from its source through the environment to rivers or lakes, and how it is altered over time. Chemical risk agents generally become diluted and may degrade after they are released. The aim of exposure assessment in this case is to determine the concentration of toxic materials in space and time as they interact with target populations. Direct contact exposure to non-drinking surface water resources contaminated with chemical compounds may occur in some cities where the population uses surface water to irrigate gardens or small farming parcels. Indirect ingestion pathway exposure may occur by fish harvest on seacoasts, in river deltas, along estuaries and river mouths contaminated with chemical compounds, and by ingestion of agricultural products irrigated with wastewater (WHO, 2006a).

Therefore, ingestion of food is a relevant route of entry for some chemical compounds in water resources.

While water can be a major source of hazardous chemicals, it is by no means the only source. Attention must be given to other sources of hazards, such as food and air, as well as to the role of poor sanitation and personal hygiene. Thus, the appropriate pathways for exposure depend on the risk agent and the level of polluted medium.

The exposure to contaminant mixtures is another factor of risk. Exposure to risk agents that act synergistically greatly complicates risk estimation because most assessments are conducted on individual risk agents. The exposure to trihalomethanes (chloroform, bromoform, dichlorobromomethane and bromodichloromethane) in drinking water is an example of exposure to contaminant mixtures (Health Canada, 2006).

Multiple sources (same as routes or pathways) may increase the risk and also complicate exposure estimation. An individual can be exposed to a single risk agent from several distinct sources. Exposures to lead, for example, can come from breathing, eating food and drinking water (in that order of health risk importance) (Covello and Merkhofer, 1993).

In general, most chemicals in water are of concern only when long-term exposure takes place; however, some hazardous chemicals that occur in drinking water are of concern because of the effects, which arise from sequences of exposures over a short period. Other hazards may arise intermittently; they are often associated with seasonal activity or seasonal conditions.

The diverse hazards that may be present in water are associated with very different adverse health outcomes. Some outcomes are acute (methemoglobinemia), and others are delayed by years (e.g., cancer); some are potentially severe (cancer, adverse birth outcomes), and others are typically mild (dental fluorosis); some especially affect certain age ranges (skeletal fluorosis in older adults often arises from exposure in childhood), and some have very specific concerns for certain vulnerable subpopulations. Moreover, any hazard may cause multiple effects.

4.3 RISK SOURCES AND RISK COMPOUNDS IN URBAN WATER CYCLE

4.3.1 Releases to water

Assuring an adequate supply of water is not the only problem faced by many cities throughout the world; they must be concerned, as well, with water quality. Surface water resources such as river and lakes, seacoasts, river deltas estuaries as well as groundwater reservoirs in urban areas are polluted by different sources containing several risk agents. In order to analyze the chemical risks, Table 4.1 summarizes five categories of sources that will be addressed in the following sections: natural releases to the aquatic environment, point sources discharges, non-point sources discharges, releases from solid and hazardous waste disposal sites, and chemical spills.

In some cases, water quality problems occur naturally, as happened with arsenic-contaminated groundwater in Bangladesh (Lenton et al., 2005). Traces of compounds such as fluoride, arsenic or uranium are present in many water sources and are associated with the type of rock that the water flows through. Under natural conditions, the highest concentrations of these compounds are usually found in groundwater as a result of

Table 4.1 Categories of risk sources

Risk agent category	Example agent	Source/description
Natural conditions	Arsenic, fluorides, mercury, uranium	Dissolution from natural deposits, local geochemical environment
Point-source discharges	Orinated products, volatile organic compounds (VOCs), solvents, persistent organic pollutants (POPs), pharmaceutical compounds, some metals	Industrial operation discharges, municipal water waste management facilities
Non-point discharges	Pesticides, organic chemicals, mercury and mining residues	Runoff from precipitation, agricultural runoff, and urban runoff, atmospheric deposition
Solid wastes and Hazardous waste disposal sites	Toxic chemicals and radioactive waste	Leaching from municipal waste management facilities, abandoned toxic waste sites, landfills, surface impoundments, mining operations
Chemical spills	Volatile organic compounds (VOCs), polycyclic aromatic hydrocarbons (PAHs), polychlorinated biphenyls (PCBs), chlorine	Chemical releases from truck, rail and ship transportation, underground petroleum storage tanks, spills from industrial production, storage facilities and pipelines

the strong influence of water–rock interactions and the physical and geochemical conditions that favour mobilization and accumulation in aquifers (WHO, 2001).

Risk sources for surface water include point-source discharges from industry, mining operations, electric power plants and municipal sewage treatment facilities. Industrial waste can include heavy metals and considerable quantities of synthetic chemicals, which are characterized by their toxicity and persistence; they are not readily degraded under natural conditions or in conventional sewage treatment plants (Maurits la Riviere, 1989).

Precipitation, agricultural and urban runoffs are non-point discharges or diffuse sources to water bodies, which are contaminated with pesticides and mining residues. Urban stormwater runoff pollution can have an acute impact on receiving waters, such as rivers, lakes or the ocean (Krejci et al., 1987; Chebbo et al., 2001; Gromaire et al., 2001).

Risk sources for groundwater include spills from underground petroleum storage tanks, frequently found in gas service stations in all cities around the world; and leaching of toxic agents from industrial hazardous waste disposal sites and municipal landfills. Sewer and stormwater pipe leakage can lead also to the degradation of urban groundwater quality. Accidental releases are another source that includes catastrophic industrial accidents, oil spills and in-plant release accidents.

Human activities leading to the depletion of groundwater reserves include agriculture, land use changes, urbanization, demand for domestic and public drinking water, industrial activities, and the rise of tourism in coastal areas. In addition, the overload of aquifers with the infiltration of agricultural pollutants, such as fertilizers and pesticides, and from the intrusion of saline water, affects groundwater quality.

Many toxic chemical compounds, such as some metals (arsenic, cadmium and chromium), polychlorinated-biphenyls (PCBs) and formaldehyde from industrial boilers and furnaces burning recycled oil or municipal waste incinerators emissions get indirectly into the water cycle through deposition from air emissions.

Many other factors influence water quality in urban cities. Different types of water treatment systems modify the composition of water in diverse manners. For example, ozone treatment, applied to remove pesticides, may lead to the formation of bromate from the bromide in the raw water (Bates, 2000). Water in the distribution system is subject to potential contamination from exposure to the elements and the nature of the materials through which the water is transported. Moreover, water use, consumption patterns and individual diet varies and affects a person's potential exposure to contaminants.

The primary consequences of the discharge of chemicals into receiving waters are the degradation of environmental quality and the loss of recreational values. Even though this problem has received increased attention over the last years (Chocat and Desbordes, 2004), more than half of the world's rivers are seriously depleted and polluted. The degradation and poisoning of the surrounding ecosystems is threatening, in threatening people's health and livelihood (World Commission on Water, 1999).

The discharge of chemicals into receiving waters also affects drinking-water quality. Drinking water in many cities exhibits some toxic compounds often at concentration levels below the drinking-water guideline values of the country's regulations. Chloroform, 1,3-butadiene, chromium, 1,2-dichloroethane, dichloromethane, lead, nickel and toluene, are examples of toxic compounds found in drinking water (Sofuoglu et al., 2003).

Some banned compounds, such as the organochlorines pesticides, e.g., DDT, have been found in water samples of cities from developing countries (Rissato et al., 2004). In addition, there are many other unregulated compounds detected in domestic and public well samples of potential human-health concern for which there is a lack of available toxicity information; that is the case of Methyl tert-butyl ether (MTBE), a common compound in oxygenated gasoline formulation (Zogorski et al., 2006).

Ongoing efforts to control or avoid wastewater discharges and runoffs of toxic compounds to rivers or lakes have resulted in current lower concentration levels in many countries. However, the accumulation process in sediments and the release of some metabolites or toxic compounds from contaminated sediment have been a major source of fish contamination during recent years. This affects inland fisheries, which are a major source of protein and other nutrients for a large proportion of the world's population. For example, the population of Cambodia gets roughly 60% of its total animal protein from the fishery resources of *Tonle Sap*, a large freshwater lake (MRC, 1997). Ingestion of fish and drinking water, which will be described in the following sections, is one of the major routes of exposure to chemical compounds in urban populations.

4.3.2 Chemical compounds

Water pollutants include a wide spectrum of organic and inorganic substances. Chemical compounds of concern in this chapter are those that may represent a negative impact or a hazard to human health in urban settlements. The focus is on the compounds most frequently found in the urban water cycle, in developed and developing countries. The aim is the identification of the release sources and a brief characterization of the main health effects caused to humans through exposure to polluted water.

Toxic chemicals are a growing concern in many parts of the world (WHO, 2006a). The number of toxic chemicals being used on a daily basis is large and growing. Tens of thousands of chemicals are being used routinely in manufacturing, agricultural production and household products. A fraction of these chemicals may find their way into the wastewater collection systems. In general, industrial wastewater discharges into sanitary sewers or drains are the main source of chemical pollution, but households may also be significant sources of toxic chemicals (WHO, 2006a).

Although a great variety of toxic compounds may be present in water resources, many are difficult to detect due to the lack of analytical techniques and to the increasing number of compounds that are being produced and discharged to sewers. However, some cities measure health-related compounds, such as a number of metals and a small group of organic compounds (including pesticides), on a regular basis. Moreover, the identification of new compounds in water resources and in drinking water has become an important task for water suppliers in the last years (van Dijk-Looijaard and van Genderen, 2000).

Some chemicals can be present in groundwater, with considerable stability and long-lasting environmental persistence. Such compounds include metals, polycyclic aromatic hydrocarbons (PAHs), other aromatic hydrocarbons, chlorinated hydrocarbons, polychlorinated biphenyls (PCBs), pesticides, dioxins and pharmaceuticals (Aertgeerts and Angelakis, 2003).

Irrigation practices with untreated or partially treated wastewater also affect the quality and safety of groundwater in shallow aquifers and surface waters that supply drinking water (WHO, 2006a). Many chemicals undergo reactive decay or change, especially over long periods in groundwater reservoirs. Chlorinated hydrocarbons, such as trichloroethylene used in industrial metal degreasing and tetrachloroethylene used in the dry cleaning industry, are examples. Both of these chemicals, which are carcinogens themselves, undergo partial decomposition reactions leading to new hazardous chemicals, such as dichloroethylene and vinyl chloride (VC), with VC more toxic than the parent contaminant (Swindoll et al., 2000).

Chemical contaminants of potential health concern are shown in Table 4.2, based on their toxicity to humans and animals, and their reported presence in wastewater, sewage sludge or in drinking water. Chemicals in Table 4.2 exemplify the potential health effects from exposure above guideline values for drinking water, and the common sources of contaminant in water resources and drinking water. Compounds in the list include toxic metals (arsenic, cadmium, chromium, lead and mercury), pesticides (2,4-D and endrin), disinfection by-products (THMs), polychlorinated biphenyl (PCBs) and petroleum components (benzene, benzo(a)pyrene, toluene and xylenes).

The following sections contain the description of the main sources responsible for the increased levels of inorganic and organic compounds in water resources and drinking water. Six inorganic compounds (arsenic, fluoride, lead, mercury, and nitrates and nitrites) and eight groups of organic compounds (hydrocarbons, chlorinated compounds, volatile organic compounds (VOCs), solvents, trihalomethanes (THMs), pesticides, persistent organic pollutants (POPs) and new chemicals) were chosen to demonstrate the diversity of risk sources. The role that the ingestion of drinking water

Table 4.2 Potential health effects, sources of risk and drinking-water guidelines for some chemical compounds

Contaminant	WHO (mg/L) (3)	MCL (mg/L) (4)	EU (mg/L) (5)	Potential health effects from exposure above guidelines	Common sources of contaminant in drinking water
Arsenic	0.010	0.010	0.010	Skin damage or problems with circulatory systems; increased risk of cancer	Erosion of natural deposits; runoff from orchards, runoff from glass & electronics wastes
Benzene	0.010	0.005	0.001	Anaemia; decrease in blood platelets; increased risk of cancer	Discharge from factories; leaching from gas storage tanks and landfills
Benzo(a)pyrene	0.0007	0.0002	0.00001	Reproductive difficulties; increased risk of cancer	Leaching from linings of water storage tanks and distribution lines
Bromodichloromethane	0.060	(1)	(1)	Genotoxic effects; kidney and large intestinal problems; risk of cancer	By-product of drinking-water disinfection
Bromoform	0.100	(1)	(1)	Tumours of the large intestine	By-product of drinking-water disinfection
Cadmiun	0.003	0.005	0.005	Kidney damage	Corrosion of galvanized pipes; erosion of natural deposits; discharge from metal refineries; runoff from waste batteries
Carbon tetrachloride	0.004	0.005	(1)	Liver problems; increased risk of cancer	Discharge from chemical plants and other industrial activities
Chloroform	0.300	(1)	(1)	Damage to the centrilobular region of the liver; secondary carcinogen	By-product of drinking-water disinfection
Chromium	0.050	0.100	0.05	Allergic dermatitis	Discharge from steel and pulp mills; erosion of natural deposits
2,4-D	0.030	0.070	(1)	Kidney, liver or adrenal gland problems	Runoff from herbicide used on row crops
Dibromochloromethane	0.100	(1)	(1)	Induced hepatic tumours	By-product of drinking-water disinfection
1,2-Dichloroethane	0.030	0.005	0.003	Increased risk of cancer	Discharge from industrial chemical factories
Dioxin (2,3,7,8-TCCD)	(1)	0.00003 g/L	(1)	Reproductive difficulties; increased risk of cancer	Emissions from waste incineration and other combustion; discharge from chemical factories
Fluoride	1.5	4.0	1.5	Bone disease (pain and tenderness of the bones); children may get mottled teeth	Water additive to promote strong teeth; erosion of natural deposits; discharge from fertilizer and aluminium factories

Contaminant	WHO (2004a) (3)	US EPA MCL (2003) (4)	EU (1998) (5)	Health effects	Source
Lead	0.010	0.015	0.010	Infants and children: delays in physical or mental development; slight deficits in attention span and learning abilities	Corrosion of household plumbing systems; erosion of natural deposits
Mercury (inorganic)	0.006	0.002	0.001	Kidney damage	Erosion of natural deposits; discharge from refineries and factories; runoff from landfills and croplands
Nitrate (as Nitrogen)	50 (NO$_3^-$)	10	50 (NO$_3^-$)	Infants symptoms below the age of 6 months include shortness of breath and blue-baby syndrome	Runoff from fertilizer use; leaching from septic tanks, sewage; erosion of natural deposits
Nitrite (as Nitrogen)	3 (NO$_2^-$)	1	0.50 (NO$_2^-$)	Infants symptoms below the age of 6 months include shortness of breath and blue-baby syndrome	Runoff from fertilizer use; leaching from septic tanks, sewage; erosion of natural deposits
Polychlorinated biphenyls (PCBs)	(1)	0.0005	(1)	Skin changes; thymus gland problems; immune deficiencies; reproductive effects; risk of cancer	Runoff from landfills; discharge of waste chemicals
Tetrachloroethylene	(1)	0.005	0.010	Liver problems; increased risk of cancer	Discharge from factories and dry cleaners
Total THMs	(1)	0.080	0.100	Liver, kidney or central nervous system problems; risk of cancer	By-product of drinking-water disinfection
1,1,1-Trichloroethane (TCA)	(2)	0.200	(1)	Liver, nervous systems or circulatory problems	Discharge from metal degreasing sites and other factories
Trichloroethylene	(1)	0.005	0.010	Liver problems; increased risk of cancer	Discharge from metal degreasing sites and other factories
Vinyl chloride plastic factories	0.0003	0.002	0.0005	Increased risk of cancer	Leaching from PVC pipes; discharge from plastic factories

(1) No available data
(2) WHO does not establish a guideline value
(3) WHO Guidelines for Drinking Water (2004a)
(4) Maximum contaminant level (MCL), US EPA National Primary Drinking Water Standards (2003)
(5) European Union, The Quality of Water Intended for Human Consumption (1998)

plays in exposure to chemical compounds and the identification of the main adverse health effects that each compound can cause are also described.

The reader is directed to recent reviews and additional information on the hazards and risks of many chemicals, such as WHO Environmental Health Criteria monographs (EHCs) and Concise International Chemical Assessment Documents (CICADs) (www.who.int/pcs/index.htm); reports by the Joint FAO/WHO Meeting on Pesticide Residues (JMPR) and Joint FAO/WHO Expert Committee on Food Additives (JECFA); information from competent national authorities, such as the US Environmental Protection Agency (US EPA) (www.epa.gov/waterscience). Information about substances of highest concern at European Union Community level may be found at http://ec.europa.eu/environment/water/index.html. These information sources have been peer reviewed and provide readily accessible information on toxicology, hazards and risks of many contaminants.

4.4 INORGANIC CHEMICAL RISK AGENTS: SOURCES AND HUMAN DISEASES OF CONCERN

Inorganic substances that are soluble in water include minerals, metals and gases. For example, carbonate, bicarbonate, phosphates, nitrate and fluoride may result from the dissolution of mineral substances in the soil matrix; hydrogen sulphide and ammonia may be products of microbial decomposition of organic material.

All metals are soluble in water to some extent. While excessive amounts of any metal may present health hazards, only those metals that are harmful in relatively small amounts are commonly labelled as toxic; other metals fall into the non-toxic group. In addition to calcium and magnesium, other non-toxic metals commonly found in water include sodium, iron, manganese, aluminium, copper and zinc. For instance, iron, sodium, sulphate and zinc are excluded from WHO drinking-water guidelines because they are not of health concern at levels found in drinking water, but they may affect acceptability due appearance, taste or odour. Toxic metals, such as arsenic, barium, cadmium, chromium, lead, mercury and silver, may also be found in water.

In this chapter, to illustrate the broad range of human-health risks associated with the exposure to inorganic compounds in the urban water cycle, we selected six compounds – arsenic, fluoride, lead, mercury, and nitrates and nitrites – as representative of toxico-logical risks through drinking water and fish consumption pathways. The following description explains the role of natural releases, mining and industrial activities as sources of risks; it also identifies different types of human diseases. In Sections 4.6 and 4.7, examples of the magnitude of the populations affected, with particular emphasis to public health in developing countries, are presented.

4.4.1 Nitrates and nitrites

Nitrate and nitrite are naturally occurring ions that are part of the nitrogen cycle (WHO, 2004). Nitrate is used mainly in inorganic fertilizers, and sodium nitrite is used as a food preservative, especially in cured meats. Nitrate concentrations in groundwater and surface water are normally low, but can reach high levels as a result of leaching or runoff from agricultural land or contamination from human or animal wastes as a consequence of the oxidation of ammonia and similar sources (WHO, 2004).

Natural concentrations of nitrate in groundwater reach a few milligrams per litre and depend strongly on soil type and geological situation (US EPA, 1987). For example, in the USA, naturally occurring levels do not exceed 4–9 mg/L for nitrate and 0.3 mg/L for nitrite (US EPA, 1987). The increasing use of artificial fertilizers, waste disposal (particularly from animal farming), and changes in land use are the main factors responsible for the progressive increase in nitrate levels in groundwater supplies over the last 20 years (WHO, 1985). For example, concentrations of up to 1,500 mg/L were found in groundwater in an agricultural region in India (Jacks and Sharma, 1983).

Nitrate concentration in surface water is normally low (0–18 mg/L), but can reach high levels as a result of agricultural runoff, refuse dump runoff, or contamination with human or animal waste. Nitrate concentrations often fluctuate with the season and may increase when rivers are fed by nitrate-rich aquifers.

Chloramination may give rise to the formation of nitrite in the water distribution system if the formation of chloramines is not sufficiently controlled. Nitrification in distribution systems can increase nitrite levels, usually by 0.2–1.5 mg/L (WHO, 2004).

In most countries, nitrate levels in drinking water derived from surface water do not exceed 10 mg/L. In 15 European countries, 0.5 to 10% of the population is exposed to nitrate levels in drinking water above the drinking-water standard (WHO, 1985; European Chemical Industry Ecology and Toxicology Center, 1988). This is equivalent to nearly 10 million people.

The primary health, concern regarding nitrate and nitrite is the formation of methemoglobinemia, also known as 'blue-baby syndrome'. Nitrate is reduced to nitrite in the gastrointestinal system of infants, and nitrite oxidizes hemoglobin (Hb) to methemoglobin (metHb), stopping it from transporting oxygen throughout the body. The reduced oxygen mobilization becomes clinically manifested when metHb concentrations reach 10% or more above normal Hb concentrations; the condition, known as methemoglobinemia, causes cyanosis and, at higher concentrations, asphyxia. The Hb of young infants is more susceptible to metHb formation than that of older children and adults; this is believed to be the result of the large amount of fetal Hb, which is more easily oxidized to metHb, still present in the blood of infants. In addition, there is a deficiency in infants of metHb reductase, the enzyme responsible for the reduction of metHb to Hb. The reduction of nitrate to nitrite by gastric bacteria is also higher in infants because of low gastric acidity.

Nitrate levels in breast milk are relatively low; when bottle-fed, however, young infants are at risk because of the potential of exposure to nitrate/nitrite in drinking water and the relatively high intake of water in relation to body weight (WHO, 2004).

4.4.2 Fluoride

Traces of fluoride are present in many water sources; seawater, for example, has been reported to have a total fluoride concentration of 1.3 mg/L (Sloof, 1988). Higher concentrations are often associated with underground sources although fluoride can also enter a river as a result of industrial discharges (Sloof, 1988). In groundwater, fluoride concentrations vary with the type of rock that the water flows through, but do not usually exceed 10 mg/L (US EPA, 1985). In fluoride-rich containing areas, well waters may contain up to about 10 mg of fluoride per litre. The highest natural fluoride level ever reported is 2,800 mg/L (WHO, 2006b).

High fluoride concentrations are found in groundwater in many areas of the world, including large parts of Africa, China, the Eastern Mediterranean and southern Asia (India, Sri Lanka) (WHO, 2004). One of the best-known high fluoride belts on land extends along the East African Rift from Eritrea to Malawi. Another belt is present from Turkey through Iraq, Iran, Afghanistan, India, northern Thailand and China. The Americas and Japan have similar belts.

Although fluoride has beneficial health effects on teeth at low concentrations, in drinking water, excessive exposure to fluoride in drinking water, or exposure to fluoride from other sources, can have a number of adverse effects. Several studies have reported the acute effects of fluoride exposure following fluoridation overdosing (Fawell et al., 2006). However, the effects resulting from long-term exposure are the major human-health concern. A large number of epidemiological studies have been conducted in many countries concerning the effects of long-term exposure to fluoride.

Adverse effects range from mild dental fluorosis to crippling skeletal fluorosis as the level and period of exposure increases. Fluorosis is endemic in several parts of the world because of the excess of natural fluoride content in drinking water. It has also been found that crippling skeletal fluorosis is a significant cause of morbidity in a number of regions of the world (Fawell et al., 2006). Some examples are presented in Sections 4.6 and 4.7

The use of fluoridated dental products can increase the total daily fluoride exposure in children. In hot climates, the higher consumption of water will often significantly increase the intake of fluoride. Even though fluoride in drinking water is the largest contributor to daily intake, other sources may occasionally be significant. Other sources of fluoride include air pollution from burning fluoride-rich coal, certain foods or drinks (such as brick tea) and fluoride supplements.

In 1984, the WHO conducted an extensive review finding that the mottling of teeth or dental fluorosis is sometimes associated with fluoride levels in drinking water above 1.5 mg/L, and that crippling skeletal fluorosis can arise when fluoride levels exceed 10 mg/L. The WHO therefore recommended a guideline value of 1.5 mg/L to minimize the risk of dental fluorosis. This guideline value was subsequently re-evaluated by the WHO, concluding this value is still valid (WHO, 2004).

4.4.3 Toxic metals

Toxic metals (arsenic, barium, cadmium, chromium, lead, mercury and silver) can be present in water sources and in drinking water in forms and levels considered to be harmful to people. Although natural resources of all the toxic metals exist, significant concentration in water can usually be traced to mining, industrial or agricultural sources. Furthermore, water in pipes can leach out accumulated metals from the plumbing.

Cumulative toxins, such as arsenic, cadmium, lead and mercury, are particularly hazardous. Indeed, several environmental processes, including aqueous speciation, adsorption or desorption, chemical transformation, precipitation and dissolution, and biological alteration, could alter the fate, bioavailability and toxicity of metals (Swindoll et al., 2000). Once in the human body, metals create adverse effects at cellular level. Some metals disrupt chemical reactions, while others block essential biological processes. Others bind to nutrients in the stomach; consequently, they prevent the absorption of nutrients into the body (Pounds, 1985). Brief contact with high concentrations of metals can cause lung damage, skin reactions and gastrointestinal symptoms.

The following description shows that some metals (arsenic, lead and mercury) accumulate in the body over time, reaching toxic concentrations after years of exposure. Many metals, including arsenic, are carcinogens; and others (such as mercury) are concentrated in the food chain. Consequently, they pose the greatest danger to organisms near the top of the chain and to human beings.

4.4.3.1 Arsenic

Arsenic (As) is a well-known poison. It has been featured in history as a poison for killing people both in fact and in fiction. It was once a popular poison because it conveniently left no easily detectable trace. The analysis was difficult, and remains so.

Arsenic is a natural component of the earth's crust in some areas and as a consequence may be found in water. The range of arsenic concentrations found in natural waters is unusually large; it goes from less than $0.5\,\mu g/L$ to more than $5,000\,\mu g/L$ under natural conditions (Page, 1981; Robertson, 1989; Welch et al., 1988). Typical concentrations of arsenic in freshwater are less than $10\,\mu g/L$ and often less than $1\,\mu g/L$. However, concentrations of arsenic in fresh water vary depending on the source of arsenic, the amount available and the local geochemical environment. Under natural conditions, the highest concentrations of arsenic are found in groundwater due to the strong influence of water–rock interactions and the physical and geochemical conditions that favour arsenic mobilization and accumulation in aquifers. Therefore, groundwater, often used as a source of drinking water, represents a particular risk (WHO 2001).

Baseline concentrations of arsenic in river waters are low ($0.1\,\mu g/L$ to $2\,\mu g/L$) according to the composition of the surface recharge, the contribution from base flow and the bedrock lithology. As with river waters, increased concentrations are found in lake waters affected by geothermal water and by mining activity. Higher arsenic concentrations have been found in sediments and soils contaminated by the products of mining activity, including mine tailings piles and effluents.

Localized groundwater arsenic problems are now being reported from an increasing number of countries and many new cases are likely to be discovered. Until recently, arsenic was not traditionally on the list of elements routinely tested by water quality testing laboratories and so many arsenic-rich sources undoubtedly remain to be identified. The recent discovery of arsenic contamination on a large scale in Bangladesh has highlighted the need for a rapid assessment of the situation in alluvial aquifers worldwide.

A number of large aquifers in various parts of the world have been identified with problems from arsenic occurring at concentrations above $50\,\mu g/L$. The most noteworthy occurrences are Argentina, Bangladesh, Chile, northern China, Hungary, India (West Bengal), Mexico, Romania, Taiwan and many parts of the US, particularly the southwest. Arsenic associated with geothermal waters has also been reported in several areas, including hot springs from Argentina, Japan, New Zealand, Chile, Kamchatka, Iceland, France, Dominica and the US (Welch et al., 1988). Mining-related arsenic problems in water have been identified for instance in Austria, Ghana, Greece, India (Madhya Pradesh), South Africa, Thailand and the US.

Available evidence suggests that arsenic non-occupational exposure occurs primarily through the ingestion of food and water, with the inhalation pathway playing only a minor role. Intake via dermal absorption is believed to be negligible and thus, hand-washing, bathing, laundry, etc. with water containing arsenic do not pose human-health risks. Depending upon the bioavailability, soil may be a potentially significant source of arsenic intake in children, particularly in areas near industrial and hazardous waste sites.

Food is more commonly the main contributor to total intake, but in areas where drinking water contains relatively high levels of arsenic, it may be the most important source of arsenic intake. Total arsenic concentrations in food from various countries vary widely, depending on the type of food, growing conditions (type of soil, water, geochemical activity, use of arsenical pesticides) and processing techniques. In general, however, the highest concentrations of total arsenic are found in seafood, followed by meats and grain; fruit, vegetables and dairy products tend to have lower total concentrations (Gunderson, 1995; Yost et al., 1998; NRC, 1999; MAFF, 1997; Dabeka et al., 1993; ANZFA, 1994).

Drinking arsenic-rich water over a long period is unsafe and the health effects are well known in some countries around the world. However, delayed effects from arsenic poisoning, the lack of common definitions, poor reporting and local awareness in affected areas are major problems in determining the extent of the problem or arsenic in drinking water and affect the development of adequate solutions.

Well-known cases of arsenic poisoning from natural arsenic in drinking water have been found in Taiwan, Chile, Argentina, Mexico and China and more recently in West Bengal (India) and Bangladesh (see Section 4.7.2). The Bengal Basin has the greatest problem in terms of population exposed to high arsenic concentrations, with perhaps 40 million people drinking water containing 'excessive' amounts of arsenic (Bailey et al., 2006). Many of the worst problems occur in poor countries lacking the necessary infrastructure to be able to respond to them rapidly.

Acute and chronic arsenic exposure can result in a wide variety of adverse health outcomes. Acute arsenic poisoning occurs usually as an acute gastrointestinal syndrome. Indirect effects include renal failure, bone marrow suppression, respiratory failure and polyneuropathy. Drinking arsenic-rich water over a long period – between 5 to 20 years – leads to arsenic poisoning or Arsenicosis and various health effects. These effects include skin problems (such as colour changes on the skin and hard patches on the palms and soles of the feet); skin, bladder, kidney and lung cancer; diseases of blood vessels in legs and feet; and possibly diabetes, high blood pressure and reproductive disorders.

In Taiwan, exposure to arsenic via drinking water has been shown to cause a severe disease of the blood vessels, known as 'black foot disease', which leads to gangrene. This disease has not been observed in other parts of the world, and it is possible that malnutrition contributes to its development. Studies in several countries have demonstrated that arsenic causes other, less severe forms of peripheral vascular disease (WHO, 2001).

4.4.3.2 Mercury

Mercury (Hg) is present in inorganic form in surface water and groundwater at concentrations usually below 0.5 mg/L, although local mineral deposits may produce higher levels in groundwater (WHO, 2003). Natural weathering of mercury-bearing minerals in igneous rocks is estimated to directly release about 800 metric tons of mercury per year to surface waters of the earth (Gavis and Ferguson 1972). Mercury releases can be in several different inorganic forms (e.g., elemental mercury vapour, gas-phase ionic mercury and particulate-phase mercury). Mercury is released from natural and anthropogenic sources. It is known that mercury is used in the electrolytic production of chlorine, in electrical appliances, in dental amalgams and as a raw material for various mercury compounds. It exhibits a complicated chemistry, and proceeds via several different pathways to humans and wildlife. As mercury is a natural element, it does not degrade to simpler compounds. Once released, mercury continues to cycle through the environment until it is sequestered (i.e. trapped in lake or ocean sediments) (US EPA, 1999).

Atmospheric deposition of elemental mercury from both natural and anthropogenic sources is identified as an indirect source of mercury to surface waters (WHO, 2003). The greatest releases of anthropogenic mercury to the environment are from combustion of fuel containing trace amounts of mercury; industrial processes that use mercury; and disposal of products that contain mercury either as an intentional constituent or as an impurity. Most of this waste was either incinerated or placed in landfills. It is estimated that one-third of anthropogenic emissions are deposited through wet and dry sedimentation around the releases sites; while the remaining two-thirds are transported outside the sites and enter the global mercury cycle (US EPA, 1999).

Mercury associated with soils can be directly washed into surface waters during rain events. Therefore, surface runoff is an important mechanism for transporting mercury from soil into surface waters, particularly for soils with high humic content.

Mercury may also be released to surface waters in effluents from several industrial processes, including chloralkali production, mining operations and ore processing, metallurgy and electroplating, chemical manufacturing, ink manufacturing, pulp and paper mills, leather tanning, pharmaceutical production, and textile manufacture. These effluents contaminate the sediments of rivers, swamps and watersheds, and bays adjacent to discharges sites with mercury concentrations (Davis et al., 1992).

The release of mercury in water discharges is believed to be small compared to atmospheric emissions, but it can have significant local effects. Mercury discharges to surface waters from abandoned gold and mercury mines are believed to be the cause of fish death as result of the presence of methylmercury in several streams and lakes.

Although the major environmental risks are associated with organic forms of mercury (i.e. methylmercury), environmental releases are usually in the inorganic form. However, methylation of inorganic mercury has been shown to occur in fresh water and in seawater (WHO, 2003). That is, inorganic mercury emitted and deposited in soil and water is, in a very complex transformation, biologically converted to methylmercury. Mercury accumulates most efficiently in the aquatic food web as methylmercury. While inorganic mercury, which is less efficiently absorbed and more readily eliminated from the body than methylmercury, does not tend to bioaccumulate (US EPA, 1999). In fact, nearly all of the mercury that accumulates in fish tissue is methylmercury.

Potential sources of general population exposure to mercury include inhalation of elemental mercury vapours in ambient air, ingestion of drinking water and foodstuffs contaminated with elemental mercury or various mercury compounds (i.e. methylmercury), and exposures through dental and medical treatments (ATSDR, 1999). Dietary intake is the most important source of non-occupational human exposure to mercury, with fish and fish products being the dominant sources of methylmercury in the diet (US EPA, 1984), and it is estimated that the mean dietary intake of mercury in various countries ranges from 2 to 20 mg/day per person (WHO, 2003). It is consistent with an international study of heavy metals detected in foodstuffs from 12 different countries (Toro et al., 1994).

The exposure route of methylmercury production in aquatic systems indicates that animals that primarily feed on fish and those that prey on these fish eaters have the greatest risk of toxic effects associated with methylmercury exposure. EPA (1999) identified the mink, river otter, kingfisher, loon, osprey and bald eagle as examples of species with increased risk. Depending on the methylmercury concentration in the fish, women may be at increased risk and may be putting their foetuses at risk of the subtle neurological and developmental effects associated with methylmercury exposure. In addition to women of childbearing age and their foetuses, young children whose nervous systems continue to develop postnatal, and children under six years of age are of particular concern.

Environmental contamination from mercury has been recognized for decades as a problem for humans and wildlife. Human epidemics of methylmercury poisoning, occurring in multiple countries, have established the toxicity of this chemical to the nervous system. Moreover, as the quantity of available mercury in the environment has increased, so have the risks of neurological and reproductive problems for humans and wildlife, making it a pollutant of considerable concern (US EPA, 1999).

Methylmercury toxic effects known in humans include permanent damage to the brain and kidney. The nervous system is particularly sensitive to mercury and methylmercury (WHO, 2003). Generally, the subtlest indicators of methylmercury toxicity are neurological changes. The neurotoxic effects include subtle decrements in motor skills and sensory ability at comparatively low doses, to tremors, inability to walk, convulsions and death at extremely high exposures.

4.4.3.3 Lead

Lead (Pb) is used principally in the production of lead-acid batteries, solder and alloys. The organolead compounds, tetraethyl and tetramethyl lead, have also been used extensively as antiknock in petrol, although their use for this purpose in many countries is being phased out. The principal route of exposure for the general population is food. Environmental sources include lead-based indoor paint in old dwellings; lead in contaminated drinking water; lead in air from the combustion of lead-containing industrial emissions; hand-to-mouth activities of young children living in polluted environments; lead-glazed pottery; and lead dust brought home by industrial workers on their shoes and clothes (Klaassen and Watkins III, 2003).

Owing to the decreasing use of lead-containing additives in petrol and of lead-containing solder in the food processing industry, concentrations in air and food are declining, and intake from drinking water constitutes a greater proportion of total intake. Lead is rarely present in tap water as a result of its dissolution from natural sources. Concentrations in drinking water are generally below 5 mg/L, although much higher concentrations (above 100 mg/L) have been measured where lead fittings are present. Its presence in tap water is primarily from household plumbing systems containing lead in pipes, solder, fittings or the service connections to homes. Polyvinyl chloride (PVC) pipes also contain lead compounds that can be leached from them and result in high lead concentrations in drinking water. The amount of lead dissolved from the plumbing system depends on several factors, including pH, temperature, water hardness and standing time of the water, with soft, acidic water being the most plumb solvent (Schock, 1989, 1990).

Lead is a general toxicant that accumulates in the skeleton. Infants, children up to six years of age and pregnant women are most susceptible to its adverse health effects. Results from prospective (longitudinal) epidemiological studies suggest that prenatal exposure to lead may have early effects on mental development that persist to the age of four years. Placental transfer of lead occurs in humans as early as the 12th week of gestation and continues throughout development. Young children absorb 4–5 times as much lead as adults, and the biological half-life may be considerably longer in children than in adults.

Lead is toxic to both the central and peripheral nervous systems, inducing subencephalopathic neurological and behavioural effects. There is electrophysiological evidence of effects on the nervous system in children, with blood lead levels well below 30 mg/dL. The balance of evidence from cross-sectional epidemiological studies indicates that there are statistically significant associations between blood lead levels of 30 mg/dL and more, and intelligence quotient deficits of about four points in children.

4.5 ORGANIC CHEMICAL RISK AGENTS: SOURCES AND HUMAN DISEASES OF CONCERN

The organic compounds comprise a large list of compounds with a complex diversity of chemical and physical properties. They have been used for a variety of purposes for different industrial processes, agricultural production and household products. Many of them are widespread pollutants found in air, water, soil and sediments. Once they are in the water sources, several types of processes (oxidation–reduction reactions, sorption, volatilization, precipitation, hydrolysis and biodegradation) may result in a change of their physical or chemical properties, which can affect their potential environmental risks by altering their mobility, bioavailability or toxicity. As a result, organic compounds differ in their toxicity, persistence or bioaccumulation potential. For the purposes of this chapter, some organic compounds have been selected, based on their occurrence in the urban water cycle, to summarize the risk sources and to draw attention to the potential adverse health effects that may arise. The groups of compounds selected are hydrocarbons, chlorinated organic compounds, two subsets of the chlorinated hydrocarbons – volatile organic compounds (VOCs) and solvents – trihalomethanes (THMs), pesticides, the persistent organic pollutants (POPs) group and emerging pollutants.

4.5.1 Hydrocarbons compounds

Petroleum contains thousands of different chemicals. Most of these chemicals are hydrocarbons, of which benzene, toluene, ethylbenzene, and xylene (BTEX) and polynuclear aromatic hydrocarbons (PAHs) are a small subset of aromatic compounds.

A majority of underground storage tanks leak some form of petroleum products, initially as non-aqueous phase liquids (NAPLs). When groundwater comes into contact with these NAPLs, it typically becomes contaminated with BTEX compounds, which are the aromatic compounds most soluble in water. Thus, the dissolved fraction of BTEX compounds are typically the most common groundwater contaminants associated with spilled or leaked fuels, such as gasoline, diesel and jet fuel. As an additional concern, methyl tert-butyl ether (MTBE), a volatile gasoline oxygenate added to substitute lead has been detected at many sites. MTBE is very soluble, slowly biodegraded and potentially carcinogenic; thus, for more recently fuel-contaminated sites, MTBE as well as BTEX compounds are the contaminants of concern (Swindoll et al., 2000). Additionally, MTBE exemplifies the case of chemicals for which there are very limited toxicological data (WHO, 2004).

Major sources of PAHs in sediments, soil, surface water and groundwater include spills of petroleum or direct discharge of petroleum-containing wastes in industrial effluents or domestic wastewaters. The behaviour of PAHs in the marine food web has been extensively reviewed (Meador et al., 1995). Marine organisms at all trophic levels are capable of accumulating PAHs from their diet, but an increase in PAHs tissue concentrations in organisms at higher trophic levels does not occur because the food web transfer generally does not occur. (Swindoll et al., 2000). Many studies have demonstrated that fish do not accumulate higher tissue PAH concentrations than their prey (Eisler, 1987; Suedel et al., 1994). Recent research has demonstrated that hydrophobic organic compounds, such as PAHs, can become sequestered in soils over time, resulting in limited bioavailability to most ecological receptors (Alexander, 1995).

Chronic exposure to benzene may manifest initially as anaemia, leukopenia, thrombocytopenia, or a combination of these diseases. There is strong evidence from epidemiologic studies that high-level benzene exposure results in an increased risk of acute myelogenous leukaemia in humans. Evidence for increased risk of other cancers is less compelling. Toluene is well absorbed from the lungs and gastrointestinal tract, and it accumulates rapidly in the brain. Manifestations of exposure range from slight dizziness and headaches to unconsciousness, respiratory depression and death. Xylenes and ethylbenzene have a limited capacity to adversely affect organs other than the central nervous system. Mild, transient liver and/or kidney toxicity have been reported occasionally in humans (Klaassen and Watkins III, 2003).

4.5.2 Chlorinated organic compounds

The ubiquity of chlorinated hydrocarbons in the environment continues to be the basis for public and regulatory concern because of their recalcitrance and potential for bioaccumulation. Currently, more than 15,000 different chlorinated organic compounds are in commercial use (Hileman, 1993). Large volumes of chlorinated compounds are produced each year for a variety of industrial and commercial uses, including solvents, lubricants, intermediates in the chemical industry, plasticizers, crop protection, pharmaceuticals and medical equipment. Because of the large-scale industrial production and use of chlorinated compounds, these chemicals have a potential for introduction into the urban water cycle (Swindoll et al., 2000). More than 100 organic compounds are identified in the US EPA list of priority pollutants in water, including 27 chlorinated compounds, 14 PAHs, 9 PCBs and 17 pesticides and metabolites.

Chlorinated organic compounds are a heterogeneous chemical family with a wide range of physical and chemical properties. The environmental fate of chlorinated organic compounds is affected primarily by the environment where the chemical is released and by the physical and chemical properties of the chemical. For example, organic compounds denser than water will sink to the bottom of an aquifer or other body of water, compounds with high water solubility will dissolve into the water and compounds with a high Henry's law constant will preferentially partition into air rather than water.

The transport and partitioning of a chemical can effectively reduce the overall concentration of a contaminant, thereby reducing or increasing the potential for exposure. For example, dilution and dispersion can reduce the overall concentration of a chemical in aqueous environments. Several types of process, including oxidation–reduction reactions, sorption, volatilization, precipitation, hydrolysis and biodegradation, result in a change in a chemical's physical or chemical properties, which can reduce potential environmental risks by altering contaminant mobility, bioavailability or toxicity. As a result, the various chlorinated organic compounds differ in their toxicity, persistence, or bioaccumulation potential (Swindoll et al., 2000).

Soils and sediments often act as sinks for many of the highly chlorinated hydrocarbons. Numerous studies have reported chlorinated aromatics in sediments, freshwater and marine environments (Oliver and Nicol, 1982; Golden et al., 1993; Palm and Lammi, 1995). Conversely, the lower chlorinated congeners of chlorobenzene are more water-soluble, thus, their residence time in the water column is longer. Many chlorinated solvents have relatively low sorptive capacity and are soluble in water at concentrations higher than the US maximum contaminant levels (MCLs) for drinking-water standards.

Hydrolysis has been identified as an important abiotic mechanism responsible for the transformation of a range of chlorinated organic compounds (1,1,1-trichloroethane, and chlorinated methanes, ethanes, ethenes and propanes). However, many chlorinated organic compounds resist both chemical and biological attacks, which contribute to the environmental persistence of these materials. Some compounds, such as PCBs and tetrachloroethene (TCE), are produced specifically for their persistence and stability.

A chemical's hydrophobicity or tendency to partition into lipids is often used to estimate bioconcentration. Thus, many chlorinated organic compounds have been identified as being bioaccumulative chemicals because of their lipophilic nature; hence, the uptake and sequestration in body lipids reflects the hydrophobic characteristics of these compounds. Many of the chlorinated aromatic compounds are classified as having a moderate to high bioconcentration potential and therefore are of potential ecological and human-health concern. For example, polychlorinated biphenyls (PCBs), dioxins, furans and dichlorodiphenyltrichloroethane (DDT), tend to accumulate because they have a very low rate of biodegradation, are very toxic in nature and are bioaccumulative (Swindoll et al., 2000).

Some chlorinated organic compounds are associated with specific risks to health, because of their persistence and lipophilicity, or as result of their genotoxic or carcinogenic effects. For example, chloromethanes have a central anaesthetic effect, some chlorinated ethanes, chlorinated ethenes and propenes are reported to be carcinogenic (Henschler, 1994), and some chlorinated ethenes are genotoxic. While many chlorinated aromatic hydrocarbons, including hexachlorobenzene, PCBs and dioxins, have been classified as nongenotoxic carcinogens (Swindoll et al., 2000).

4.5.2.1 Volatile organic compounds

Volatile organic compounds (VOCs) are organic chemical compounds that have a high vapour pressure relative to their solubility in water. VOCs include components of gasoline, fuel oils, lubricants, organic solvents, fumigants, some adjuvant in pesticides, as well as some by-products of chlorine disinfection. They are a concern in groundwater because many are mobile, persistent and toxic. Because VOCs are used in numerous domestic, commercial and industrial products and applications, they can be released to groundwater through sources, such as sewage systems, leaking water mains landfills and dumping sites, leaking storage tanks and stormwater runoff.

VOCs frequently detected in aquifer samples in concentrations generally less than 1 μg/L include solvents (chloromethane, methylene chloride, 1,1,1-trichloroethane, trichlorethene and perchloroethene); trihalomethanes (bromodichloromethane and chloroform); and gasoline compounds (toluene, 1,2,4-trimethylbenzene and methyl tert-butyl ether) (Squillace and Moran, 2006). Examples of VOCs concentrations range detected in water sources of some developed cities are included in Section 4.6.4.

VOCs can be transported through the unsaturated zone during groundwater recharge, in soil vapour, or as non-aqueous-phase liquid. A shallow water table and abundant groundwater recharge would favour rapid transport through the unsaturated zone and increase the likelihood of VOCs reaching groundwater. In fact, VOCs as non-aqueous phase liquids could migrate to the water table by gravity without being transported during water recharge or any other transport mechanism.

The transport of VOCs dissolved in groundwater may be affected by sorption, advection and/or dispersion process. Sorption of VOCs to organic carbon in the soil may slow transport. It is dependent on the solubility of the VOC, the organic carbon content the density and porosity of soil. Some very soluble VOCs, such as MTBE, have a small sorption tendency, and consequently move as quickly as water does; whereas other less soluble VOCs, such as carbon tetrachloride, have a larger sorption tendency and move very slowly relative to the rate of groundwater flow (Wiedemeier et al., 1999).

VOCs that are persistent in water are more likely to be detected in groundwater because they can travel greater distances from their source before degradation and dilution occurs. In groundwater, VOCs undergo selective abiotic and biotic transformation. Hydrolysis is an example of abiotic transformation in water; for example, 1,1,1-trichloroethane (TCA) is the only commonly used chlorinated solvent that can be transformed to 1,1-dichloroethene and acetic acid by hydrolysis (Wiedemeier et al., 1999).

The transport, behaviour and fate of VOCs in streams depend on the effects of various chemical, physical and biological processes. The relative importance of each of these processes depends on the characteristics of the VOC and the stream. For example, volatilization is likely to be the dominant process affecting the concentrations of VOCs in streams. However, it is controlled by the mixing conditions within the water phase. Dry deposition of VOCs with particles is not expected to be important for most of the organic compounds commonly found in water. However, wet deposition of VOCs can occur with precipitation. Microbial degradation could be important in deep, low-velocity streams, with lesser importance in intermediate-depth, high-velocity streams; while microbial degradation in shallow, medium-velocity streams is not expected to be important, compared to volatilization (Rathbun, 2000).

Analysis of aquatic toxicity values indicates that the concentrations of VOCs necessary to cause acute adverse effects in aquatic organisms are not likely to be routinely present in streams. Though, spills could result in short-term exposures to high concentrations. The effect of long-term chronic exposure to low concentrations of VOCs is largely unknown. However, the persistence of VOCs in streams for long periods is unlikely because of the volatile nature of these compounds; while the concentration of some VOCs in sediments may be expected.

4.5.2.2 Solvents

Chlorinated solvents are used in a variety of commercial, industrial, manufacturing and domestic applications. They are also used in the aerospace and electronics industries; dry cleaning; manufacture of foam; paint removal/stripping; manufacture of pharmaceuticals; metal cleaning and degreasing; and wood manufacturing. Solvents also can be found in a variety of household consumer products including drain, oven and pipe cleaners; shoe polish; household degreasers; typewriter correction fluid; deodorizers; leather dyes; photographic supplies; tar remover; waxes; and pesticides (US EPA, 1980).

The most commonly used chlorinated solvents are methylene chloride, 1,1,1-trichloroethane (TCA), trichloroethylene, tetrachloroethylene and perchloroethene (PCE). Carbon tetrachloride was the first solvent produced in the United States and was the main solvent used for the first half of the twentieth century. Production of trichloroethene (TCE) and perchloroethene (PCE) began in the 1920s. After World War II, production and use of these two solvents in industry increased markedly, and they became the most commonly used solvents. Two other commonly used solvents during this period were methylene chloride and 1,1,1-trichloroethane. The production of methylene chloride, PCE, TCA and TCE peaked in the 1970s and has been declining since then due mostly to the human-health and environmental concerns associated with these compounds (Pankow and Cherry, 1996). Although production has been declining recently, large quantities of these solvents continue to be produced and used by many commercial and industrial sectors of society.

In general, chlorinated solvents have low water solubilities and high volatilities and densities relative to other VOCs. For the solvents, three types of abiotic transformation processes are important – hydrolysis, reduction and photolysis. Chlorinated solvents are often persistent in soil and water, leading to bioconcentration and groundwater pollution. For example, trichloroethane degrades in soil to form the more toxic chemical vinyl chloride (Lave and Upton, 1987). Many of the solvents have high water solubilities. This means that even small spills of some solvents can result in substantial groundwater contamination problems with respect to human health (Moran, 2006). Relative to other VOCs, Moran (2006) reported that solvents were ranked high in all data sets in terms of the frequency of concentrations higher than Maximum Contaminant Levels (MCLs). Mixtures have been a common mode of occurrence of solvents. The probability of occurrence of solvents in groundwater has also been strongly associated with urban land use and population density and with variables that represented the transport of solvents through the soil zone to groundwater.

The health effects of solvents can be different depending on the route of exposure. There are numerous exposure routes for solvents in the human body. For solvents in groundwater, exposure occurs mainly through drinking water containing solvents and through breathing solvents that have been transferred from water to air. Solvents can get into the air by volatilization from contaminated drinking water in showers or taps.

The effects of acute inhalation of methylene chloride in humans include decreased visual, auditory and psychomotor functions and effects on the central nervous system, but these effects are reversible once exposure ceases. The major effects from chronic inhalation exposure to methylene chloride in humans are effects on the central nervous system, headaches, dizziness, nausea and memory loss. No studies have indicated developmental or reproductive effects in humans from inhalation or oral exposure (ATSDR, 2000). According to the US Department of Health and Human Services (2005) methylene chloride is reasonably anticipated to be a human carcinogen, based on sufficient evidence of carcinogenicity in experimental animals.

The effects of acute inhalation of TCA in humans include mild hepatic effects, mild motor impairment (for example, increased reaction time), light-headedness, impaired balance, dizziness, nausea, vomiting, diarrhoea, loss of consciousness, decreased blood pressure and ataxia. Cardiac arrhythmia and respiratory arrest may result from the depression of the central nervous system (ATSDR, 2004). Most studies have not reported adverse effects from chronic oral or inhalation exposure to low levels of TCA in humans (ATSDR, 2004). Epidemiologic studies have found no relation between adverse pregnancy outcomes and exposure of mothers or fathers to TCA, and information is not available on the carcinogenic effects of TCA in humans.

The effects of acute inhalation of PCE in humans include irritation of the upper respiratory tract and eyes, kidney dysfunction, and, at lower concentrations, neurological effects, such as reversible mood and behavioural changes, impairment of coordination, dizziness, headaches, sleepiness and unconsciousness. The major non-cancer effects from chronic inhalation exposure to PCE in humans include neurological effects, such as sensory symptoms and headaches, impairments in cognitive and motor neurobehavioural functioning, colour vision decrements, cardiac arrhythmia, liver damage and possible kidney effects (ATSDR, 1997). The US EPA's Science Advisory Board placed PCE on a continuum between probable human carcinogen and possible human carcinogen.

TCE has been linked to potential human-health effects including anaemia, arthritis, cancer, birth defects and damage to the liver, kidneys, immune system and nervous system. Acute inhalation of TCE in humans primarily produces effects on the central nervous systems with symptoms, such as sleepiness, fatigue, headaches, confusion and feelings of euphoria. Effects on the liver, kidneys, gastrointestinal system and skin also have been noted. The major non-cancer effects from chronic inhalation exposure to TCE in humans include dizziness, headaches, sleepiness, nausea, confusion, blurred vision, facial numbness and weakness (ATSDR, 1997). Effects to the liver, kidneys and immune and endocrine systems have also been observed in humans exposed to TCE from contaminated drinking water (US EPA, 2001). Some studies have indicated reproductive or developmental effects from exposure to TCE (US EPA, 2000). An analysis of available epidemiological studies reports TCE exposure to be associated with several types of cancers in humans, especially kidney, liver, cervix and lymphatic system (US EPA, 2001).

4.5.2.3 Trihalomethanes

The use of chlorine in the treatment of drinking water has virtually eliminated water-borne microbial diseases, because of its ability to kill or inactivate essentially all enteric pathogenic microorganisms. Chlorine is the most convenient and easily controlled disinfectant; it is a strong oxidant, its residue can be maintained in the distribution system to prevent bacterial re-growth, although, the use of chlorine can lead to the formation of disinfection by-products (DBPs).

The DBPs most commonly found in drinking water are trihalomethanes (THMs), which include chloroform, bromodichloromethane (BDCM), dibromochloromethane (DBCM) and bromoform. Haloacetic acids (HAAs) are also disinfection by-products, which were detected in chlorinated drinking waters (Christman et al., 1983; Miller and Uden, 1983) nine years after the discovery of THMs (Bellar et al., 1974; Rook, 1974). Over 500 DBPs have been identified to date (Richardson, 1998); haloacetic acids (HAAs), after THMs, represent the second most abundant DBPs species in water.

The formation of DBPs is a complex process that occurs when chlorine reacts with naturally present organic matter. This process depends on naturally occurring organic precursor concentration, chlorine dose, contact time, water pH and temperature (seasonal variation), bromide ion concentration, water sampling location within the distribution system (contact time, spatial variation), and on the type of disinfection process (chlorination, chloramination, ozonation) used.

Levels of THMs, are generally higher in treated surface water than in treated groundwater because of the higher concentrations of precursor organic materials in lakes and river waters and especially because the rate of formation of disinfection by-products increases at higher temperatures in warmer months (LeBel et al., 1997). It must be noted that the formation of brominated by-products, such as BDCM, depends on the presence of bromine in the source water.

After drinking-water exposure, the secondary exposure pathways to THMs, particularly to chloroform and BDCM, are via inhalation and dermal absorption from tap water during showering and bathing. Moreover, most of the chloroform in indoor air is present as a result of volatilization from drinking water.

The occurrence and the fate of THMs in the water supply system of many cities of developed and developing countries have been widely assessed. The chlorination practice as a disinfection process and the natural bromide and dissolved organic carbon concentrations of groundwater resources varies from site to site. In some cases, the total THMs formation is always below WHO, European Union and US EPA drinking-water standards, as illustrated in the examples described in Sections 4.6 and 4.7.

THMs are of great concern to public health, owing to their potential reproductive, carcinogenic and mutagenic effects (IPCS, 2000; Nieuwenhuijsen et al., 2000). Numerous epidemiological studies have been conducted to investigate the correlation between chlorination of drinking water and cancer mortality (Flaten, 1992; Morris, 1995; Pilotto, 1995; Gallagher et al., 1998; Yang et al., 1998, 2000). The few analytic studies suggest a clear link between exposure to chlorinated drinking water and the development of urinary bladder cancer (McGeehin et al., 1993; Pilotto, 1995; Cantor et al., 1998; Yang et al., 1998). They also suggest a possible link with rectal cancer. Some studies have also demonstrated that exposure to chlorination by-products in water is related to spontaneous abortion and other adverse reproductive outcomes (Kramer et al., 1992; Bove et al., 1995; Waller et al., 1998, Health Canada, 2006).

It must be pointed out that the health risks from disinfection by-products, are much less than the risks from consuming water that has not been disinfected. Public utilities should make every effort to maintain concentrations of all disinfection by-products as low as reasonably achievable without compromising the effectiveness of disinfection.

4.5.3 Pesticides

Large quantities of pesticides, such as DDT, were released into rivers and coastal waters before these chemicals were banned (mostly in developed countries) in the 1970s, resulting in contaminated sediments with pesticides and their metabolites, and the food web pathways by which they may be transferred from sediments to higher trophic levels in food chain. Actually, the surface water of rivers and lakes inside cities or in its surroundings may be contaminated by pesticides concentrations above the allowable levels due to agriculture activities. In some cities, these pesticides could hardly be determined in drinking-water samples (wells) and their concentrations rarely exceeded detection limits (Hung and Thiemann, 2002). Although many of them were banned some time ago, they are still observed in developing countries (see examples in Section 4.7). Pesticides commonly found in rivers and lakes inside developing cities or in the surroundings are: alpha-HCH, beta-HCH, gamma-HCH o lindane, delta-HCH, 4,4'-DDE; 4,4'-DDD o TDE; 4,4'-DDT, aldrin, dieldrin, endrin and heptachlor (Hung and Thiemann, 2002).

It should not be surprising that pesticides are important risk agents, considering that they are designed to kill. Because of their extensive use and the manner in which they are applied, pesticides are found everywhere – in drinking water, food, air and soils. The main types of pesticides are insecticides and herbicides. Insecticides include organochlorines compounds (DDT, polychlorinated biphenyls (PCBs), lindane, aldrin/dieldrin, heptachlor, dioxin), organophosphates compounds (diazinon, malathion, parathion) and carbamates (aldicarb, carbaryl, methomyl).

Organochlorine compounds act as neurotoxins, that is, they tend to simulate the nervous system in insects and mammals, causing tremors and convulsions. Therefore, they have a greater potential for chronic toxicity to humans. Most organochlorine pesticides are now banned or restricted because of their persistence in the environment and their tendency to accumulate in the tissues of living organisms. However, some of them (for example DDT) are still applied in several developing countries, under legal or illegal circumstances. DDT poisoning affects the central nervous system function in humans, but major pathologic changes are observed in the liver and reproductive organs (Klaassen and Watkins III, 2003).

Today there are some 200 different organophosphates in the marketplace. Organophosphate pesticides break down rapidly and do not accumulate in tissues but are often extremely toxic and non-selective. Organophosphates operate by interfering with the nervous system in a complex way that ultimately affects the respiratory and circulatory systems by binding to and, thereby, blocking the activity of acetylcholinesterase, the enzyme responsible for the destruction and termination of the biological activity of the neurotransmitter acetylcholine, which controls the stimulus to nerve ends (Klaassen and Watkins III, 2003).

Carbamate pesticides, the most widely used pesticides, are similar to organophosphates in that they operate by inactivating acetylcholinesterase. However, the bond formed is not as stable, so the effect is relatively brief. Herbicides, such as 2,4-D and 2,4,5-T, atrazine, dicamba, paraquat and linuron, vary greatly in their selectivity, their persistence in tissues and the environment, and in their ability to be absorbed by plants.

4.5.4 Persistent organic pollutants

During the last half of the twentieth century, the global environment has become contaminated with a number of persistent, fat-soluble chemical contaminants, commonly referred to as persistent organic pollutants (POPs).

POPs include the organochlorine pesticides (e.g., DDT, chlordane and toxaphene), the polyhalogenated–biphenyls (PHBs; including polychlorinated biphenyl (PCBs)), dibenzo-p-dioxins (PHDDs; including polychlorinated dibenzo-p-dioxin (PCDDs)), dibenzofurans (PHDFs; including polychlorinated dibenzofuran (PCDDFs)), and polychlorinated naphthalenes (PCNs) (Ross and Birnbaum, 2001).

Their introduction into the environment through various processes, such as leakage, discharge, combustion, incineration and agricultural application, and their physicochemical properties have led to a contamination of aquatic food chains in particular. Aquatic food chains are vulnerable to contamination by POPs as a result of the lipophilic characteristics of these chemicals as well as their resistance to breakdown. As a result of biomagnification, organisms occupying high trophic levels are often exposed to high concentrations of such chemicals. Therefore, the primary sources of these contaminants to humans and top predators in the environment are food or prey items from the freshwater and marine environment food chain. Certain human consumer groups, including sports fishing families, immigrant communities and aboriginal populations, can be at increased risk, because of exposure to environmental contaminants through their consumption of fish and other aquatic foods (Dewailly et al., 1989, 1994, 2000; Jacobson and Jacobson, 1996).

The Inuit from the Canadian Arctic, for example, are exposed to high levels of dioxins through the consumption of food, exceeding the average daily intake by up to ten times (Kuhnlein et al., 1995). Their food pattern is an important part of their cultural heritage; the elimination of beluga whale skin and blubber alone was estimated to reduce dietary exposure by approximately 50%, because of the high degree of contamination of these particular lipid-rich products (Dewailly et al., 1996).

Many of the effects of the toxic POPs were first identified in fish-eating wildlife, but the precise mechanisms of action and chemicals involved were subsequently elucidated in carefully controlled experimental designs in laboratory animals. Increasing evidence of POP-related toxicities in humans exposed through their diets has been strengthened by evidence from wildlife. Certain wildlife species can therefore serve as 'sentinels' for human-health risks for complex mixtures of POPs.

The accumulating weight of evidence strongly implicates POPs, in incidents of endocrine and immune dysfunction, reproductive impairment, developmental abnormalities and neurological function in vertebrate species. Dioxin and related compounds are reproductive and developmental toxicants, immunotoxic, neurotoxic and carcinogenic in multiple species, including domestic and laboratory animals, wildlife and people. Acute human exposures are rare and often poorly documented, but several incidences of poisoning have occurred after accidental mixing of PCB/PCDD/PCDF. Symptoms have included weight loss, chloracne and immunotoxicity (Nakanishi et al., 1985; Takayama et al., 1991).

Chronic exposure to low or moderate levels represents a more insidious health risk to high trophic level organisms and humans, mainly because of the persistence and ubiquity of POPs in the global environment. An example is the fish consumption at the remote Lake Laberge in the Yukon Territory in northern Canada, where high POP concentrations in fish appear to be due to an influx of atmospheric pollutants and altered trophodynamic structure (AMAP, 1998). Much work remains to be done on the links between exposures, burden and effects, although laboratory animal models provide a basis for a mechanistic understanding in this area.

4.5.5 Emerging pollutants

A new list of compounds, such as pharmaceutical compounds, natural and synthetic hormones and polycyclic musk residues, have been recognized as one of the emerging issues in urban and industrial water pollution.

The earliest report on human hormones in water was published in 1965, showing that steroids were not completely eliminated during wastewater treatment (Stumm-Zollinger and Fair, 1965). While other reports demonstrating the presence of human hormones were published in the 1970s and 1980s, little attention was focused on these trace pollutants until their occurrence became linked to toxicological impacts in fish (Kramer et al., 1998; Renner, 1998; Snyder et al., 2001). The first reports on pharmaceuticals in wastewater effluents and surface waters were published in the United States in the 1970s (Tabak and Bunch, 1970; Hignite and Azarnoff, 1977). Similar to the steroid hormones, pharmaceuticals as environmental contaminants did not receive a great deal of attention until the link was established between a synthetic birth control pharmaceutical (ethynylestradiol) and impacts on fish (Purdom et al., 1994; Desbrow et al., 1998; Jobling et al., 1998).

The effect of these substances on the ecosystem and other animals is not yet well known. Some naturally occurring and man-made chemicals, including certain pharmaceuticals, pesticides, industrial chemicals, combustion by-products, phytoestrogens and hormones excreted by animals and humans are widely considered to be endocrine disruptors. The evidence of potential endocrine activity is limited, incomplete or controversial for many other chemicals. An even greater number of chemicals have not yet been tested for potential endocrine activity using any of the available methods. Additionally, therefore, any list of chemicals defined as endocrine disruptors is speculative (Kim et al., 2007). More research is needed to determine which chemicals are likely to persist in the environment, which of these may be harmful at the concentrations present in wastewater and what treatment techniques are most effective at removing them.

It is now well established that pharmaceuticals and natural and synthetic hormones are ubiquitous contaminants of wastewater effluents. In recent years, polycyclic musk compounds, frequently used as synthetic fragrances in perfumes, and various household products have also been recognized as important organic residues in the aquatic environment, especially, in urbanized areas. These persistent contaminants are primarily discharged via municipal sewage effluents into the receiving surface waters. Most often, all these compounds have been found at concentrations up to the μg/L-level in sewage influent and effluent samples and also in several surface waters located downstream from municipal sewage treatment plants.

4.6 CHEMICAL RISKS IN URBAN CITIES IN DEVELOPED COUNTRIES

The following examples illustrate the compounds most frequently reported in researches published in scientific journals and performed in cities of developed countries: fluoride, arsenic, mercury, volatile organic compounds (VOCs), trihalomethanes (THMs) and new chemicals. Some examples reflect national water quality status and the concern on monitoring, analysis and identification of compounds at risk level. Others focus on chronic diseases associate to natural releases to water resources.

4.6.1 Fluoride

The following examples address naturally occurring fluoride in drinking water in levels greater than WHO guideline value of 1.5 mg/L (WHO, 2004). A detailed description of these and other cases, as well as specific references of original researches, may be found in *Fluoride in Drinking-water* (Bailey et al., 2006).

4.6.1.1 China

In China, fluorosis results from consumption of drinking water containing elevated fluoride levels, pollution caused by burning fluoride-rich coal and high levels of consumption of brick tea. Endemic fluorosis is prevalent in China, occurring in 29 provinces, municipalities and autonomous regions. High fluoride levels in drinking water are seen mainly in the arid and semi-arid regions, such as Helongjiang, Jilin, Liaoning, Inner Mongolia, Hebei, Henan, Shanxi, Shaanxi, Ningxia, Gansu and Xinjiang provinces or autonomous regions. Fluoride levels in drinking water can be high, exceeding 3 mg/L. For example, in shallow well water in Pei county (Jiangsu province) levels in excess of 13 mg/L have been reported, while hot spring water has been shown to have fluoride levels up to 17 mg/L.

The main type of endemic fluorosis in China results from high levels of fluoride in drinking water. It has been estimated that over 26 million people in China suffer from dental fluorosis due to elevated fluoride in their drinking water, with a further 16.5 million cases of dental fluorosis resulting from coal smoke pollution. It has been demonstrated that people living in high fluoride drinking-water regions (>4 mg/L) and consuming a nutrient deficient diet have the highest incidence of dental and skeletal fluorosis. Over one million cases of skeletal fluorosis are thought to be attributable to drinking water, with a further million cases due to coal smoke pollution.

4.6.1.2 Japan

Dental fluorosis was observed in children of the Ikeno district in the Aichi prefecture in the 1970s, following the unintentional supply of drinking water containing up to

7.8 mg/L fluorides since 1960. On discovery of the high fluoride levels, an alternative supply with a fluoride level of 0.2 mg/L was provided. A total of 1,060, 10–12 year-old, lifetime residents were examined to estimate the prevalence of dental fluorosis in Japanese communities exposed to naturally occurring fluoride up to 1.4 mg/L. The prevalence of dental fluorosis was found to increase as fluoride levels increased, ranging from 1.7% at 0.2–0.4 mg/L up to 15.4% in the group exposed to 1.1–1.4 mg/L.

4.6.1.3 United States of America

Historically, dental fluorosis was quite widespread in the US. Originally, the problem was termed 'mottled enamel' or, local to Colorado Springs, as 'Colorado brown strain'. In 1930, the link was made between mottled enamel and high levels of fluoride in drinking-water supplies (2.0–13.7 mg/L) and the term fluorosis was adopted. Deans Index was developed as a system for estimating the severity of dental fluorosis (Dean, 1933). Arizona, Arkansas, California, Colorado, Idaho, Illinois, Iowa, Kansas, Minnesota, Nevada, New Mexico, North Carolina, North Dakota, Oklahoma, Oregon, South Carolina, South Dakota, Texas, Utah and Virginia were reported to have areas with endemic fluorosis. These studies also led to the realization of the link between low levels of fluoride and high prevalence of caries, which culminated in the recommendation that levels of fluoride in water be adjusted to between 0.7 and 1.2 mg/L and the adoption of widespread fluoridation in places where water and food characteristics lead to a low F consumption. Fifty years later, it was observed that more than 700 communities in the US were thought to have water supplies that contained at least twice the recommended optimum level of fluoride (i.e. 2.4 mg/L and above). It was found, for example, that the mean fluoride concentrations in Illinois were between 1.06 and 4.07 mg/L. Moreover, at the highest fluoride concentration, only 5.2% of children were considered to have normal teeth or questionable mottling.

4.6.2 Arsenic

Few examples of exposure to arsenic are found in developed cities. However, Canada, China and US cases represent a good sample of groundwater supply with arsenic concentrations greater than WHO guideline value of 10 µg/L (WHO, 2004). A detailed description of these and other cases, as well as the original references, may found in the *United Nations Synthesis Report on Arsenic in Drinking Water* (WHO, 2001) or may be consulted at the website www.who.int/water_sanitation_health/dwq/arsenic3/en/index.html.

4.6.2.1 Canada

Water quality monitoring data obtained from six Canadian Provinces, over a four-year period (1985–1988), have been compiled and analyzed. It was found that the percentage of drinking-water samples having total arsenic concentrations greater than 5 µg/L varied from 0 to 32%. However, most of the high arsenic concentrations were identified in samples from groundwater sources in Nova Scotia. In areas dominated by naturally occurring high arsenic-containing ores, or where gold mining had previously occurred, arsenic concentrations in drinking-water supplies of between 150 and up to 500 µg/L were reported.

4.6.2.2 China

Arsenic has been found at high concentrations (in excess of the Chinese national standard of 50 µg/L) in groundwater from Inner Mongolia as well as Xinjiang and Shanxi Provinces. The first cases of arsenic poisoning were recognized in Xinjiang Province in the early 1980s when arsenic concentrations in groundwater were found at up to 1,200 µg/L. Ten years later arsenic concentrations were found between 40 µg/L and 750 µg/L in deep artesian groundwater from the Dzungaria Basin on the north side of the Tianshan Mountains (from Aibi Lake in the west to Mamas River in the east).

4.6.2.3 United States of America

A review of water quality monitoring data collected during the period 1976–1993 revealed that concentrations of arsenic in drinking waters in the US lie between less than 2.5 and 28 µg/L in surface waters, and between less than 5 to 48 µg/L in ground-water sources. Based on these data, it was estimated that approximately 2% of the US population is exposed to drinking water containing more than 10 µg/L of arsenic.

Many areas have been identified in the US with arsenic problems in groundwater. Most of the worst affected and best-documented cases occur in the south-western states (Nevada, California and Arizona). However, within the last decade, aquifers in Maine, Michigan, Minnesota, South Dakota, Oklahoma and Wisconsin have been found with concentrations of arsenic exceeding 10 µg/L and smaller areas of high arsenic ground-water have been found in many other states. For example, in Iowa, Missouri and Ohio, arsenic, apparently of natural origin, was found in groundwater at concentrations between 34 and 490 µg/L. The city of Fallon, Nevada (population 8,000) is an example of groundwater supply with 100 µg/L of arsenic concentration, which for many years has been supplied without treatment other than chlorination. In groundwater from the Tulare Basin of the San Joaquin Valley, California, a large range of groundwater arsenic concentrations from less than 1 µg/L to 2,600 µg/L have been found.

Arsenic contamination from mining activities has been also identified in numerous areas of the US. Groundwater from some areas has been locally reported to have very high arsenic concentrations (up to 48,000 µg/L).

4.6.3 Mercury

Brief information is presented to show the risks of drinking water, the fish consumption advisories resulting from mercury discharges to surface waters as well as the identification of the vulnerable population in US. The other examples are regarding contaminated water–fish consumption pathways in other countries.

4.6.3.1 Canada Arctic

Beluga whales have been hunted for food by native people in the Canadian Arctic since prehistoric time. Mercury accumulates in the liver of the whales over time so that the whale ages are usually linked statistically to their levels of mercury in liver. Lockhart et al. (2005) reported that the levels of total mercury in the most analyzed whale organs fell in the order of liver (highest levels), kidney, muscle and muktuk (lowest level) from samples collected over the period 1981–2002. They found that virtually all the samples of 566 animals analyzed contained mercury in the liver at concentrations higher than the Canadian consumption guideline of 0.5 µg/g (wet weight) for fish. However, while muktuk had the lowest level of the organs most frequently analyzed, it is the preferred food item from these whales and it still exceeded the consumption guideline in most instances. They also reported that samples from locations in the Mackenzie Delta in the western Canadian Arctic and from Pangnirtung in the eastern Canadian Arctic were obtained more often than from other location. Previous analyses of data from geographically distinct groups had suggested that whales in the western Canadian Arctic had higher levels of mercury than those from the eastern Canadian Arctic. However, the results of their study suggest that such regional differences have diminished and are no longer statistically significant.

4.6.3.2 China

An assessment of exposure to mercury performed in Changchun City, China (Zhibo et al., 2006) is an example of the impact of mercury emissions in the local water cycle. Changchun City is located in the centre of north-east China and has an annual average temperature of 4.9°C. Therefore, heating systems are used from November to March, with coal as the main energy source for heating. Control of contaminant diffusion is difficult because most boilers lack desulphurization equipment and most of the mercury in coal is emitted to the atmosphere during coal combustion. It then circulates among various environmental compartments. Zhibo et al. (2006) found that diet was the most important path, with fish and shellfish consumption being the important exposure route. The range of concentrations in fish and shellfish was 0.010–0.126 µg/g (mean = 0.041 µg/g fresh wt). The range of mercury concentrations in water sources of drinking water reported was 0.002 to 0.085 µg/L with a mean value of 0.018 µg/L. They estimated adult intake of all mercury species via all routes of 6.78 µg/day (excluding dental amalgam). Therefore, the dietary intake of Hg in Changchun City was higher than in some other areas (e.g., Australia, Finland and Sweden), but it was lower than in polluted areas, such as Brazil and Japan.

4.6.3.3 Japan

Minamata disease: Minamata disease (methylmercury poisoning) was first discovered in 1956 around Minamata Bay, Kumamoto Prefecture, Japan. A second epidemic in Japan occurred in 1965 along the Agano River, Niigata Prefecture (Eto, 2000).

From 1932 to 1968, Chisso Corporation, a company located in Kumamoto Japan, dumped an estimated 27 tons of mercury compounds into Minamata Bay. Kumamoto is a small town about 570 miles south-west of Tokyo. The town consists of mostly farmers and fishermen. When Chisso Corporation dumped this massive amount of mercury into the bay, thousands of people, whose normal diet included fish from the bay, unexpectedly developed symptoms of methylmercury poisoning. The illness became known as 'Minamata Disease'. Victims were diagnosed as having a degeneration of their nervous system. Numbness occurred in their limbs and lips. Their speech became slurred and their vision constricted. Some people had serious brain damage, while others lapsed into unconsciousness or suffered from involuntary movements. Finally, in July 1959, researchers from Kumamoto University concluded that organic mercury was the cause of 'Minamata Disease'. By 1974, only 798 victims had been officially recognized as having 'Minamata Disease'. Approximately 3,000 more people were waiting verification from the board of physicians in Kumamoto Prefecture. In 1993, almost forty years later, victims were still being compensated for damages.

Mercury in cetacean products: The concentrations of total mercury were determined in 61 whale meat products (bacon, blubber, red meat, liver, intestine and tongue) purchased from retail outlets across Japan by Simmonds et al. (2002). They found that mean (range) concentration in all samples were total mercury 4.17 µg/g (0.01–204 µg/g). They concluded that for sensitive consumers and those with high-level consumption (e.g., whaling communities), exposure to mercury from certain whale blubber and bacon, and striped dolphin liver products could lead to chronic health effects. In addition, they recommended that the Japanese community should exercise a precautionary approach to the consumption of such foods in excess, particularly by high-risk members of the population.

Cetacean products sold for human consumption in Japan originate from a wide range of whale, dolphin and porpoise species caught off several areas of the Japanese coast, and Antarctic and North Pacific Oceans. Endo et al. (2003) surveyed the total mercury (T-Hg) levels in red meat from the most popular cetacean products in Japan. The red meats originating from nine species of odontocete and six species of mystecete were sold in Japanese markets. T-Hg concentrations in all odontocete red meats (0.52–81.0 µg/wet g, n = 137) exceeded the provisional permitted level of T-Hg in marine foods set by the Japanese government (0.4 µg/wet g). The highest and second highest levels of T-Hg in the red meats were found in the false killer whale (81.0 µg/wet g) and striped dolphin (63.4 µg/wet g), respectively. The result of the study is that these concentrations of T-Hg exceeded the permitted level of T-Hg by about 200 and 160 times, respectively, suggesting the possibility of chronic intoxication by methylmercury, due to frequent consumption of odontocete red meats. The conclusion of the study is that the higher concentration levels in odontocete species caught off southern areas than in those caught off northern areas are probably reflecting a higher mercury concentration in the seawater and/or their diet (squid and fish) in the southern area.

4.6.3.4 United States of America

Mercury discharges to surface waters from abandoned gold and mercury mines in the western United States are believed to be the cause of fish advisories for methylmercury in a number of streams and lakes. An example is the contamination of Clear Lake in California by the Sulphur Mercury Mine Superfund Site. Releases of methylmercury from sediments have not been well quantified, but high concentrations of methylmercury in sediments often coincide with high concentrations of methylmercury in fish tissue and result in mercury fish consumption advisories (US EPA, 1999). Thirty-nine states have some form of mercury fish advisory for their water bodies. State-wide advisories for mercury occur consistently across the north-eastern states and Gulf Coast states also have advisories in all coastal waters.

Based on its analysis of dietary surveys, US EPA concluded that typical fish consumers in the United States are not in danger of ingesting harmful levels of methylmercury. This is a result of the relatively low levels of fish consumed by the typical US citizen. However, US EPA's risk assessment also concluded that between one and three per cent of women of childbearing age (e.g., between the ages of 15 and 44) eat sufficient amounts of fish for their foetuses to be at risk from methylmercury exposure, depending on the methylmercury concentrations in the fish. Also, based on the same analysis of United States dietary survey data from the mid-1990s, US EPA estimated the percentage of people from different sub-populations who consume methylmercury in excess of the Reference Dose (RfD). Among white/non-Hispanic sub-populations the fraction above the RfD was 9.0%, among black/non-Hispanics, 12.7%, and among people of Asian/Pacific Islander ethnicity, Native American tribal members and non-Mexican Hispanics (e.g., people from Puerto Rico and other Caribbean islands) 16.6%. Among women of childbearing age (i.e. 15 through 44 years) 7% of the more than 58 million women in the group (i.e. more than 4 million women) are exposed to methylmercury from fish at levels in excess of the RfD, using month-long exposures as the basis for calculation (US EPA, 1999).

4.6.4 Volatile organic compounds

Volatile organic compounds (VOCs) in water resources and in drinking-water samples have been evaluated in many developed cities. The following examples show that such evaluations require analytical resources and detailed sampling programmes over time to find out range concentrations and detection frequencies for these compounds, mostly detected at low concentrations.

4.6.4.1 Netherlands

In the drinking water of three cities, of the ten cities of The Netherlands that were evaluated, trichloroethane, trichloroethene and tetrachloroethene were detected (Kool and van Kreyl, 1998). One conclusion in the study is that in these water-catchment's areas organic pollutants have reached groundwater. Another survey carried out in The Netherlands (2001) investigating the presence of methyl tert-butyl ether (MTBE) in drinking water found concentrations ranged between 10 ng/L and 420 ng/L. However, at one location MTBE was found at a level of 2,900 ng/L caused by point source contamination of the groundwater (11,900 ng/L).

4.6.4.2 United States of America

The US Geological Survey's National Water-Quality Assessment Program recently completed a national assessment of volatile organic compounds (VOCs) in groundwater (Moran et al., 2006; Zogorski et al., 2006). Water samples from 3,497 domestic and public wells were analyzed for 55 VOCs. Eight of the 55 VOCs had concentrations greater than the Maximum Contaminant Levels (MCLs), in 45 well samples (1% of all well samples): trichloroethene (TCE), ethylene dibromide, methylene chloride, dibromochloropropane, perchloroethene (PCE), 1,1-dichloroethene, 1,2-dichloropropane and vinyl chloride; these concentrations may be of potential human-health concern if the water were to be ingested without treatment for many years. Most of the volatile compounds found are carcinogenic and the liver is the major target organ.

In California, the most frequently detected VOCs in sampled drinking-water sources (groundwater) during a 7-year period (1995–2001) were chloroform, PCE and TCE (Williams et al., 2002). However, after adjusting for the occurrence of each VOC in drinking water, chloroform and PCE were found to pose the greatest relative cancer risk. Williams et al., also explain, 'Despite media reports about significant MTBE contamination of drinking-water supplies in California, MTBE detections were infrequent, and this chemical was found to pose the least cancer risk relative to the other VOCs'.

Concerning MTBE, Moran et al. (2005) found that only 13 groundwater samples from all study types (0.3%), had concentrations of MTBE that exceeded the lower limit of the US EPA Drinking-Water Advisory. The detection frequency of MTBE was highest in monitoring wells located in urban areas and in public supply wells. Groundwater in areas with high population density, in areas where MTBE is used as a gasoline oxygenate, and in areas with high recharge rates had a greater probability of MTBE occurrence.

4.6.5 Trihalomethanes

The total trihalomethanes (THMs) concentrations range in the following chlorinated drinking-water examples complies in some cases with WHO guidelines value and with the European Commission rule set at 100 μg/L. Even though in some cases, the mean and maximum concentrations are close to the US EPA standard of 80 μg/L.

4.6.5.1 Alaska

In Alaska, most of the population is concentrated in three cities with the remainder spread between approximately 225 towns and villages. Shallow lakes and streams are the sole source of drinking water, since more than half of Alaska land is on continuous or discontinuous permafrost (i.e. permanently frozen ground). Natural organic matter (NOM) is ubiquitous in northern surface waters making the water difficult to treat and disinfect. Consequently, higher total trihalomethanes (TTHMs) concentrations ranged from 37 to 1,050 μg/L and haloacetic acids (HAAs) from 37 to 1,200 μg/L were found (White et al., 2003) in 17 different drinking-water systems. Both the median and the mean TTHMs and HAAs values would exceed levels set by the Stage 1 D/DBP Rule (80 and 60 μg/L respectively) of the Safe Drinking Water Act. The results also showed that NOM contributing to disinfection by-products (DBPs) were primarily phenolic compounds, a finding consistent with previous studies.

4.6.5.2 Canada

The majority of drinking-water treatment plants in Canada use some form of chlorine to disinfect drinking water. Concentrations of THMs have been determined in drinking-water supplies at a considerable number of locations across Canada (Water Quality Issues Sub-Group, 2003). Eight provinces provided THMs data (1994–2000) for just over 1,200 water systems serving a sampled population of over 15 million Canadians with a mean THMs level of about 66 μg/L in drinking-water samples from all systems. Some systems had average values in the 400 μg/L ranges, and some systems had maximum or peak values in the 800 μg/L range.

Results from Health Canada studies (Williams et al., 1995, 1997), including a national survey of disinfection by-products (DBPs) in Canadian drinking water

(53 systems) and a 1-year monthly survey of three systems using different disinfection processes, indicated that THMs and HAAs were the major DBPs found in all facilities for all treatment processes including chlorine disinfectant, and that HAAs levels often equalled or exceeded THMs concentrations.

Recent data indicate that, in general, average trihalomethanes levels in Canadian drinking-water supplies are well below the guideline. However, some systems show average levels well above the guidelines; these systems serve only a small proportion of Canadians (less than 4%) and are generally smaller treatment systems with limited ability to remove organic matter before adding the chlorine disinfectant (Health Canada, 2006).

4.6.5.3 United Kingdom

A study was conducted in the UK to establish whether total THMs is a satisfactory indicator for other DBPs. The study was conducted in three water regions covering a wide range of water types and thus, qualities. Final disinfection at water treatment plants invariably involves the use of chlorine. Tap water samples in each region were obtained for analysis of THMs and HAAs (haloacetic acids) as well as temperature, pH, and free and total chlorine. They found that there are ranges of HAAs levels in drinking water with the means ranging from 35–95 µg/L and a maximum concentration of 244 µg/L, concentrations in excess of the current US standard of 60 µg/L for HAAs (US EPA, 1998). For THMs levels in drinking water, they found the means ranging from 27 to 51 µg/L and a maximum concentration of 76 µg/L (Toledano et al., 2005). The ratio between total THMs and total HAAs levels was significantly correlated with the other measured parameters indicating that these are important determinants in the formation of HAAs and THMs, but may affect the formation differently. Toledano et al. (2005) concluded that total THMs levels could not be assumed to be a good indicator for HAAs levels and possibly for other DBPs.

4.6.5.4 United States of America

As part of the US Geological Survey's National Water-Quality Assessment (NAWQA) Program, samples of untreated groundwater from drinking-water supply wells (1,096 public and 2,400 domestic wells) were analyzed for THMs and other volatile organic compounds (VOCs) during 1986–2001, or compiled from other studies. The report (Ivahnenko and Zogosrski, 2006) indicates that chloroform was the most frequently detected in public and domestic wells samples collected. Statistical analyses indicate that population density, percentage of urban land and the number of Resource Conservation and Recovery Act hazardous-waste facilities near the sampled wells have a significant association with the detection of chloroform. The other THMs (bromodichloromethane, dibromochloromethane and bromoform) were among the 10 most frequently detected VOCs in the NAWQA data sets for public well samples. Although chloroform was detected frequently in public and domestic well samples and other THMs were present in some samples, no total THM concentrations in samples from either well type exceeded the US Environmental Protection Agency's Maximum Contaminant Level of 80 µg/L.

4.6.6 Emerging pollutants

There is evidence that some pharmaceutical compounds originating from human therapy are not completely eliminated during wastewater treatment and are, thus, discharged as contaminants into the receiving waters (Heberer, 2002). In some investigations carried out in Austria, Brazil, Canada, Croatia, England, Germany (Heberer et al., 2002), Greece, Italy, Spain, Switzerland, The Netherlands and the US (Kolpin et al., 2002), more than 80 compounds, pharmaceuticals and several drug metabolites, were detected in the aquatic environment. Positive findings of pharmaceutical compounds have also been reported in groundwater contaminated by landfill leaching or manufacturing residues. However, to date, only in a few cases have pharmaceutical compounds been detected at trace-levels in drinking-water samples.

A German study, conducted by Heberer et al. (1997), found that two pharmaceutically active compounds, the drug metabolite clofibric acid (2-(4)-chlorophenoxy-2-methyl propionic acid) and N-(phenylsulfonyl)-sarcosine, were detected as organic contaminants in drinking-water samples from Berlin with the maximum concentrations of 270 ng/L for clofibric acid and 105 ng/L for N-(phenylsulfonyl)-sarcosine. They concluded that both polar contaminants are not eliminated by the drinking-water treatment used by the Berlin waterworks. Furthermore, they advise that the occurrence of clofibric acid and N-(phenylsulfonyl)-sarcosine in environmental water samples is not just a local phenomenon, since there are positive findings in screening analyses of surface water samples from the river Danube in Germany and the river Po in Italy.

The results of another German study (Adler et al., 2001), indicate that in raw sewage the total concentrations of natural steroid hormones (estradiol and estrone) and the synthetic oestrogen ethinylestradiol vary between 1.5 to 13 ng/L. However, in surface waters, the median concentrations of the unconjugated analytes were generally below the detection limit (0.05 to 0.5 ng/L). In groundwater, only the median for the synthetic oestrogen ethinylestradiol was above the detection limit. In raw water, only the median of estrone was above the detection limit. In the drinking-water samples, the medians of all steroids were below the detection limit. An important conclusion of the research is that the measured ranges do not support the assumption of harmful effects to aquatic organisms or human health.

Other types of hormone have been detected in the aquatic environment in The Netherlands. The occurrence of four oestrogenic hormones (17 beta-estradiol, 17 alpha-estradiol, estrone and 17 alpha-ethinylestradiol) could be detected at low concentrations, up to 6 ng/L, in surface water and in effluents of wastewater treatment plants (Belfroid et al., 1999).

To provide a national occurrence of pharmaceutical, hormones and other organic wastewater contaminants in surface water resources in the US, the US Geological Survey performed a study of water samples from more than 135 streams across 30 states in the country (1999–2000). The most frequently detected compounds were coprostanol (fecal steroid), cholesterol (plant and animal steroid), N,N-diethyltoluamide (insect repellent), caffeine (stimulant), triclosan (antimicrobial disinfectant), tri(2-chloroethyl)phosphate (fire retardant) and 4-nonylphenol (non-ionic detergent metabolite) (Kolpin et al., 2002). The measured concentrations were generally low and rarely exceeded drinking-water guidelines, drinking-water health advisories, or aquatic-life criteria. Many compounds, however, do not have such guidelines established.

4.7 CHEMICAL RISKS IN URBAN CITIES IN DEVELOPING COUNTRIES

To illustrate the chemical health risks to which the population in cities from developing countries is exposed, the following examples were chosen. They focus on the hundreds and even thousands of people with some type of disease because of the exposure to inorganic compounds naturally found in water resources. They also address the diversity of diseases a chemical agent may cause, the severity and endemic characteristics that some diseases have, and the poverty, the low sanitation level achieved, and the poor regulation conditions under which such exposures occur.

4.7.1 Fluoride

Endemic fluorosis diseases have been reported in 20 developing countries. Almost 50% of them are in Africa, where, in four countries that share land in the Rift Valley (Tanzania, Uganda, Kenya and Ethiopia) in the East part of Africa, it is a public health problem. The other 25% are in Asia and 15% are in Latin America and in the Middle East, respectively. A brief description of these examples is presented in this section. A detailed description of these and the other cases (Eritrea, Niger, Senegal, South Africa and Sudan); as well as specific references to original researches, may be found in *Fluoride in Drinking-water* (Bailey et al., 2006).

4.7.1.1 Brazil

In Brazil, elevated fluoride levels in drinking-water supplies have been observed. In Paraiba State, which is located in the north-east region, for example, well-derived drinking-water supplies contain 0.1–2.3 mg/L fluoride. Although the concentration varies according to the time of the year and depends on rainfall levels, in Olho D'Agua (Ceara State), levels of fluoride in drinking-water supplies of 2–3 mg/L have been recorded. Over 62% of children examined in Olho D'Agua had fluorosis scores of 3 or more, based on the Thylstrup–Fejerkov Index.

4.7.1.2 Ethiopia

In the Rift Valley, dental fluorosis was reported (1979) as 99% of 239, 6–7 year-old children examined living in Wonji and Awassa with a fluoride concentration of 12.4 and 3.5 mg/L, respectively. A study conducted ten years later (1987) found dental fluorosis in more than 80% of sampled children resident in the Rift Valley since birth (1,221 out of 1,456). The maximum prevalence was seen in the 10–14 year-old age group and 32% of the children showed severe dental mottling. Males were affected more than females.

Skeletal fluorosis was first reported in Ethiopia in 1973 in the Wonji-Shoa sugar estates in the Ethiopian Rift Valley. Three areas, Wonji-Shoa, Alemtena and Samiberta, have been identified as having cases of skeletal fluorosis. The highest incidence was found at the Wonji-Shoa sugar estates, where a linear relationship was observed between the development of crippling fluorosis, fluoride concentration in drinking-water supplies and period of exposure. The first cases of skeletal fluorosis appeared among workers (98% males) in the estates who had been consuming water with fluoride content of more than 8 mg/L for over 10 years. Between 1976 and 1984, 530 workers were retired from Wonji-Shoa at the age of 45–50 years because of the inability to perform their physically strenuous jobs. About 46% of these workers were found to have skeletal fluorosis. In August 1984, radiological evidence of skeletal fluorosis was found in 65% of the 300 persons from Wonji-Shoa examined and 30 (10%) had crippling fluorosis.

4.7.1.3 India

At least 17 (out of 32) states are affected by elevated fluoride levels in drinking water, namely: Andhra Pradesh, Assam, Bihar, Delhi, Gujarat, Haryana, Jammu and Kashmir, Karnataka, Kerala, Madhya Pradesh, Maharashtra, Orissa, Punjab, Rajasthan, Tamil Nadu, Uttar Pradesh and West Bengal. They have been progressively identified since the first report in 1937, with Assam being the most recently identified state. It has been estimated that the total population consuming drinking water containing elevated levels of fluoride is over 66 million. In Rajasthan, for example, fluoride concentrations have been found to vary between 0.6 mg/L and 69.7 mg/L. In Haryana, the highest fluoride concentration was found in the village of Karoli and it was recorded at 48 mg/L.

The prevalence of dental fluorosis was investigated in 15 tribal villages in Rajasthan. At mean fluoride concentrations of 1.4 and 6 mg/L, dental fluorosis was seen in 25.6% and 84.4% of schoolchildren (<16 years) and 23.9% and 96.9% of adults respectively.

Endemic skeletal fluorosis was reported in India in the 1930s. It was observed first in Andhra Pradesh bullocks used for ploughing, when farmers noticed the bullocks' inability to walk, apparently due to painful and stiff joints. Several years later, the same disease was observed in humans. For example, the prevalence of skeletal fluorosis in Rajasthan was examined in adults exposed to mean fluoride levels of 1.4 and 6 mg/L. At 1.4 mg/L over 4% of adults were reported to be affected, while at 6 mg/L, 63% of adults were reported to be affected. A recent (2003) study conducted in Andhra Pradesh, found skeletal fluorosis affecting just between 0.2 and 1% of the population examined, where the maximum drinking water fluoride concentration was 2.1 mg/L.

4.7.1.4 Kenya

The highest concentrations of fluoride in groundwater are reported to occur in the peri-urban areas of Nairobi, in the Rift Valley around Nakuru, Naivasha and Mount Kenya, and near the northern frontier. Local pockets of intermediate concentrations of 2–20 mg/L have been reported throughout the country. In a study of over 1,000 ground-water samples taken nationally, over 60% exceeded 1 mg/L, 20% exceeded 5 mg/L and 12% exceeded 8 mg/L. The volcanic areas of the Nairobi Rift Valley and Central Provinces had the highest concentrations, with maximum groundwater fluoride concentrations reaching 50 mg/L. The excess fluoride in surface water occurs in Lakes Rudolph (12 mg/L), Hannington (1,100 mg/L), Baringo (6 mg/L), Nakuru (2,400 mg/L), Magadi and elsewhere. Furthermore, concentrations of fluoride up to 1,640 mg/L and 2,800 mg/L have been reported in lakes Elmentaita and Nukuru respectively.

Dental fluorosis is reported in the Northern Frontier (Turkana), north-west Kenya, South Rift Valley and Central and Eastern Regions. A survey of 1,307 Asian and African schoolchildren found that 67% of Asian children and 47% of African children were affected. The degree of dental fluorosis was reported to be more severe among Asian children, and it was speculated that this could have been related to their vegetarian diet. All of the 110 children living in an area of Kenya with 2 mg/L fluoride in the water exhibited dental fluorosis. Altitude may have an effect on the level of dental fluorosis. In low fluoride drinking-water zones (<0.5 mg/L) 36% of the children at sea level had dental fluorosis, compared with 78% at 1,500 m and 100% at 2,400 m. In higher fluoride zones (0.5–1.0 mg/L), 71% had dental fluorosis at sea level as compared to almost 94% at 1,500 m.

4.7.1.5 Mexico

Fluorosis is considered a largely unrecognized environmental health problem in Mexico. The exposure to elevated levels of fluoride began in the late 1960s. Díaz-Barriga et al. (1997) estimated that approximately 5 million people might be exposed to elevated levels of fluoride in their drinking-water supplies. Mean fluoride concentrations in urban locations ranged from 1.5 to 2.8 mg/L, although individual sources were recorded as having concentrations up to 7.8 mg/L (Hermosillo in Sonora State). In rural locations, a similar pattern occurred with mean levels between 0.9 and 4.5 mg/L and the highest recorded concentration of 8 mg/L (Abasolo in Guanajuato State). States with elevated fluoride levels included Aguascalientes, Chihuahua, Coahuila, Durango, Guanajuato, San Luis Potosi and Sonora. In the city of Durango, it has been estimated that almost 95% of the residents were exposed to fluoride concentrations in drinking water greater than 2 mg/L. In the city of San Luis Potosi, for example, 98% of children exposed to fluoride drinking-water concentrations of greater than 2 mg/L were reported to exhibit signs of dental fluorosis. Bottled juice and bottled water containing high levels of fluoride were also found to contribute to the levels of dental fluorosis in San Luis Potosi, and help explain the high levels of fluorosis in children not exposed to elevated levels in their drinking-water supply.

4.7.1.6 Saudi Arabia

In the Hail region, over 90% of 2,355 rural children examined and aged 12–15 years were reported to show dental fluorosis, and a strong association was seen between fluoride level (0.5–2.8 mg/L) in well water used for drinking and the severity of dental fluorosis. Mecca (with a fluoride concentration up to 2.5 mg/L) was also reported to be an urban area with endemic fluorosis.

4.7.1.7 South Africa

There are 803 endemic fluorosis areas in South Africa, mostly the north-western, western and Karoo Regions of Cape Province, Western and Central Free State, and northern, eastern and western areas of Transvaal. For example, high fluoride groundwater is found inside the Pilanesberg Alkaline Igneous Complex (mean 3.7 mg/L) and very high fluoride concentrations (mean 57 mg/L) are found around the perimeter. High fluoride concentrations in groundwater are also found in the Nebo Granite and the mineralized Lebowa Granite. It is suggested that the cause of most high fluoride concentrations in groundwater is the dissolution of fluoride-bearing minerals in bedrock and soil. In the Western Bushveld areas, which are known to have endemic dental fluorosis, about 300,000 people drink water with fluoride concentrations above 0.7 mg/L. Dental fluorosis in both children and adults is clearly manifested in many villages.

4.7.1.8 Turkey

There are a number of areas in Turkey where drinking-water fluoride concentrations can be very high, especially in the middle and eastern part of Turkey. In Denizli-Sarayköy and Çaldiran Plain levels can reach 13.7 mg/L, while in Eskioehir and Isparta levels from 1.9 to 7.5 mg/L and from 3.8 to 4.9 mg/L respectively, have been reported.

4.7.1.9 United Republic of Tanzania

Endemic fluorosis is a public health problem in some parts of the United Republic of Tanzania, namely: Arusha, Moshi, Singida and Shinyanga regions. It is particularly severe around Arusha, which is situated in the Rift Valley (on the foothills of Mount Meru, which has approximately 135,000 inhabitants), and Moshi (on the foothills of Mount Kilimanjaro). In the African Rift System, fluoride-rich waters are associated with volcanic activity. Due to high temperatures and high pH levels, surface waters (as well as groundwater) contain high fluoride concentrations. It was suggested that the high level of fluorosis seen in the low fluoride area might be at least partly explained by the use of high fluoride foods, such as magadi, in weaning preparation.

All the 119 children examined in Maji ya Chai aged between 9 and 13 years were found to have dental fluorosis; 87.4% of them to a severe degree. They had lived their entire lives in this area, consuming only the local spring or river water (18.6 mg/L). A similar prevalence level was found around Arusha and Moshi, where between 83 and 95% of people examined exhibited dental fluorosis. In Arusha, the prevalence and severity of fluorosis was greater in non-vegetarians (fluorosis – 95%, severe fluorosis – 35%) than in vegetarians (fluorosis – 67%, severe fluorosis – 21%). In addition, skeletal manifestations have been reported around Arusha.

4.7.2 Arsenic

The well-known eight cases of arsenic poisoning from natural arsenic in drinking water in developing countries is presented. A detailed description of these and other cases may be found in the *United Nations Synthesis Report on Arsenic in Drinking Water* (WHO, 2001) at the website www.who.int/water_sanitation_health/dwq/arsenic3/en/index.html.

4.7.2.1 Argentina

The Chaco-Pampean Plain of central Argentina constitutes perhaps one of the largest regions of high-arsenic groundwater known, covering around 1 million km². High concentrations of arsenic have been documented from Córdoba, La Pampa, Santa Fe and

Buenos Aires Provinces in particular. Symptoms typical of chronic arsenic poisoning, including skin lesions and some internal cancers, have been recorded in these areas.

4.7.2.2 Bangladesh – West Bengal, India

In terms of the population exposed, arsenic problems in groundwater from the alluvial and deltaic aquifers of Bangladesh and West Bengal represent the most serious occurrences identified globally. The problem in Bangladesh was undetected for many years because of the lack of suitable arsenic testing laboratories within the country. The problem is made worse by the very large number of wells present (6–11 million wells), most of them operated by hand-pumps and obtaining their water from depths of 10–70 metres.

Concentrations in groundwater from the affected areas have a very large range from less than $0.5\,\mu g/L$ to $3200\,\mu g/L$. Resultant health problems were first identified in West Bengal in the late 1980s. But the first diagnosis in Bangladesh was not made until 1993. Between 30 and 36 million people in Bangladesh are exposed to arsenic in drinking water at concentrations above $50\,\mu g/L$, and up to 6 million of these are in West Bengal. Chatterjee et al. (1995) analyzed groundwater from six districts of West Bengal, India. Mean total arsenic levels ranged from 193 to $737\,\mu g/L$ with a maximum value of $3,700\,\mu g/L$. A year later, Mandal et al. (1996) reported that 44% of groundwater samples collected in West Bengal contained total arsenic levels greater than $50\,\mu g/L$. In other studies, levels as high as $3.7\,mg/L$ have been recorded at some locations. Similarly, in areas of Bangladesh bordering India, 38% of groundwater samples in 27 districts were found to contain arsenic at levels greater than $50\,\mu g/L$ (Dhar et al., 1997).

Skin disorders including hyper/hypopigmentation changes and keratosis are the most common external manifestations, although skin cancer has also been identified. Around 5,000 patients have been identified with arsenic-related health problems in West Bengal (including skin pigmentation changes), although some estimates put the number of patients with Arsenicosis at more than 200,000. The number in Bangladesh is not known, but it must be many times greater than in West Bengal. The occurrence of internal arsenic-related health problems is not known, but it may be appreciable.

4.7.2.3 Chile

Health problems related to arsenic in drinking water were first recognized in northern Chile in 1962, where 100,000 people had consumed drinking water containing $800\,\mu g/L$ of arsenic for up to 12 years. Typical symptoms included skin-pigmentation changes, keratosis, squamouscell carcinoma (skin cancer), cardiovascular problems and respiratory disease. More recently, arsenic ingestion has been linked to lung and bladder cancer. It has been estimated that around 7% of all deaths occurring in Antofagasta between 1989 and 1993 were due to past exposure to arsenic in drinking water at concentrations around $500\,\mu g/L$. Since exposure was chiefly in the period 1955–1970, this pointed to a long latency period of cancer mortality. Other reported symptoms include impaired resistance to viral infection and lip herpes.

4.7.2.4 Mexico

The Laguna Region of north central Mexico has a well-documented arsenic problem in groundwater with significant resulting chronic health problems. The region is arid and groundwater is an important resource for drinking-water supply. Arsenic concentrations were found in the range 8 µg/L to 624 µg/L, with half the samples having concentrations greater than 50 µg/L. Groundwater from the region also has high concentrations of fluoride (up to 3.7 mg/L). The estimated population exposed to arsenic concentrations in drinking water greater than 50 µg/L is around 400,000 in the Laguna Region. High arsenic concentrations have also been identified in groundwater from the state of Sonora in north-west Mexico in the range 2–305 µg/L with highest concentrations in groundwater from the cities of Hermosillo, Etchojoa, Magdalena and Caborca. The arsenic concentrations were also positively correlated with fluoride. It is also believed that high arsenic groundwater have been found in other parts of northern Mexico.

4.7.2.5 Taiwan

The south-west coastal zone of Taiwan was perhaps the first area to be identified as a problem area for health effects arising from chronic arsenic exposure. Chen et al. (1994) reported mean arsenic levels in groundwater of south-west Taiwan of 671 µg/L. Arsenic problems are also documented in north-eastern parts of the island. Awareness of the arsenic problem began during the 1960s and arsenic-related health problems have been well documented by several workers since then. Taiwan is the classic area for the identification of black-foot disease, but a number of other typical health problems, including internal cancers, have been described. In south-west Taiwan, about 100,000 people have been exposed (i.e. prior to 1970) to high concentrations of arsenic in drinking water (range 10 to 1800 µg/L, mean 500 µg/L).

4.7.2.6 Thailand

Probably the worst recorded case of arsenic poisoning related to mining activity is that of the Ron Phibun District in Nakhon Si Thammarat Province of southern Thailand. Health problems were first recognized in the area in 1987. Around 1,000 people have been diagnosed with arsenic-related skin disorders, particularly in and close to Ron Phibun town. The affected area lies within the south-east Asian Tin Belt. Arsenic concentrations have been found at up to 5,000 µg/L in shallow groundwater from Quaternary alluvial sediment that has been extensively dredged during tin-mining operations. Deeper groundwater from older limestone aquifers have been found to be less contaminated, although a few high arsenic concentrations occur, presumably also as a result of contamination from the mine workings.

4.7.2.7 Vietnam

Little was known about the arsenic concentrations in groundwater in Vietnam until recently. UNICEF and EAWAG/CEC (Hanoi National University) are now carrying out extensive investigations to assess the scale of the problem. Preliminary results from Hanoi indicate that there is a significant arsenic problem in shallow tube wells in the city, particularly in the south. There appears to be a seasonal pattern, with significantly higher concentrations in the rainy season. This could be related to the local hydrology, since there are significant interactions between the aquifer and the adjacent Red River. Little is known about the arsenic concentrations in groundwater from the middle and upper parts of the Mekong Delta (and into adjacent Cambodia and Laos) and other smaller alluvial aquifers in Vietnam, but investigations are presently taking place.

4.7.3 Mercury

The following cases address the continuing exposure among the population in gold and mercury-mining regions in developing countries regarding contaminated water–fish consumption pathways; they are examples of exposure scenarios that could be taking place in other countries.

4.7.3.1 Brazil

The Madeira River Basin was the second most important gold mining region in the Amazon. The gold rush in this basin started around 1975 as an individual operation at riverbanks and islands during the dry season. Between 1979 and 1990, about 87 tons of mercury were discharged (45% directly to rivers as metallic mercury). Bastos et al. (2006) reported mercury concentrations in different environmental (river sediments, forest soils, river suspended matter and fish) samples from the lower Madeira River, Amazon, sampled between 2001 and 2003, about 15–20 years after the almost complete cessation of gold mining activities in the region, which reached its peak in the late 1980s. Their results indicate that reduction in mercury concentrations seems to be restricted to areas close to old point sources, and only for abiotic compartments. They found that fish mercury concentrations are very similar between the late 1980s survey of Malm et al. (1990) [0.92 µg/g ± 0.75 (0.10–2.10 µg/g)] and their values [0.34 µg/g ± 0.36 (0.01–2.52 µg/g)], notwithstanding the large reduction of the activity witnessed by 2000. These results suggest no reduction in mercury levels in biological compartments of the Madeira River Basin. Moreover, for the Amazon riverside populations, fish is their main protein source with daily consumption of up to 500 g of fish; therefore, they concluded that even relatively low mercury concentration in fish might result in high exposure doses for these human groups.

4.7.3.2 Philippines

A mine of mercury located south-east of Manila in the Philippines, was in operation from 1955 to 1976. During this time, approximately 2,000,000 tons of mine-waste calcines (retorted ore) were produced during mining and roughly 1,000,000 tons of these calcines were dumped into nearby Honda Bay to construct a jetty to facilitate mine operations where about 2,000 people reside in the nearby three barangays. In October 1994, the Department of Health received a request from the Provincial Health Office for technical assistance relative to the investigation of increasing complaints of unusual symptoms (e.g., miscarriages, tooth loss, muscle weakness, paralysis, anaemia and tremors) among residents of three barangays. Initial health reports revealed significant elevation of blood mercury levels exceeding the recommended exposure level of 20 ppb in 12 out of the 43 (27.9%) residents examined. The majority of the volunteers were former mine workers. In 2002, Maramba et al. (2006) performed an assessment of human exposure levels and environmental mercury contamination in selected communities near the abandoned mercury mine. The results of their study show that there is a continuing mercury exposure among the residents from different exposure pathways in the environment: fish, air, water and soil, in that order of importance. Four species of fish (ibis, tabas, lapu-lapu and torsillo) exceeded the recommended total mercury and methylmercury levels in fish (N > 0.5 µg/g f.w., N > 0.3 µg/g f.w., respectively). Surface water quality at the mining area, Honda Bay and during some monitoring periods at Palawan Bay exceeded total mercury standards (N > 0.002 ng/mL). Soil samples in two sites, Tagburos and Honda Bay, exceeded the US EPA Region 9 Primary Remediation Goal recommended values for total mercury for residential purposes (N > 23 mg/kg). They conclude that the assessment of the risk of mercury exposure among residents in the area should consider the contribution of these different exposure pathways to the overall mercury burden of residents in the area. Likewise, the maternal contribution to prenatal exposure of the foetus to mercury and the hand-to-mouth activity among infants and children are other significant routes for mercury exposure that cannot be underestimated.

4.7.3.3 South Africa

An international example of mercury pollution from an industrial source exists in Natal, South Africa, where the Thor Chemical Plant houses large quantities of mercury wastes that have leaked/leached to the nearby environment and groundwater. The South African government has assigned a high priority to addressing this situation.

4.7.4 Trihalomethanes

The trihalomethanes (THMs) concentrations reported in developing countries and presented in this section have a similar pattern to that observed for developed countries. However, the management and control of these risks associated with drinking-water chlorination might have a different focus, since avoiding microbial diseases seems to be the priority policy in developing cities.

4.7.4.1 Greece

A wide variety of disinfection by-products (DBPs) was determined in the finished drinking water of 15 important cities of Greece, Athens included, distributed in the southern, northern, west continental land and in the islands of the Aegean and Ionian Seas (Kampioti and Stephanou, 2002). Trihalomethanes, haloacetonitriles, haloacetic acids, chloropicrin, halogenated ketons and chloral hydrate were the main DBPs determined. In most cities, the total trihalomethanes concentrations were below 60 µg/L. However, in Athens drinking water, total trihalomethanes (TTHMs) ranged from 0.75 up to 126 µg/L. In chlorinated drinking water of coastal cities, brominated DBPs were more abundant than their chlorinated homologues due to the higher bromide concentration in the raw waters of these areas. Thus, in Greek cities TTHMs concentrations comply with the US EPA standards (80 µg/L; US EPA, 1997b) and European Commission rules (100 µg/L; 1998), except in Athens.

4.7.4.2 Malaysia

In two districts of Malaysia (Tampin and Sabak Bernam) water samples were collected from water treatment plants and its distributive system during 2001. Only chloroform, dichlorobromomethane and dibromochloromethane could be quantified in all samples. Bromoform was not detected for all the samples under review. For the districts of Tampin and Sabak Bernam, the total THMs concentration varies from 3 μg/L up to 135 μg/L, but all the maximum concentrations at Sabak Bernam were greater than 80 μg/L up to 135 μg/L (Pauzi et al., 2003).

4.7.4.3 Mexico

Trihalomethane levels below the Mexican standard (200 μg/L) were found in chlorinated water samples from México City (Mazari-Hiriat et al., 2003) before entering the distribution system that supplies dwellings.

In other studies, performed in four communities located in the central zone of the country, total THMs concentrations varying from 5 to 22 μg/L were found in tap water. However, free residual chlorine in drinking water fell below the permissible limit in 50% of measurements and it represents a microbial health risk for the population of concern. The conclusion of the research is that the risks to health from these by-products at the levels at which they occur in drinking water are small in comparison with risks associated with inadequate disinfection (Navarro et al., 2006).

4.7.4.4 Turkey

The occurrence (in 1999–2000) of total trihalomethanes (TTHMs) in the water distribution system in Ankara, the capital of Turkey and the second largest city in the country, ranged from 25 to 74 μg/L, from 28 to 73 μg/L, and from 25 to 110 μg/L in winter, spring and summer, respectively (Tokmak et al., 2004). The TTHMs concentrations were highest in summer for all 22 districts sampled.

Uyak (2006) investigated the occurrence of THMs within the water distribution systems in the city of Istanbul between April and September of 2004. Total concentrations of THMs in tap water samples measured vary from 33 up to 100 μg/L values. However, the total concentrations of THMs in tap water samples from four districts (Gurpinar, Avcilar, Esenyurt and Sariyer) are higher than EU and US EPA guideline values of 100, and 80 μg/L, respectively. The risk assessment results are consistent with other studies that indicate a higher cancer risk through oral ingestion than through the other two pathways, inhalation and dermal exposure.

4.7.5 Pesticides

The following cases illustrate the presence of banned pesticides, such as DDT and its metabolites, measured in water resources and in drinking water in some developing cities, they are examples of circumstances that could be taking place in other developing cities.

4.7.5.1 Brazil

The presence of organochlorinated pesticides in the water supply system of the city of Bauru, State of São Paulo, Brazil, was reported by Rissato et al. (2004) in concentrations less than 0.035 μg/L for BHC, dieldrin, endosulfan, aldrin, and heptachlor and for DDT, which are within the values established in Brazilian regulations.

4.7.5.2 Egypt

Among water samples, groundwater samples had the highest residues of gamma-HCH and DDT, followed by Nile River water and then tap water. However, the organochlorine pesticide residues were found at concentrations below the maximum allowable limits set by the World Health Organization for drinking water (Dogheim et al., 1996).

4.7.5.3 South Africa

Persistent organochlorine pesticides, such as DDT and its metabolites (DDDs and DDEs), chlordane, hexachlorobenzene (HCB), heptachlor and endosulfan, were determined in drinking, ground, surface and marine waters from the Eastern Cape Province of South Africa (Fatok and Awofolu, 2004). The study reports levels of persistent organochlorine pesticides ranged from 5.5 ng/L (2,4-DDD) to 160 ng/L (HCB) in the water samples. Some endocrine disrupting pesticides, such as DDT, DDE, heptachlor, endosulfan and chlordane, were also detected.

4.8 CHEMICAL RISK MANAGEMENT IN URBAN WATER CYCLE

The review of the major chemical health risks analyzed in the last sections, to which the population is exposed in urban cities, illustrate the complexity of the problem. Many findings, opinions, judgements and conclusions may follow from the cases described. For example, human diseases due to exposure to inorganic compounds found naturally in water resources for drinking water are widely and accurately addressed in both developed and developing countries. Meanwhile, information on exposure to organic compounds is mainly addressed for cities in developed countries; little is known about their occurrence in developing ones. They are just a sample from the cases published to highlight the topics that most influence the risk to hazard chemical compounds in the water cycle. They outline some relevant exposure factors the urban population could be facing in the short or long term that should be taken into account in the understanding and management of the chemical risks, to prevent and reduce them.

4.8.1 Chemical risks identification in urban water cycle

4.8.1.1 Drinking water

The implementation of a successful risk management strategy requires the development of an understanding of those hazards that may affect the quality of water being provided to an urban community. A wide range of chemicals in drinking water could potentially cause adverse human-health effects. The detection of these chemicals in both raw water and in water delivered to consumers is often slow, complex and costly, which means that detection is too impractical and expensive to serve as an early warning system. Moreover, the identification of the risk sources is not sufficient to face chemical risks. Actually, the monitoring of water resources and drinking-water systems does not ensure safe water in urban cities, except for a few cases. Water quality monitoring networks have been used extensively to assess global trends in water quality, (Meybeck et al., 1990; UNEP, 1995) and there is a growing interest in the monitoring of, and in the indicators for, major environmental health risks, particularly in the developing world. As it is neither physically nor economically feasible to test for all chemical constituents in drinking water on an equal basis, monitoring efforts and resources should be carefully planned and directed at significant or key parameters. Surveillance of drinking-water quality also contributes to the protection of public health by promoting improvements of the quality, quantity, access, affordability and continuity of water supplies.

A comprehensive assessment of the water supply is essential in the development of a preventive approach to the management of chemical risks in drinking water. Although many chemicals can be of health concern, the true nature and severity of their impact often remains uncertain (Howard, 2001). When assessing the chemical constituents of drinking water, the following factors should be carefully considered before undertaking more extensive, and often expensive, analysis of the water: what is the extent of the problem – is there strong evidence that the chemicals in water sources are present, or are likely to be present? What is the relevant contribution from drinking-water sources compared with other sources (e.g., food)? How severe is the potential health concern in the context of other health problems?

Unless there is strong evidence that particular chemicals are currently found or will be found in the near future, at levels that may compromise the health of a significant proportion of the urban population, the inclusion of those chemicals in drinking-water monitoring programmes is not justified, particularly where resources are limited. It is often more effective to maintain pollution control and risk assessment programmes at the basic level. In that sense, a relevant issue that should be taken into consideration is the existing databases. Even though they are of a widely varying quality and quantity, information on many cities is still lacking, especially in the developing world. Furthermore, current large-scale data collection efforts are generally disconnected from micro-scale studies in health sciences, which in turn have insufficiently investigated the behavioural and socio-economic factors that influence exposure.

Drinking-water quality can vary significantly throughout the system; consequently, the assessment of the drinking-water supply should aim to determine whether the final quality of water delivered to the consumer routinely meets established health-based targets. The assessment needs to take into consideration the behaviour of individual constituents, or groups of constituents, that may influence water quality. If the assessment indicates that the system is unlikely to be able to meet the targets, this means that the targets are unrealistic under current operating conditions.

Many different chemicals may occur in drinking water; however, only a few are important in any given circumstance. Of particular importance are adverse health outcomes relating to chemical constituents of drinking water arising primarily from prolonged exposure. It is extremely unlikely that all the chemicals included in the WHO *Guidelines for Drinking-water Quality* (WHO, 2004) will be present in a drinking-water supply system. Consequently, it is important that cities identify those chemicals of concern according to local circumstances. Chemical contaminants in drinking water should be prioritized to ensure that scarce resources are not unnecessarily directed towards management of chemicals that pose no threat to health and do not affect the acceptability of drinking water.

One chemical may cause several different types of adverse health effects (ranging from relatively minor skin irritation to life-threatening cancer), and these effects may manifest themselves at different exposure levels. Since there is rarely sufficient information to conduct a thorough risk assessment on every health effect associated with a chemical, it is generally necessary to focus on the most important specific forms of toxicity (e.g., neurotoxicity, or carcinogenicity) that can be caused by a chemical, and then evaluate whether these forms of toxicity might be expressed in exposed humans. The magnitude, duration and timing of the doses that people receive because of chemical exposures are critical to evaluating potential health risks. Unfortunately, detailed data on local concentrations of toxic chemicals are rarely available.

The presence of a limited number of chemical contaminants in drinking water is usually already known in many cities and in many local systems. Significant problems, even crises, can occur, however, when chemicals posing a high health risk are widespread but their presence is unknown because their long-term health effect is caused by chronic exposure as opposed to acute exposure. Such has been the case of arsenic in groundwater in Bangladesh and West Bengal. Therefore, the probability that any particular chemical may occur in significant concentrations in any particular setting must be assessed on a case-by-case basis.

Chemical contaminants in drinking water may be categorized in various ways; however, the most appropriate is to consider the primary source of the contaminant – i.e. to group chemicals according to where control may be effectively exercised. This aids in the development of approaches that are designed to prevent or minimize contamination, rather than those that rely primarily on the measurement of contaminant levels in final waters. In general, approaches to the management of chemical hazards in drinking water vary between those where the source water is a significant contributor (with control achieved, for example, through source water selection, pollution control or treatment) and those from materials and chemicals used in the production and distribution of drinking water (controlled by process optimization or product specification).

Once priority chemicals within a particular drinking-water system have been identified, a management policy should be established and implemented to provide a framework for the prevention and reduction of these chemicals. Appropriate monitoring programmes should be established to ensure the chemical quality of drinking water remains within appropriate standards.

Every city should have a policy on drinking-water quality. Effective local programmes to control drinking-water quality depend ideally on the existence of adequate legislation, standards and codes. Cities that face similar patterns of chemical risks, often experience widely differing human-health impacts. Therefore, the nature and form of drinking-water standards may vary between cities – no single approach is universally applicable. It is essential in the development and implementation of standards to take into account current and planned legislation relating to the water, health and local government sectors and to assess the capacity of potential regulators in the city. Approaches that may have worked in one city do not necessarily transfer to other cities. It is essential that each city undertakes a review of its needs and capacity for drinking-water standards before embarking on the development of a regulatory framework.

4.8.1.2 Other water-related chemical risks

For some contaminants, there will be exposure from sources other than drinking water, such as food and air, and this may need to be taken into account when setting risk management policies and considering the need for standards. It may also be important when considering the need for monitoring. In some cases, drinking water will be a minor source of exposure, and controlling levels in water will have little impact on overall exposure. In other cases, controlling a contaminant in water may be the most cost-effective way of reducing exposure. Drinking-water monitoring strategies, therefore, should not be considered in isolation from other potential routes of exposure to chemicals in the environment.

It is recognized that exposure from various media may vary with local circumstances. In those areas where relevant exposure data are available, authorities are encouraged to develop context-specific guideline values that are tailored to local circumstances and conditions. In addition, in cases in which guideline values are exceeded, efforts should be made to assess the contribution of other sources to total intake, to interpret the health significance of the risks and to orient remedial measures to sources of exposure that are most relevant.

Special attention should be paid to critical chemical pollutants of concern in lakes, rivers and coastal areas, such as organochlorides and metals compounds, known as persistent organic pollutants (POPs) or persistent bioaccumulative toxic chemicals. Food is the primary route of human exposure to these chemicals, and consumption of fish is the most important source of exposure originating directly from that water resources.

It is extremely difficult to estimate critical pollutant loadings entering urban water resources via rivers, precipitation, sewage treatment plants, waste sites, agricultural areas and other sources. The levels of contaminants entering the urban water resources are constantly changing in response to many known and unknown factors. As a result, loadings data are often limited and rely on numerous assumptions. Although quantitative loadings information may be difficult to obtain, qualitative indicators provided by the environmental monitoring of water, sediment and aquatic organisms can often provide sufficient information to identify those contaminant sources that need to be controlled. Improving the database on sources and loadings of critical pollutants is a high priority, as is determining effective ways to virtually eliminate these critical pollutants from urban water resources. Public health advisories and other guidelines should be followed to minimize contaminant exposures, as has been done in some regions.

4.8.2 Vulnerability and variability

The individual behaviour and the social conditions are important factors of risk. For a given individual, chemical exposure via drinking water is determined by the chemical level in the water and by the daily water consumption (litres per day). Water consumption data are most readily available for countries, such as Canada, the US and the UK. However, national consumption figures, especially for developing countries, may be of limited use for the purpose of the management of the urban water cycle, because there are likely to be major differences between urban communities with fully piped supplies and rural communities using wells and boreholes with hand pumps. Furthermore, for a given individual, water consumption increases with temperature, humidity levels, exercise and health status, and it is modified by other factors including diet. Thus, it follows that total daily chemical exposure can vary markedly from one region to another.

The scarcity of water and the low quality of drinking water have provoked an increase in the consumption of bottled water all around the world, particularly in cities. In developed countries, a high percentage of the population consumes bottled water. In developing countries, this tendency is not very clear; this behaviour is highly variable from one city to another and even within urban areas. The population that consumes bottled water has less risk than those that drink tap water without any household treatment. However, the quality of this water is not under frequent control; risks of chemical compounds in bottled water could be advised, and it sets a new exposure scenario requiring appropriate environmental health assessment and surveillance.

Vulnerability to chemical risks is dynamic and influenced by multiple factors, including population growth and regional shifts in population, urbanization, technology, urban government policies, land use and other natural resource management practices, water use trends, and increasing environmental awareness.

Factors, such as low income, poor housing and public services, unplanned urbanization, and environmental mismanagement, increase the risk. The chemical risks associated with the drinking-water pathway must be dealt with caution in some developing countries, where less than 60% of the urban population have access to a water supply, which may be through a household connection, public standpipe, borehole, protected dug well, protected spring or rainwater collection. Examples of selected countries are Afghanistan, Angola, Chad, Guinea-Bissau, Mauritania and Sudan with less than 35% of urban population with access to a water supply (UN, 2003).

The water quality problem is severe in many cities, particularly in developing countries with high growth rates and sometimes only rudimentary environmental legislation. The appropriate regulatory response in developing countries with large populations exposed to chemical risks in drinking water, often exceeding the WHO guidelines, is complex (Smith and Smith, 2004). Meeting a standard in drinking water can represent significant financial implications for treatment plants. For instance, to assure that drinking water is microbiologically safe is a priority topic on their agenda; meanwhile programmes to reduce disinfection by-products formation in the short term, in most of those cities, are not feasible. Moreover, for cities in most developing countries, expanding sanitation coverage poses a far greater challenge than expanding water supply coverage.

Risk and vulnerability assessments, are a prerequisite to identify the areas at greatest risk and the most appropriate risk management measures for a given urban community. It is vital to understand the interplay between chemical risks and the development process, to ascertain the way in which current and future development planning and implementation leads to, or has the potential to, increase vulnerability and risk.

4.8.3 Urban water policy

Urban water government policy, in general, is reduced to provide water rather than to apply an integral service of water. Urban water problems are growing more complex and acute all over the globe. Widespread mismanagement of water resources, growing competition for the use of freshwater and degraded sources heightens the depth of these problems. These problems can only be addressed properly through a concerted effort, which involves scientific, social and institutional approaches. New paradigms for improved urban water management are emerging – reflecting integrated management of all components, and solutions adapted to the particular physical and socio-economic settings.

To ensure an effective integration of chemical risk management concerns into urban water government, particular focus in chemical risk reduction and management policies should take highest priority in development plans at all levels, including prevention, mitigation, response and public health advisories. This can most appropriately be provided through multidisciplinary and inter-sector interactions, taking into consideration socio-economic and cultural factors. It involves the adoption of suitable regulatory and other legal measures, institutional reform, improved analytical and methodological capabilities, appropriate technologies, capacity building, financial planning and public education. A preventive management strategy, operating from the water resources to the tap, should be implemented to ensure drinking-water quality. The extent to which a risk may become a disastrous event has a lot to do with the planning, early warning and protective measures taken.

Lack of coordination among institutions at urban or metropolitan levels is a major constraint to implement effectively chemical risk reduction, as it results in narrow, segmented approaches and poor planning. It is increasingly recognized that reducing the vulnerability to water-related diseases involves far more groups than just the water sector.

Urban water consumers, for example, are frequently unaware of the potential health risks associated with exposure to water-borne contaminants and often consult practising physicians who are unfamiliar with water pollution issues and their subsequent impact on human health. Because of the time delay in exposure and symptom onset, as well as the complexity of the processes involved (individual behaviour and health status), these illnesses are often overlooked, diagnosed incorrectly – or denied. Misdiagnosis and under-diagnosis of water-borne disease by the medical community may result in significant morbidity and mortality, particularly in vulnerable populations at increased risk of disease as a result of exposure to water-borne pathogens and chemical contaminants. Therefore, any future strategic plan to maintain and protect water quality and safety in urban cities must include physicians as stakeholders and active participants in this ongoing public health challenge.

REFERENCES

Aertgeerts, R. and Angelakis, A. (eds) 2003. *State of the Art Report. Health Risks in Aquifer Recharge Using Reclaimed Water.* Who Regional Office for Europe, Dk–2100, Copenhagen, Denmark, EUR/03/5041122.

Adler, P., Steger-Hartmann, T. and Kalbfus, W. 2001. Distribution of Natural and Synthetic Estrogenic Steroid Hormones in Water Samples from Southern and Middle Germany, *Acta Hydrochimica et Hydrobiologica*, Vol. 29, No. 4, pp. 227–241.

Alexander, M. 1995. How Toxic are Toxic Chemicals in Soil? *Environmental Science and Technology*, Vol. 29, pp. 2713–2717.

AMAP. 1998. Arctic Marine Assessment Program (AMAP) Report: *Arctic Pollution Issues.* AMAP. 1-859. 1998. Oslo, AMAP Secretariat.

ANZFA. 1994. *The 1994 Australian Market Basket Survey – A Total Diet Survey of Pesticides and Contaminants.* Australian New Zealand Food Authority, Canberra.

ATSDR. 1997. *Toxicological Profile for Tetrachloroethylene.* Available at: www.atsdr.cdc.gov/toxprofiles/tp18.html.

ATSDR. 1999. *Toxicological Profile for Mercury*, Atlanta, GA, US Department of Health and Human Services, Public Health Service. Available at: www.atsdr.cdc.gov/toxprofiles/tp46.html.

ATSDR. 2000. *Toxicological Profile for Methylene Chloride*. Available at: www.atsdr. cdc.gov/toxprofiles/tp14.html.

ATSDR. 2004. *Toxicological Profile for 1,1,1-Trichloroethane*. Available at: www.atsdr. cdc.gov/toxprofiles/tp70.html.

Bailey, K., Chilton, J., Dahi, E., Lennon, M., Jackson, P. and Farell, J. (eds) 2006. *Fluoride in Drinking-water*. WHO Drinking-water Quality Series, Printed by TJ International (Ltd), Padstow, Cornwall, UK.

Bastos, W.R., Oliveira, J.P., Cavalcante, R., Almeida, R., Nascimento, E.L., Bernardi, J.V., Drude de Lacerda, L., da Silveira, E.G. and Pfeiffer, W.C. 2006. Mercury in the Environment and Riverside Population in the Madeira River Basin, Amazon, Brazil, *Science of the Total Environment*, Vol. 368, pp. 344–351.

Bates, A.J. 2000. Water as Consumed and its Impact on the Consumer – do we Understand the Variables? *Food and Chemical Toxicology*, Vol. 38, pp. S29–S36.

Belfroid, A.C., Van der Horst, A., Vethaak, A.D., Schafer, A.J., Rijs, G.B.J., Wegener, J. and Cofino, W.P. 1999. Analysis and Occurrence of Estrogenic Hormones and Their Glucuronides in Surface Water and Wastewater in The Netherlands, *Science of the Total Environment*, Vol. 225, No. 1–2, pp.101–108.

Bellar, T.A., Lichtenberg, J.J. and Kroner, R.C. 1974. The Occurrence of Organohalides in Chlorinated Drinking Waters, *Journal of the American Water Works Association*, Vol. 66, No. 12, pp. 703–706.

Bove, F.J., Fulcomer, M.C., Klotz, J.B., Esmart, J., Dufficy, E.M. and Savrin, J.E. 1995. Public Drinking Water Contamination and Birth Outcome, *American Journal of Epidemiology*, Vol. 141, pp. 850–862.

Cantor, K.P., Lynch, C.F., Hildesheim, M.E., Dosemeci, M., Lubin, J., Alavanja, M. and Craun, G. 1998. Drinking Water Source and Chlorination Byproducts. I. Risk of Bladder Cancer, *Epidemiology*, Vol. 9, pp. 21–28.

Chatterjee, A., Das, D., Mandal, B.K., Chowdhury, TR., Samanta, G. and Chakraborti, D. 1995. Arsenic in Groundwater in Six Districts of West Bengal, India: The Biggest Arsenic Calamity in The World. Part 1. Arsenic Species in Drinking Water and Urine of the Affected People, *Analyst*, Vol. 120, pp. 643–650.

Chebbo, G., Gromaire, M.C., Ahyerre, M. and Garnaud, S. 2001. Production and Transport of Urban Wet weather Pollution in Combined Sewer Systems: The 'Le Marais' Experimental Urban Catchment in Paris, *Urban Water*, Vol. 3, pp. 3–15.

Chen, S.L., Dzeng, S.R. and Yang, M.H. 1994. Arsenic Species in Groundwater of the Blackfoot Disease Area, Taiwan, *Environmental Science and Technology*, Vol. 28, pp. 877–881.

Chocat, B. and Desbordes, M. 2004. Proceedings of Novatech 2004: Sustainable Techniques and Strategies in Urban Water Management, *Water Science and Technology*, Vol. 51, No. 2, pp. 258.

Christman, R.F., Norwood, D.L., Millington, D.S. and Johnson, J.D. 1983. Identity and Yields of Major Halogenated Products of Aquatic Fulvic Acid Chlorination, *Environmental Science and Technology*, Vol. 17, No. 10, pp. 625–628.

Covello, V.T. and Merkhofer, M.W. 1993. *Risk Assessment Methods. Approaches for Assessing Health and Environmental Risks*. Plenum Press, New York.

Dabeka, R.W., McKenzie, A.D., Lacroix, G.M.A., Cleroux, C., Bowe, S., Graham, R.A. and Conacher, H.B.S. 1993. Survey of Arsenic in Total Diet Food Composites and Estimation of the Dietary Intake of Arsenic by Canadian Adults and children, *JAOAC International*, Vol. 76, pp. 14–25.

Davis, J.A., Gunther, A.J. and O'Connor, J.M. 1992. Priority Pollutant Loads from Effluent Discharges to the San Francisco Estuary, *Water Environment Research*, Vol. 64, pp. 134–140.

Dean, H.T. 1933. Distribution of Mottled Enamel in the United States, *Public Health Reports*, Vol. 48, No. 25, pp. 703–734.

Desbrow, C., Routledge, E.J., Brighty, G.C., Sumpter, J.P. and Waldock, M. 1998. Identification of Estrogenic Chemicals in STW Effluent. 1. Chemical Fractionation and in Vitro Biological Screening? *Environmental Science and Technology*, Vol. 32, No. 11, pp. 1549–1558.

Dewailly, E., Nantel, A., Weber, J. P. and Meyer, F. 1989. High Levels of PCBs in Breast Milk of Inuit Women from Arctic Québec, *Bulletin of Environmental Contamination and Toxicology*, Vol. 43, pp. 641–646.

Dewailly, E., Ryan, J.J., Laliberte, C., Bruneau, S., Weber, J.P., Gingras, S. and Carrier, G. 1994. Exposure of Remote Maritime Populations to Coplanar PCBs, *Environmental Health Perspectives*, Vol. 102, pp. 205–209.

Dewailly, E., Ayotte, P., Blanchet, C., Grondin, J., Bruneau, S., Holub, B. and Carrier, G. 1996. Weighing Contaminant Risks and Nutrient Benefits of Country Food in Nunavik, *Archives of Medical Research*, Vol. 55, pp. 13–19.

Dewailly, E., Ayotte, P., Bruneau, S., Gingras, S., Belles-Iles, M. and Roy, R. 2000. Susceptibility to Infections and Immune Status in Inuit Infants Exposed to Organochlorines, *Environmental Health Perspectives*, Vol. 108, pp. 205–211.

Dhar, R.K., Biswas, B.K., Samanta, G., Mandal, B.K., Chakraborti, D., Roy, S., Fafar, A., Islam, A., Ara, G., Kabir, S., Khan, A.W., Ahmed, S.A. and Hadi, S.A. 1997. Groundwater Arsenic Calamity in Bangladesh, *Current Science*, Vol. 73, No. 1, pp. 48–59.

Díaz-Barriga, F., Navarro-Quezada, A., Grijalva, M.I., Grimaldo, M., Loyola-Rodriguez, J.P. and Ortiz, M.D. 1997. Endemic Fluorosis in Mexico, *Fluoride*, Vol. 30, No. 4, pp. 233–239.

Dogheim, S.M., Mohamed, E.Z., Alla, S.A.G, ElSaied, S., Emel, S.Y., Mohsen, A.M. and Fahmy, S.M. 1996. Monitoring of Pesticide Residues in Human Milk, Soil, Water, and Food Samples Collected from Kafr El-Zayat Governorate, *Journal of AOAC International*, Vol. 79, No. 1, pp. 111–116.

European Chemical Industry Ecology and Toxicology Center. 1988. *Nitrate and drinking water. Technical Report No. 27*. Brussels.

Eisler, R. 1987. *Polycyclic Aromatic Hydrocarbon Hazards to Fish, Wildlife, and Invertebrates: a Synoptic Review*. U.S. Fish & Wildlife Service, Washington DC. Biological Report No. 85(1.17), p. 72.

Endo, T., Hotta, Y., Haraguchi, K. and Sakata, M. 2003. Mercury Contamination in the Red Meat of Whales and Dolphins Marketed for Human Consumption in Japan, *Environmental Science and Technology*, Vol. 37, No. 12, pp. 2681–2685.

Environment Canada and Health Canada 2001. *Canadian Environmental Protection Act, 1999*. Priority Substances List assessment report. Chloroform. Ottawa.

Eto, K. 2000. Minamata Disease, *Neuropathology*, Suppl:S1, pp. 4–9.

European Union Commission. 1998. *The Council Directive 1998/EC on the Quality of Water Intended for Human Consumption*. DGI 12767/97, Brussels.

Fatok, O.S. and Awofolu, O.R. 2004. Levels of Organochlorine Pesticide Residues in Marine-, Surface-, Ground- and Drinking Waters from the Eastern Cape Province of South Africa, *Journal of Environmental Science and Health Part B-Pesticides Food Contaminants and Agricultural Wastes*, Vol. 39, No. 1, pp. 101–114.

Fawell, J., Bailey, K., Chilton, J., Dahi, E., Fewtrell, L. and Magara, Y. (eds) 2006. *Fluoride in Drinking Water*. World Health Organization (WHO), Geneva, Switzerland.

Flaten, T.P. 1992. Chlorination of Drinking Water and Cancer Incidence in Norway, *International Journal of Epidemiology*, Vol. 21, pp. 6–15.

Gallagher, M.D., Nuckols, J.R., Stallones, L. and Savitz, D.A. 1998. Exposure to Trihalomethanes and Ddverse Pregnancy Outcomes, *Epidemiology*, Vol. 9, No. 5, pp. 484–489.

Gavis J. and Ferguson J.F. 1972. The Cycling of Mercury Through the Environment, *Water Research*, Vol. 6, pp. 986–1008.

Golden, K.A, Wong, C.S., Jeremiason, J.D., Eisenreich, S.J., Hallgren, J., Swackhamer, D.L., Engstrom, D.R. and Long, D.T. 1993. Accumulation and Preliminary Inventory of Organochlorines in Great Lakes Sediments, *Water Science and Technology*, Vol. 28, pp. 19–31.

Gromaire, M.C., Garnaud, S., Saad, M. and Chebbo, G. 2001. Contribution of Different Sources to Pollution of Wet Weather Flows in Combined Sewers, *Water Research*, Vol. 35, No. 2, pp. 521–533.

Gunderson, E.L. 1995. FDA Total Diet Study – 1986–1991 – Dietary Intakes of Pesticides, Selected Elements, and other Chemicals, *JAOAC International*, pp. 1353–1363.

Health Canada 2006. *Guidelines for Canadian Drinking Water Quality: Guideline Technical Document — Trihalomethanes*. Water Quality and Health Bureau, Healthy Environments and Consumer Safety Branch, Health Canada, Ottawa, Ontario.

Heberer, T. 2002. Occurrence, Fate, and Removal of Pharmaceutical Residues in the Aquatic Environment: a Review of Recent Research Data, *Toxicology Letters*, Vol. 131, No. 1–2, pp. 5–17.

Heberer, T., Dünnbier, U., Reilich, C. and Stan, H.J. 1997. Detection of Drugs and Drug Metabolites in Groundwater Samples of a Drinking Water Treatment Plant, *Fresenius' Environmental Bulletin*, Vol. 6, pp. 438–443.

Heberer, T., Reddersen, K. and Mechlinski, A. 2002. From Municipal Sewage to Drinking Water Fate and Removal of Pharmaceutical Residues in the Aquatic Environment in Urban Areas, *Water Science and Technology*, Vol. 46, No. 3, pp. 81–88.

Henschler, D. 1994. Toxicity of Chlorinated Organic Compounds: Effects of the Introduction of Chlorine in Organic Molecules, *Angewandte Chemie International Edition in English*, Vol. 33, pp. 1920–1935.

Hignite, C. and Azarnoff, D.L. 1977. Drugs and Drug Metabolites as Environmental Contaminants: Chlorophenoxyisobutyrate and Salicylic Acid in Sewage Water Effluent, *Life Science*, Vol. 20, No. 2, pp. 337–341.

Hileman, B. 1993. Concerns Broaden Over Chlorine and Chlorinated Hydrocarbons, *Chemical Engineering News*, Vol. 71, pp. 11–20.

Howard, G. 2001. *Water Supply Surveillance: A Reference Manual*. Water, Engineering and Development Centre (WEDC), Loughborough University, Loughborough, United Kingdom.

Hung, D.Q. and Thiemann, W. 2002. Contamination by Selected Chlorinated Pesticides in Surface Waters in Hanoi, Vietnam, *Chemosphere*, Vol. 47, No. 4, pp. 357–367.

IPCS. 2000. *Disinfectants and disinfectant by-products. Environmental Health Criteria 216*. World Health Organization, International Programme on Chemical Safety, Geneva.

IPCS. 2004. IPCS Glossary of Key Exposure Assessment Terminology. *IPCS Risk Assessment Terminology*. World Health Organization, International Programme on Chemical Safety (Harmonization Project Document No. 1), Geneva.

Ivahnenko, T. and Zogorski, J.S. 2006. *Sources and Occurrence of Chloroform and other Trihalomethanes in Drinking-Water Supply Wells in the United States, 1986–2001*. U.S. Geological Survey Scientific Investigations Report 2006 – 5015, 13 pp.

Jacks, G. and Sharma, V.P. 1983. Nitrogen Circulation and Nitrate in Groundwater in an Agricultural Catchment in Southern India, *Environmental Geology*, Vol. 5, pp. 61–64.

Jacobson, J.L. and Jacobson, S.W. 1996. Intellectual Impairment in Children Exposed to Polychlorinated Biphenyls in Utero, *New England Journal of Medicine*, Vol. 335, pp. 783–789.

Jobling, S., Noylan, M., Tyler, C.R., Brighty, G. and Sumpter, J.P. 1998. Widespread Sexual Disruption in Wild Fish, *Environmental Science and Technology*, Vol. 32, pp. 2498–2506.

Kampioti, A.A. and Stephanou, E.G. 2002. The Impact of Bromide on the Formation of Neutral and Acidic Disinfection by-Products (DBPs) in Mediterranean Chlorinated Drinking water, *Water Research*, Vol. 36, pp. 2596–2606.

Kim, S.D., Cho, J., Kim, I.S., Vanderford, B.J. and Snyder, S.A. 2007. Occurrence and Removal of Pharmaceuticals and Endocrine Disruptors in South Korean Surface, Drinking, and Wastewaters, *Water Research*, Vol. 41, pp. 1013–1021.

Klaassen, C. D. and Watkins III, J.B. (eds) 2003. *Casarett and Doull's Essentials of Toxicology*. McGraw-Hill Companies, Inc., US

Kolpin, D.W., Furlong, E.T., Meyer, M.T., Thurman, E.M, Zauff, S.D., Barber, L.B. and Buston, H.T. 2002. Pharmaceuticals, Hormones, and other Organic Wastewater Compounds in US

Streams, 1999–2000: A National Reconnaissance, *Environmental Science and Technology*, Vol. 36, No. 6, pp. 1202–1211.

Kool, H.J. and van Kreyl, C.F. 1998. Mutagenic Activity in Drinking Water Prepared from Groundwater: A Survey of Ten Cities in The Netherlands, *Science of the Total Environment*, Vol. 77, No. 1, pp. 51–60.

Kramer, M.D., Lynch, C.F., Isacson, P. and Hanson, J.W. 1992. The Association of Waterborne Chloroform with Intrauterine Growth Retardation, *Epidemiology*, Vol. 3, pp. 407–413.

Kramer, V.J., Miles-Richardson, S., Pierens, S.L. and Giesy, J.P. 1998. Reproductive Impairment and Induction of Alkaline-Labile Phosphate, A Biomarker of Estrogen Exposure, in Fathead Minnows (Pimephales Promelas) Exposed to Waterborne 17bestradiol, *Aquatic Toxicology*, Vol. 40, pp. 335–360.

Krejci, V., Dauber, L., Novak, B. and Gujer, W. 1987. Contribution of Different Sources to Pollutant Loads in Combined Sewers. *Proc. 4th Int. Conf. on Urban Storm Drainage, Lausanne, Swizerland*, pp. 34–39.

Kuhnlein, H.V., Receveur, O., Muir, D.C.G., Chan, H.M. and Soueida, R. 1995. Arctic Indigenous Women Consume Greater than Acceptable Levels of Organochlorines, *The Journal of Nutrition*, Vol. 125, pp. 2501–2510.

Lave, L.B. and Upton, A.C. (eds) 1987. *Toxic Chemicals, Health and the Environment*. Johns Hopkins University Press, Baltimore.

LeBel, G.L., Benoit, F.M. and Williams, D.T. 1997. A One-year Survey of Halogenated Disinfection by-Products in the Distribution System of Treatment Plants Using Three Different Disinfection Processes, *Chemosphere*, Vol. 34, pp. 2301–2317.

Lenton, R., Wright, A.M. and Lewis, K. 2005. *Health, Dignity and Development: What Will It Take?* UN Millennium Project Task Force on Water and Sanitation, Earthscan, London, ISBN-1-84407-219-3.

Lockhart, W.L., Stern, G.A., Wagemann, R., Hunt, R.V., Metner, D.A., DeLaronde, J., Dunn, B., Stewart, R.E., Hyatt, C.K., Harwood, L. and Mount, K. 2005. Concentrations of Mercury in Tissues of Beluga Whales (Delphinapterus leucas) from Several Communities in the Canadian Arctic from 1981 to 2002, *Science of the Total Environment*, Vol. 351–352, pp. 391–412.

McGeehin, M.A., Reif, J.S., Becher, J.C. and Mangione, E.J. 1993. Case-control Study of Bladder Cancer and Water Disinfection Methods in Colorado, *American Journal of Epidemiology*, Vol. 138, pp. 492–501.

MAFF. 1997. Ministry of Agriculture, Fisheries and Food. *Lead, arsenic and other metals in food. Food Surveillance Paper No 52*. Her Majesty's Stationery Office, 'London, UK.

Malm, O., Pfeiffer, W.C., Souza, C.M.M. and Reuther, R. 1990. Mercury Pollution Due to Gold Mining in the Madeira River Basin, Brazil, *Ambio*, Vol. 19, pp. 11–15.

Mandal, B.K., Chowdhury, T.R., Samanta, G., Basu, G.K., Chowdhury, P.P., Chanda, C.R., Lodh, D., Karan, N.K., Dhar, R.K., Tamili, D.K., Das, D., Saha, K.C. and Chakraborti, D. 1996. Arsenic in Groundwater in Seven Districts of West Bengal, India – The Biggest Arsenic Calamity in the World, *Current Science*, Vol. 70, No. 11, pp. 976–986.

Maramba, N.P.C, Reyes, J.P., Francisco-Rivera, A.T., Panganiban, L.C.R., Dioquino, C., Dando, N., Timbang, R., Akagi, H., Castillo, M.T., Quitoriano, C., Afuang, M., Matsuyama, A., Eguchi, T. and Fuchigami Y. 2006. Environmental and Human Exposure Assessment Monitoring of Communities Near an Abandoned Mercury Mine in the Philippines: A Toxic Legacy, *Journal of Environmental Management*, Vol. 81, pp. 135–145.

Mazari-Hiriart, M., Lopez-Vidal, Y., De Leon, S.P., Castillo-Rojas, G., Hernandez, E.C. and Rojo, F. 2003. Bacteria and Disinfection Byproducts in Water from Southern Mexico City, *Archives of Environmental Health*, Vol. 58, No. 4, pp. 233–237.

Maurits la Riviere, J.W. 1989. Threats to the World's Water, *Scientific American*, September 1989, pp. 80–94.

Meador, J.P., Stein, J.E., Reichert, W.L. and Varanasi, U. 1995. Bioaccumulation of Polycyclic Aromatic Hydrocarbons by Marine Organisms, *Reviews of Environmental Contamination and Toxicology*, Vol. 143, pp. 79–163.

Meybeck, M., Chapman, D.V. and Helmer, R. 1990. *Global Freshwater Quality*. Blackwell, Oxford, UK.

Miller, J.W. and Uden, P.C. 1983. Characterization of Nonvolatile Aqueous Chlorination Products of Humic Substances, *Environmental Science and Technology*, Vol. 17, No. 3, pp. 150–157.

Moran, M.J. 2006. *Occurrence and Implications of Selected Chlorinated Solvents in Groundwater and Source Water in the United States and in Drinking Water in 12 Northeast and Mid-Atlantic States, 1993–2002*. US Geological Survey Scientific Investigations Report 2005-5268, p. 70.

Moran, M.J., Hamilton, P.A. and Zogorski, J.S. 2006. *Volatile Organic Compounds in the Nation's Groundwater and Drinking-water Supply Wells – A Summary*. US Geological Survey Fact Sheet 2006–3048, 6 pp.

Moran, M.J., Zogorski, J.S. and Squillace, P.J. 2005. MTBE and Gasoline Hydrocarbons in Groundwater of the United States, 2005, *Ground Water*, Vol. 43, No. 4, pp. 615–627.

Morgenstern, R.D., Shih, J.S. and Sessions, S.L. 2000. Comparative Risk Assessment: An International Comparison of Methodologies and Results, *Journal of Hazardous Materials*, Vol. 78, No. 1–3, pp. 19–39.

Morris, R.D. 1995. Drinking Water and Cancer, *Environmental Health Perspectives*, Vol. 103, Suppl. 8, pp. 225–231.

MRC. 1997. *Greater Mekong Sub-Region: State of the Environment Report*. Mekong River Commission, Bangkok.

Nakanishi, Y., Shigematsu, N., Kurita, Y., Matsuba, K., Kanegae, H., Ishimaru, S. and Kawazoe, Y. 1985. Respiratory Involvement and Immune Status in Yusho Patients, *Environmental Health Perspectives*, Vol. 59, pp. 31–36.

Navarro, I., Jiménez, B., Maya, C. and Lucario, E.S. 2006. Assessment of Potential Cancer Risks from THMs in Water Supply at Mexican Rural Communities. Submitted to *Water Science & Technology*.

Nieuwenhuijsen, M.J., Toledano, M.B., Eaton, N.E., Elliott, P. and Fawell, J. 2000. Chlorination Disinfection by-Products in Water and their Association with Adverse Reproductive Outcomes: A Review, *Journal of Occupational and Environmental Medicine*, Vol. 57, pp. 73–85.

NRC. 1998. National Research Council. *Issues in Potable Reuse: the Viability of Augmenting Drinking Water Supplies with Recycled Water*. National Academy Press, Washington DC.

NRC. 1999 United States Nation Research Council. *Arsenic in drinking water*. National Academy Press, Washington DC, 310 pp.

Oliver, B.G. and Nicol, K.D. 1982. Chlorobezenes in Sediments, Water and Select Fish from Lakes Superior, Huron, Erie, and Ontario, *Environmental Science and Technology*, Vol. 6, pp. 532–536.

Page, G.W. 1981. Comparison of Groundwater and Surface Water for Patterns and Levels of Contamination by Toxic Substances, *Environmental Science and Technology*, Vol. 15, No. 12, pp. 1475–1481.

Palm, H. and Lammi, R. 1995. Fate of Pulp Mill Organochlorines in the Gulf of Bothnia Sediments, *Environmental Science and Technology*, Vol. 29, pp. 1722–1727.

Pankow, J.F. and Cherry, J.A. 1996. *Dense solvents: Portland, Oreg.* Waterloo Press, 522 pp.

Pauzi, A., Yew, C.H. and Ramli, M.S. 2003. Formation, Modeling and Validation of Trihalomethanes (THM) in Malaysian Drinking Water: A Case Study in the Districts of Tampin, Negeri Sembilan and Sabak Bernam, Selangor, Malaysia, *Water Research*, Vol. 37, pp. 4637–4644.

Pilotto, L.S. 1995. Disinfection of Drinking-water, Disinfection By-products and Cancer—What about Australia, *Australian Journal of Public Health*, Vol. 19, No. 1, pp. 89–93.

Pounds, J.G. 1985. The Toxic Effects of Metals. Williams, P.L. and Burson, J.L. (eds) *Industrial Toxicology*. Van Nostrand Reinhold, New York, pp. 197–210.

Purdom, C.E., Hardiman, P.A., Bye, V.J., Eno, N.C., Tyler, C.R. and Sumpter, J.P. 1994. Estrogenic Effects of Effluents from Sewage Treatment Works, *Chemistry and Ecology*, Vol. 8, pp. 275–285.

Rathbun, R. E. 2000. Transport, behavior, and Fate of Volatile Organic Compounds in Streams, *Critical Reviews in Environmental Science and Technology*, Vol. 30, No. 2, pp. 129–295.

Richardson, S.D. 1998. Drinking Water Disinfection Byproduct. Meyers, R.A. (ed.) *Encyclopedia of Environmental Analysis and Remediation*, Vol. 3. New York, Wiley, pp. 1398–1421.

Rissato, S.R., Libânio, M., Passos, G. and Gerenutti, M. 2004. Determinação De Pesticidas Organoclorados Em Água De Manancial, Água Potável E Solo Na Região De Bauru (SP), *Quimica Nova*, Vol. 27, No. 5, pp. 739–743.

Renner, R. 1998. Human Estrogens Linked to Endocrine Disruption, *Environmental Science and Technology*, Vol. 32, No. 1, p. 8A.

Robertson, F.N. 1989. Arsenic in Groundwater Under Oxidizing Conditions, South-west United States, *Environmental Geochemistry Health*, Vol. 11, pp. 171–185.

Rook, J.J. 1974. Formation of Haloforms During Chlorination of Natural Waters, *Water Treatment Examinations*, Vol. 23, pp. 234–243.

Ross, P.S. and Birnbaum, L.S. 2001. Persistent Organic Pollutants (POPs) in Humans and Wildlife. In, *Integrated Risk Assessment. Report prepared for the WHO/UNEP/ILO International Programme on Chemical Safety.* Available at: www.who.int/pcs/emerg_site/integr_ra/ ira_report.htm.

Schock, M.R. 1989. Understanding Lead Corrosion Control Strategies, *Journal of the American Water Works Association*, Vol. 81, No. 10, p. 4.

Schock, M.R. 1990. Causes of Temporal Variability of Lead in Domestic Plumbing Systems, *Environmental Monitoring Assessment*, Vol. 15, No. 1, pp. 59–82.

Simmonds, M.P., Haraguchi, K., Endo, T., Cipriano, F., Palumbi, S.R. and Troisi, G.M. 2002. Human Health Significance of Organochlorine and Mercury Contaminants in Japanese Whale Meat, *Journal of Toxicology and Environmental Health A*, Vol. 65, No. 17, pp. 1211–1235.

Sloof, W. (ed.) 1988. *Basisdocumentfluoriden, Report no. 758474005.* National Institute of Public Health and Environmental Protection, Bilthoven, Netherlands.

Smith, A.H. and Smith, M.M.H. 2004. Arsenic Drinking Water Regulations in Developing Countries with Extensive Exposure, *Toxicology*, Vol. 198, No. 1–3, pp. 39–44.

Snyder, S.A., Villeneuve, D.L., Snyder, E.M. and Giesy, J.P., 2001. Identification and Quantification of Estrogen Receptor Agonists in Wastewater Effluents, *Environmental Science and Technology*, Vol. 35, No. 18, pp. 3620–3625.

Sofuoglu, S.C., Lebowitz, M.D., O'Rourke, M.K., Robertson, G.L., Dellarco, M. and Moschandreas, D.J. 2003. Exposure and Risk Estimates for Arizona Drinking Water, *Journal American Water Works Association*, Vol. 95, No. 7, pp. 67–79.

Squillace, P.J. and Moran, M.J. 2006. *Factors Associated with Sources, Transport, and Fate of Volatile Organic Compounds in Aquifers of the United States and implications for groundwater management and assessments.* US Geological Survey Scientific Investigations Report 2005-5269, 40 pp.

Stumm-Zollinger, E. and Fair, G.M. 1965. Biodegradation of Steroid Hormones, *Journal Water Pollution Control Federation*, Vol. 37, pp. 1506–1510.

Suedel, B.C., Boraczek, J.A, Peddicord, R.K., Clifford, P.A. and Dillon, T.M. 1994. Trophic Transfer and Biomagnification Potential of Contaminants in Aquatic Ecosystems, *Reviews of Environmental Contamination and Toxicology*, Vol. 136, pp. 21–89.

Swindoll, M., Stahl Jr, R.G. and Ells, S.J. 2000. *Natural Remediation of Environmental Contaminants: Its Role in Ecological Risk Assessment and Risk Management.* Published by the Society of Environmental Toxicology and Chemistry (SETAC), 472 pp.

Tabak, H.H. and Bunch, R.L. 1970. Steroid Hormones as Water Pollutants. I. Metabolism of Natural and Synthetic Ovulation Inhibiting Hormones by Microorganisms of Activated sludge and Primary Settled Sewage, *Developments in Industrial Microbiology*, Vol. 11, pp. 367–376.

Takayama, K., Miyata, H., Mimura, M., Ohta, S. and Kashimoto, T. 1991. Evaluation of Biological Effects of Polychlorinated Compounds Found in Contaminated Cooking oil Responsible for the Disease 'Yusho', *Chemosphere*, Vol. 22, pp. 537–546.

Tokmak, B., Capar, G., Dilek, F.B. and Yetis, U. 2004. Trihalomethanes and Associated Potential Cancer Risks in the Water Supply in Ankara, Turkey, *Environmental Research*, Vol. 96, pp. 345–352.

Toledano, M.B., Nieuwenhuijsen, M.J., Bennet, J., Best, N., Whithaker, H., Cockings, S., Fawell, J., Jarup, L., Briggs, D. and Elliott, P. 2005. Chlorination Disinfection Byproducts and Adverse Birth Outcomes in Great Britain: Birthweight and Still Birth, *Environmental Health Perspectives*, Vol. 113, pp. 225–232.

Toro, E.C., Das, H.A. and Fardy, J.J. 1994. Toxic Heavy Metals and other Trace Elements in Foodstuffs from 12 Different Countries. Schrauzer G.N., *Biological Trace Element Research*. Humana Press Inc.

UN. 2003. *Critical Importance of Water Issues for the Least Developed Countries (LDCS)*. Report of The United Nations Office of The High Representative for The Least Developed Countries, Landlocked Developing Countries and Small Island Developing States Presented at The Third World Water Forum, Kyoto, Japan 16 to 23 March 2003.

UNEP. 1995. *Water Quality of World River Basins*. UNEP/GEMS, Environment Library no. 14, UNEP, Geneva, Switzerland.

US Department of Health and Human Services. 2005. *Report on Carcinogens, Eleventh Edition*. Department of Health and Human Services, Washington, DC', National Toxicology Program, 955 pp.

US EPA. 1980. *Sources of Toxic Compounds in Household Wastewater*. Office of Research and Development, US EPA, Washington DC 600/2–80–128, 39 pp.

US EPA 1984. *Mercury Health Effects Updates: Health Issue Assessment. Final Report*. U.S. Environmental Protection Agency, Office of Health and Environmental Assessment, Washington, DC. Document no. EPA 600/8-84-019F.

US EPA. 1985. *Drinking Water Criteria Document on Fluoride, TR-823-5*. Office of Drinking Water, US Environmental Protection Agency, Washington, DC.

US EPA 1987. *Estimated National Occurrence and Exposure to Nitrate and Nitrite in Public Drinking Eater Supplies*. Office of Drinking Water, US Environmental Protection Agency, Washington, DC.

US EPA 1997a. *Exposure Factors Handbook*. US Environmental Protection Agency, Washington DC.

US EPA 1997b. *Disinfectants/disinfection By-products*. Fed. Reg., 62:212:59486, 1997.

US EPA 1998. *Disinfectants and Disinfection Byproducts; Final Rule*. Federal Register 63(241), 69478.

US EPA 1999. *Mercury Research Strategy, NCEA-I-0710 Workshop Review Draft*. Office of Research and Development, US Environmental Protection Agency, Washington DC, 20460.

US EPA. 2000. *Technology Transfer Network, Air Toxics Website, Trichloroethylene*. Available at: www. epa.gov/ttnatw01/hlthef/tri-ethy.html#ref1

US EPA. 2001. *Trichloroethylene Health risk Assessment—Synthesis and Characterization*. Office of Research and Development, Washington DC, external review draft, EPA/600/P–01/002A.

US EPA 2003. *EPA National Primary Drinking Water Standards*. Office of Water (4606M). EPA 816-F-03-016. Available at: www.epa.gov/safewater.

Uyak, V. 2006. Multi-pathway Risk Assessment of Trihalomethanes Exposure in Istanbul Drinking Water Supplies, *Environment International*, Vol. 32, pp. 12–21.

van Dijk-Looijaard, A.M. and van Genderen, J. 2000. Levels of Exposure from Drinking Water, *Food and Chemical Toxicology*, Vol. 38, pp. S37–S42.

Vose, D. 2000. *Risk analysis: a quantitative guide*. 2nd ed. John Wiley & Sons Ltd., Chichester, England.

Waller, K., Swan, S.H., DeLorenze, G. and Hopkins, B. 1998. Trihalomethanes in Drinking Water and Spontaneous Abortion, *Epidemiology*, Vol. 9, pp. 134–140.

Ware, G.W. 1989. Mercury. US EPA Office of Drinking Water Health Advisories, *Reviews of Environmental Contamination and Toxicology*, Vol. 107, pp. 93–102.

Water Quality Issues Sub-Group 2003. *Water Quality Issues Sub-Group final report*. Prepared for the Chlorinated Disinfection By-Product (CDBP) Task Force. Health Canada, Ottawa.

Welch, A.H., Lico, M.S. and Hughes, J.L. 1988. Arsenic in Groundwater of the Western United States, *Ground Water*, Vol. 26, No. 3, pp. 333–347.

White, D. M., Garland, D.S., Narr, J. and Woolard, C.R. 2003. Natural Organic Matter and DBP Formation Potential in Alaskan water supplies, *Water Research*, Vol. 37. pp. 939–947.

WHO. 1985. *Health Hazards from Nitrate in Drinking Water*. Report on a WHO meeting, Environmental Health Series, No. 1, Copenhagen, 5–9 March 1984. WHO Regional Office for Europe, Copenhagen.

WHO. 2001. *United Nations Synthesis Report on Arsenic in Drinking Water (Draft)*. Available at: www.who.int/water_sanitation_health/dwq/arsenic3/en/index.html.

WHO. 2003. *Mercury in Drinking-water. Background Document for Preparation of WHO Guidelines for Drinking-water Quality*. Geneva, World Health Organization, Geneva (WHO/SDE/ WSH/03.04/10).

WHO. 2004. *Guidelines for Drinking-Water Quality, Third Edition*. World Health Organization, Geneva, Available at: www.who.int/water_sanitation_health/dwq/gdwq3rev/en/index.html.

WHO. 2006a. *Guidelines for the Safe use of Wastewater, Excreta and Greywater*. World Health Organization, Geneva.

WHO. 2006b. *Fluoride in Drinking Water*. WHO Drinking-water Quality Series, Printed by TJ International (Ltd), Padstow, Cornwall, UK.

Wiedemeier, T.H., Rifai, H.S., Newell, C.J. and Wilson, J.T. 1999. *Natural Attenuation of Fuels and Chlorinated Solvents in the Subsurface*. John Wiley & Sons, New York, 617 pp.

Williams, D.T., LeBel, G.L. and Benoit, F.M. 1995. *A National Survey of Chlorinated Disinfection By-products in Canadian Drinking Water*. Report 95-EHD-197, Environmental Health Directorate, Health Canada, Ottawa.

Williams, D.T., LeBel, G.L. and Benoit, F.M. 1997. Disinfection By-products in Canadian Drinking Water, *Chemosphere*, Vol. 34, pp. 299–316.

Williams, P., Benton, L., Warmerdam, J. and Sheehan, P. 2002. Comparative Risk Analysis of Six Volatile Organic Compounds in California Drinking water, *Environmental Science and Technology*, Vol. 36, No.22, pp. 4721–4728.

World Commission on Water. 1999. *World's Rivers in Crisis – Some are Dying; Others Could Die*. World Water Council.

Yang, C.Y., Chiu, H.F., Cheng, M.F. and Tsai, S.S. 1998. Chlorination of Drinking Water and Cancer Mortality in Taiwan, *Environment Research*, Vol. 78, pp. 1–6.

Yang, C.Y., Cheng, B.H., Tsai, S.S., Wu, T.N., Lin, M.C. and Lin, K.C. 2000. Association between Chlorination of Drinking Water and Adverse Pregnancy Outcome in Taiwan, *Environmental Health Perspectives*, Vol. 108, No. 8, pp. 765–768.

Yost, L.J., Schoof, R.A. and Aucoin, R. 1998. Intake of Inorganic Arsenic in the North American Diet, *Human and Ecological Risk Assessment*, Vol. 4, pp. 137–152.

Zhibo, Li, Wang, Q. and Luo, Y. 2006. Exposure of the Urban Population to Mercury in Changchun city, Northeast China, *Environmental Geochemistry and Health*, Vol. 28, pp. 61–66.

Zogorski, J.S., Carter, J.M., Ivahnenko, Tamara, Laphham, W.W., Moran, M.J., Rowe, B.L., Squillace, P.J. and Toccalino, P.L. 2006. *The Quality of our Nation's Waters – Volatile Organic Compounds in the Nation's Groundwater and Drinking-water Supply Wells*. US Geological Survey Circular 1292, 101 pp.

Chapter 5

Risk management in the urban water cycle: climate change risks

Claudia Sheinbaum Pardo

Instituto de Ingeniería, Universidad Nacional Autónoma de México, Mexico

ABSTRACT: This chapter presents a review of climate change impacts on the hydrological cycle and its actual and potential risks on the urban areas. The evaluation of impacts on urban settlements has to be identified on a regional scale, because there are only a few studies on impacts of climate change in cities. It is recognized that climate change, is an additional environmental challenge that will increase the already existing water problems in different regions. The negative impacts of climate change, such as the decreasing availability of water, the increase in annual river flows, or changes related to glacial melt, sea level rise and extreme events, will affect cities in almost every region; but cities in developing countries will suffer more because of challenges that are already faced, as well as a lack of economic resources for infrastructure and integrated water management programmes. To address urban water management, the concept of the urban water cycle has arisen, which provides a unifying concept for addressing climatic, hydrologic, land use, engineering and ecological issues in urban areas. Under this concept, adaptation strategies have to be addressed, integrating climate and water responses into development and poverty eradication.

5.1 INTRODUCTION

5.1.1 Global climate change

According to the Intergovernmental Panel of Climate Change (IPCC), the greenhouse gas effect can be described as follows. The earth absorbs radiation from the sun mainly at the surface. This energy is redistributed by the atmospheric and oceanic circulations and radiated back to space at longer wavelengths. For the annual mean and for the earth as a whole, the incoming solar radiation energy is balanced approximately by the out-going terrestrial radiation. Any factor that alters the radiation received from the sun or lost to space, or that alters the redistribution of energy within the atmosphere and between the atmosphere land and ocean can affect climate (IPCC, 2001a). The greenhouse gases in the atmosphere, such as water vapour, carbon dioxide and methane, absorb the outgoing terrestrial radiation and then re-emit it at higher altitudes and lower temperatures.

Greenhouse gases are necessary to life as it is known, because they keep the planet surface at a certain temperature. However, as the concentration of theses gases continues to increase in the atmosphere, the earth's temperature is climbing, producing what is known as the global climate change. Some of the greenhouse gases have a long life in the atmosphere; this means that anthropogenic climate change will persist for many centuries, even if emissions are stabilized. That is why strong mitigation policies are needed. For example, several centuries after CO_2 emissions occur, about a quarter of the increase in CO_2 concentration caused by these emissions is still present in the atmosphere (IPCC, 2001a).

Changes in climate are already occurring, and there is strong evidence that most of the warming observed in the last decades is attributable to human activities. As Karl and Trenberth (2003) pointed out, there is no doubt that the composition of the atmosphere is changing because of human activities, and today greenhouse gases are the largest human influence on global climate.

The very recent report on Climate Change: The Physical Science Basis Summary for Policymakers (IPCC, 2007) presents new findings and conclusions. Some of them are presented as follows:

Carbon dioxide is the most important anthropogenic greenhouse gas. The global atmospheric concentration of carbon dioxide has increased from a pre-industrial value of about 280 parts per million (ppm) to 379 ppm in 2005. The primary source of the increased atmospheric concentration of carbon dioxide since the pre-industrial period results from fossil fuel use, with land use change providing another significant but smaller contribution.

Eleven of the last twelve years (1995–2006) rank among the 12 warmest years in the instrumental record of global surface temperature (since 1850). The linear warming trend over the last 50 years (0.13 [0.10 to 0.16]°C per decade) is nearly twice that for the last 100 years.

Mountain glaciers and snow cover have declined on average in both hemispheres. Widespread decreases in glaciers and ice caps have contributed to sea level rise (ice caps do not include contributions from the Greenland and Antarctic ice sheets).

New data since the Third Assessment Report (TAR) now show that losses from the ice sheets of Greenland and Antarctica have very likely contributed to sea level rise over 1993 to 2003.

Global average sea level rose at an average rate of 1.8 [1.3 to 2.3] mm per year from 1961 to 2003. The rate was faster from 1993 to 2003, about 3.1 [2.4 to 3.8] mm per year. Whether the faster rate for 1993 to 2003 reflects decadal variability or an increase in the longer-term trend is unclear. There is *high confidence* that the rate of observed sea level rise increased from the 19th to the 20th century. The total 20th century rise is estimated to be 0.17 [0.12 to 0.22] m.

Based on the same report (IPCC, 2007), future impacts of climate change according to different emission scenarios show that from the end of the twentieth century to the last decade of the twenty-first century temperature will increase from 1.8°C (range from 1.1 to 2.9°C) to 4°C (range from 2.4 to 6.4°C) and the sea will rise from 0.28 m (range from 0.18 to 0.38 m) to 0.43 m (range from 0.26 to 0.59 m). Box 5.1 shows the last results from IPCC (2007).

Climate change is one of the biggest threats to human development, and since climate and water are inextricably related, shifts in the hydrological cycle and rainfall patterns and the impacts of higher surface temperature on water evaporation will affect, or are affecting already, the fresh water availability and distribution on the planet.

BOX 5.1 Comparison of observed continental-and global-scale changes in surface
temperature with results simulated by climate models using natural and
anthropogenic forcings (See also colour plate 1)

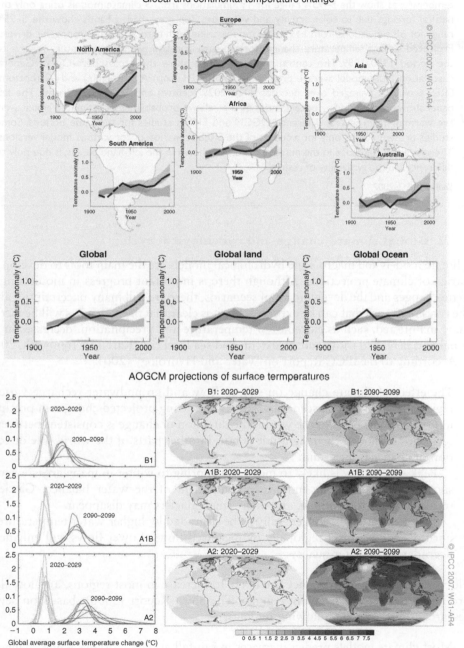

Global and continental temperature change

AOGCM projections of surface termperatures

(Continued)

BOX 5.1 (*Continued*)

Decadal averages of observations are shown for the period 1906–2005 (black line) plotted against the centre of the decade and relative to the corresponding average for 1901–1950. Lines are dashed where spatial coverage is less than 50%. The grey shaded bands that represent lower temperatures show the 5–95% range for 19 simulations from 5 climate models using only the natural forcings due to solar activity and volcanoes. Top grey shaded bands show the 5–95% range for 58 simulations from 14 climate models using both natural and anthropogenic drivers. Projected surface temperature changes for the early and late twenty-first century relative to the period 1980–1999. The central and right panels show the Atmosphere-Ocean General Circulation Multi-model average projections for the B1 (top), A1B (middle) and A2 (bottom) SRES scenarios averaged over decades 2020–2029 (centre) and 2090–2099 (right). The left panel shows corresponding uncertainties as the relative probabilities of estimated global average warming from several different AOGCM and EMICs studies for the same periods. Some studies present results only for a subset of the SRES scenarios, or for various model versions. Therefore the difference in the number of curves, shown in the left-hand panels, is due only to differences in the availability of results.

Source: IPCC, 2007.

5.1.2 Global climate change and hydrological cycle

Climate models and macro-scale hydrological models are the main tools to develop scenarios of climate projections. Although there is important progress in monitoring climate changes and the development of scenarios, there are still many uncertainties about the rates of change that can be expected, but it is clear that these changes will be increasingly manifested, such as extremes of temperature and precipitation, decreases in seasonal and perennial snow and ice extent, and sea level rise (Karl and Trenberth, 2003).

According to the IPCC Report on Water and Hydrology (2001):

The effect of climate change on stream flow and groundwater recharge varies regionally and between scenarios, largely following projected changes in precipitation. In some parts of the world, the direction of change is consistent between scenarios, although the magnitude is not. In other parts of the world, the direction of change is uncertain.

Peak streamflow is likely to move from spring to winter in many areas where snowfall currently is an important component of the water balance. Glacier retreat is likely to continue, and many small glaciers may disappear.

Water quality is likely generally to be degraded by higher water temperature, but this may be offset regionally by increased flows. Lower flows will enhance degradation of water quality

Flood magnitude and frequency are likely to increase in most regions, and low flows are likely to decrease in many regions. The Stern Report (2006), based on several researches, presents the following findings:

Most climate models predict increases in rainfall at high latitudes while changes in circulation patterns are expected to cause a drying of the subtropics.

There is more uncertainty about changes in rainfall in the tropics mainly because of complicated interactions between climate change and natural cycles like *El Niño* that dominate climate in tropics (Collins and the CMIP Modelling Groups, 2005).

Greater evaporation and more intense rainfall will increase the risk of droughts and flooding in areas already at risk.

Hurricanes and other storms are likely to become more intense in a warmer more energized world, as the water cycle intensifies (Huntington, 2006) meaning that severe floods, droughts and storms will occur more often; but changes to their location and overall numbers remain less certain.

The Fourth Assessment Report of the IPCC for policymakers, presented in early 2007, illustrates the following issues regarding precipitation and extreme events:

Sea ice is projected to shrink in both the Arctic and Antarctic under all scenarios. In some projections, Arctic late-summer sea ice disappears almost entirely by the latter part of the 21st century.

It is *very likely* that hot extremes, heat waves, and heavy precipitation events will continue to become more frequent.

Based on a range of models, it is likely that future tropical cyclones (typhoons and hurricanes) will become more intense, with larger peak wind speeds and more heavy precipitation associated with ongoing increases of tropical SSTs. There is less confidence in projections of a global decrease in numbers of tropical cyclones. The apparent increase in the proportion of very intense storms since 1970 in some regions is much larger than simulated by current models for that period.

Extra-tropical storm tracks are projected to move poleward, with consequent changes in wind, precipitation, and temperature patterns, continuing the broad pattern of observed trends over the last half-century.

Increases in the amount of precipitation are *very likely* in high-latitudes, while decreases are likely in most subtropical land regions.

Box 5.2 shows projected patterns of precipitation changes according to the last report of the IPCC.

5.1.3 Mitigation of GHG emissions

While mitigation of GHG emissions is not an issue for this chapter, it is important to mention that over a decade ago, most countries joined an international treaty – the United Nations Framework Convention on Climate Change (UNFCCC) – to begin to consider what can be done to reduce global warming and to cope with whatever temperature increases are inevitable. The Kyoto Protocol envisages a reduction in carbon dioxide emissions of 5% against 1990 level by 2012. However, the goals of the Kyoto protocol are difficult to assess. One of the major problems is that the United States has not ratified the protocol. The US emissions represents about one-fifth of the global GHG emissions. Therefore, successful mitigation of climate change will necessarily need additional commitments.

BOX 5.2 Relative changes in precipitation (in %) for the period 2090–2099, relative to 1980–1999

Projected patterns of precipitation changes

© IPCC 2007: WG1-AR4

Values are multi-model averages, based on the SRES A1B scenario for December to February (left) and June to August (right). White areas are where less than 66% of the models agree in the sign of the change and stippled areas are where more than 90% of the models agree in the sign of the change

Source: IPCC, 2007.

5.2 WATER IN AN URBANIZED WORLD

According to the UN, the twentieth century witnessed the rapid urbanization of the world's population. The global proportion of urban population increased from a mere 13% in 1900 to 29% in 1950 and, according to the *2006 Revision* of *World Urbanization Prospects*, reached 50% in 2006. Since the world is projected to continue to urbanize, 60% of the global population is expected to live in cities by 2030, meaning 4.9 billion people.

Urbanization can have positive sides for economic growth and development. However, from an environmental and social point of view, urbanization presents enormous challenges, especially in developing countries. Poverty is already a growing problem in urban areas and increasingly extending in peri-urban areas or informal settlements. As the population of cities expands, poverty is becoming urbanized. About 1 billion people live in slums, places characterized by insecurity of tenure, poor housing conditions, deficient access to safe drinking water and sanitation or severe overcrowding (UN, 2006).

In addition, the rapid urbanization that occurs in developing countries hinders the development of adequate planning and infrastructure. Urban environmental problems include air pollution; land and water pollution; conversion of land from agriculture and forest to urban, which in many cases reduces infiltration to aquifers; and settlements in unsuitable and risk locations, such as floodplains and hillsides. Urbanization affects surface and groundwater in terms of quantity and quality. It depends on the sustainable use of the resource. In developing countries, lack of resources, infrastructure and planning leads to over-exploitation and pollution. Marsalek et al. (2005) describe in detail the impacts of urbanization in the environment.

The following numbers show the proportion of urban water supply and sanitation by region, according to WHO/UNICEF (2000):

- Urban water supply:
  - World: 94%
  - North America: 100%
  - Europe: 100%
  - Oceania: 98%
  - Latin America and the Caribbean: 93%
  - Asia: 93%
  - Africa: 85%
- Urban sanitation:
  - World: 86%
  - North America: 100%
  - Europe: 99%
  - Asia: 78%
  - Latin America & the Caribbean: 87%
  - Oceania: 99%
  - Africa: 84%

According to WHO/UNICEF (2006), to achieve the Millennium Development Goal, by 2015, 961 million urban dwellers must gain access to improved water supply and over 1 billion must gain access to improved sanitation.

These figures present a huge challenge to meet water and sanitation, but they are even more outstanding if it is recognized that many urban centres already face difficulties in obtaining sufficient freshwater. Many cities have outgrown their capacity to provide adequate water supplies, as all nearby surface water sources have been tapped and/or groundwater resources are being drawn much faster than their natural rate of recharge.

5.2.1 Water scarcity

Although the world has more water than the 1,700 cubic metre per person that is established by convention as the amount needed to grow food, support industries and the environment, it is unevenly distributed on the planet, both, within regions and in time. Much of Asia receives almost its 90% of the annual flows in 100 hours a year, generating risks of floods during the peak flows and drought the rest of the year (UNDP, 2006). In China, 42% of the population in the northern region has access to only 14% of the country's water.

Availability below 1,000 cubic metres represents, according to hydrologists, water scarcity or regions under water stress. Today, about 700 million people in 43 countries live under this condition, and it has been estimated that by 2025, the share of the world's population living in water-stressed areas will be more than 3 billion (UNDP, 2006). Box 5.3 shows annual average water availability by sub-region in 2000 (UNEP, 2002), while Boxes 5.4 and 5.5 show the growth rate of the urban population

BOX 5.3 Water availability by sub-region (See also colour plate 2)

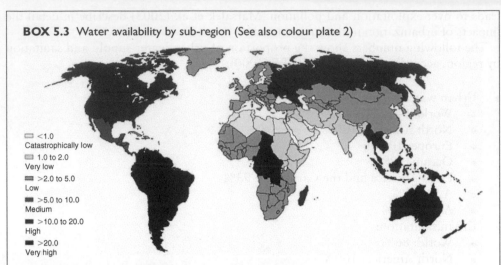

☐ <1.0
Catastrophically low
☐ 1.0 to 2.0
Very low
▬ >2.0 to 5.0
Low
▬ >5.0 to 10.0
Medium
▬ >10.0 to 20.0
High
▬ >20.0
Very high

Map shows water availability in terms of 1,000 m³ per capita/year

Source: Compiled from UNDP, UNEP, World Bank and WRI 2000 and United Nations Population Division
2001. From: Global Environmental Outlook 3, UNEP, 2002.

BOX 5.4 Urban population in major areas according to UN (2005) (See also colour plate 3)

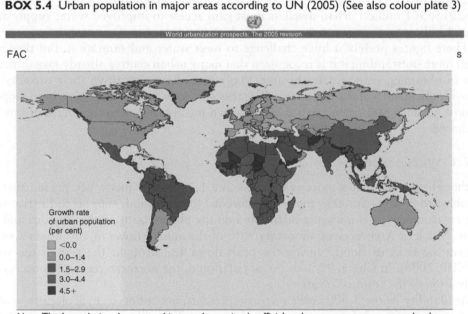

World urbanization prospects: The 2005 revision

FAC   s

Growth rate
of urban population
(per cent)
☐ <0.0
☐ 0.0–1.4
■ 1.5–2.9
■ 3.0–4.4
■ 4.5+

Note: The boundaries shown on this map do not imply official endorsement or acceptance by the
United Nations.
Source: United Nations, Department of Economic and Social Affairs, Population Division (2006).
World Urbanization Prospects: The 2005 Revision. Working Paper No. ESA/P/WP/200.

BOX 5.5 Worldwide distribution of cities by size class, above one million

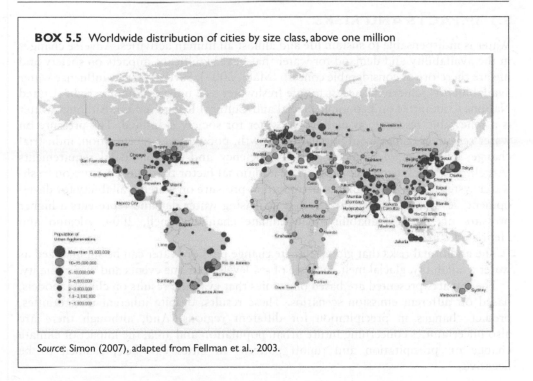

Source: Simon (2007), adapted from Fellman et al., 2003.

(UN, 2005) and worldwide distribution of cities by size class, above one million (Simon, 2007).

Comparing figures, it is clear that Sub-Saharan Africa and southern Asia both have increasing urban populations under water-stressed conditions.

According to UNDP (2006), by 2025 it is expected that the stress across Sub-Saharan Africa will intensify, with the share of the region's population in water-stressed countries rising from 20% to 85%. By 2025, average water availability is expected to be just over 500 cubic metres per person, and more than 90% of the Middle East and North Africa will be living in water-scarce countries. India and China are highly populated countries that will have water stress. An additional pressure on water availability will be the competition among consuming sectors. Globally, agriculture represents 70% of global water withdrawals, the industrial sector 22% and municipal 8%. However, according to UNDP (2006) from FAO data, in developing countries, agriculture share rises to 80%, while in high-income OECD countries agriculture represents just 40% and industrial use the other 40%. In addition, in terms of water use share, in the future, the patterns of water demand are expected to change, as urbanization and the growth of manufacturing increases, it is expected that non-agriculture water use will increase. However, the increased population will need more food production, and therefore, agricultural use is also expected to increase.

As will be shown in next part, climate change is an additional environmental challenge that will exacerbate the already existing water crisis.

5.3 IMPACTS AND RISKS

Water is indispensable to sustain life and almost all human activities. Adverse changes in the availability and demand for water that have significant impacts on society and life are therefore of considerable concern (Mata, 2003). Several factors influence water availability to achieve the access to safe freshwater as a universal human right (United Nations Committee on Economic, Social and Cultural Rights, 2003). Although water is a renewable resource, availability of water for society is limited. The pressure on water resources comes from population growth, population concentration, industrial change, expansion of irrigation, water efficiency and environmental requirements (Arnell, 1999). Climate change is just an additional factor that puts pressure on freshwater systems, and although it is an important pressure on water availability for development, several researchers affirm that increasing water demand represents a higher pressure on water availability than climate change (Arnell, 2004; Alcamo and Henrichs, 2002).

The additional risks that global climate change puts on water can be summarized as water availability, glacial melt, increase of sea levels, extreme events and water quality.

The impacts presented are based on studies that evaluate results on climate models, based on different emission scenarios. These results, despite inherent uncertainties, predict changes in precipitation for different regions. And, although there are also uncertainties concerning future urban population and area, the impact of climate change on precipitation and runoff patterns in cities can be presumed to be negative.

5.3.1 Water availability and glacial melt

There are important uncertainties in impacts on water availability related to global climate change. For the same emission scenario, different models present different results in geographical distributions of impacts and rates of growth. However, the Stern report (2006), based on different references (Arnell, 2006; Milly et al., 2002; Burke et al., 2006; Lehner et al., 2001, Warren et al., 2006, Trenberth et al., 2003; Trenberth et al., 2004) gives the following predictions:

> Differences in water availability between regions will become increasingly pronounced. Areas that are already relatively dry such as the Mediterranean basin and parts of Southern Africa and South America are likely to experience further decreases in water availability. In contrast, South Asia and parts of Northern Europe and Russia are likely to experience increases in water availability.
>
> The changes in the annual volume of water that each regions receives, masks the impacts on year to year and seasonal variability. An increase in annual river flows is not necessarily beneficial, particularly in highly seasonal climates because there may not be sufficient storage to hold extra water and rivers may flood more frequently.
>
> Some studies predict growing water shortages in Africa, the Middle East, Southern Europe and parts of South and Central America.
>
> 1–5 billion people mostly in South and East Asia may receive more water, however much of the extra water will come during the wet season.

Climate change will have serious consequences for people who depend heavily on glacier melt water to maintain supplies during the dry season, including large parts of the Indian sub-continent, over quarter of a billion in China, and tens of millions in the Andes. Water flows may increase in the spring as the glacier melts more rapidly; this can increase the risk of damaging glacial lake outburst floods, especially in the Himalayas, and also lead to shortages later in the year. In the long run dry-season water will disappear permanently once the glacier has completely melted. Parts of the developed countries in western US, Canadian prairies and Western Europe will also have their summer water supply affected unless storage capacity is increase.

Lenton (2004) estimates that many developing countries in the tropics not only face the usual technical, financial and institutional challenges that are an inherent part of the management of water, but also a set of additional problems related to climate and climate variability, especially rainfall and stream flow concentrated in a few short months and significant seasonal and annual departures from historical averages that translate into dry spells and recurrent droughts and floods.

The *Human Development Report* (UNDP, 2006) describes impacts of melting water flows for Pakistan, Nepal, China, the Andes region and Central Asia.

5.3.2 Sea level rise and extreme events

Sea level rise is a consequence of several factors. The most important is expansion due to increase in ocean temperature, changes in runoffs and ice sheet melting.

Based on the Stern report (2006) and Warren et al. (2006); Nicholls (2004), Nicholls and Tol (2006); Arnell (2006), the following predictions are related to sea level rise and extreme events.

Sea level rise will increase coastal flooding, rise costs of coastal protection, lead to loss of wet lands and coastal erosion, and increase intrusion into surface groundwater producing salinization.

Rising sea levels will increase the amount of land lost and people displaced due to permanent inundation. Coastal areas are amongst the most densely populated areas in the world. Critical infrastructure is often concentrated around coastlines, including oil refineries, power plants, port and industrial facilities. Currently, more than 200 million people live in coastal floodplains around the world. Bangladesh' s population (around 35 million) lives within the coastal floodplain. Many of the world' s major cities (22 up to 50) are at risk of flooding from coastal surges. Tokyo, Shangai, Hong Kong, Mumbai, Calcutta, Karachi, Buenos Aires, New York, Miami and London.

South and East Asia will be most vulnerable because of their large coastal populations in low-lying areas, such as Vietnam, Bangladesh and parts of China (Shangai) and India. Millions will be at risk around the coastline of Africa, particularly in the Nile Delta and along the west coast. Small island states in the Caribbean, and in the Indian and Pacific Oceans are acutely threatened, because of their high concentrations of development along the coast.

By increasing the amount of energy available to fuel storms, climate change is likely to increase the intensity of storms.

Models suggest that climate change will bring a warmer, wetter monsoon by the end of the century. This could increase water availability for around two billion people in South and East Asia. However the increase runoff would probably increase flood risk.

Confidence in projections of future rainfall variability is relatively low; however, this represents the difference between steady, predictable rainfall and destructive cycle of flooding and drought. Most models predict a modest increase in year to year variability but to differing degrees. At the heart of this are the projections of what happen with El Niño. Changes in variability within the wet season are more uncertain, but also vital to livelihoods. For example, in 2002, the monsoon rains failed during July, resulting in a seasonal rainfall deficit of 20%.

5.3.3 Water quality

According to US EPA and based on IPCC (2001b), water quality can be affected by higher water temperatures and changes in the timing, intensity and duration of precipitation. Higher temperatures reduce dissolved oxygen levels, which can have an effect on aquatic life. Decrease in stream flow and lake levels contribute to less dilution of pollutants; while, increased frequency and intensity of rainfall produce more pollution and sedimentation due to runoff.

Changes in water quality due to direct and indirect impacts may affect control facilities that have been designed to certain levels of pollution. In developing countries, where water quality is already a problem, especially to the poor population, climate change will be an additional factor to overcome (Jiménez, 2003).

Water quality is obviously related to health. Water-borne diseases have long been considered to be a major public health problem throughout the world, especially in developing countries. In general, these include such serious diseases as cholera, cyclosporiasis, cryptosporidiosis, campylobacter disease and leptospirosis (Watson et al., 2005).

5.3.4 Changes in the past decades related to global climate change

There are still important discussions in the scientific community about the impacts of anthropogenic climate change in recent drought and floods as well as other natural disasters. The 2007 IPCC report brings out the partial consensus of the scientific community on these issues. However, in this chapter we mention some reports that suggest changes in climate patterns related to global climate change.

According to Kron and Bertz, (2007) the number of great inland flood catastrophes during the last 10 years (between 1996 and 2005) is twice as large, per decade, as between 1950 and 1980, while the economic losses have increased by a factor of five. According to the World Disaster Report (2003 and 2004), floods have been the most reported natural disaster events in Africa, Asia and Europe, and have affected more people across the globe (140 million per year on average) than all other natural disasters. Mirza (2003) and Mirza et al. (2003) report that in Bangladesh, three extreme floods occurred in the last two decades, and in 1998 about 70% of the country area was inundated.

The IPCC report of the meeting on 'A multi millennia perspective on drought and implications for the future' (2003), reported that, although the relation between anthropogenic climate change and drought in the last century is not clear, scientists report that globally, as summarized by Dai et al (2004), the drought area increased by more than 50% during the twentieth century, largely due to drought conditions over the Sahel and southern Africa during the later part of the century.

However, Trenberth (2005) reports that:

> Although variability is large, trends associated with human influences are evident in the environment in which hurricanes form, and our physical understanding suggests that the intensity of and rainfalls from hurricanes are probably increasing, even if this increase cannot yet be proven with a formal statistical test. Model results suggest a shift in hurricane intensities toward extreme hurricanes. The fact that the numbers of hurricanes have increased in the Atlantic is no guarantee that this trend will continue, owing to the need for favorable conditions to allow a vortex to form while limiting stabilization of the atmosphere by convection. The ability to predict these aspects requires improved understanding and projections of regional climate change. In particular, the tropical ocean basins appear to compete to be most favorable for hurricanes to develop; more activity in the Pacific associated with El Niño is a recipe for less activity in the Atlantic. Moreover, the thermohaline circulation and other climate factors will continue to vary naturally.
>
> Trends in human-influenced environmental changes are now evident in hurricane regions. These changes are expected to affect hurricane intensity and rainfall, but the effect on hurricane numbers remains unclear. The key scientific question is not whether there is a trend in hurricane numbers and tracks, but rather how hurricanes are changing.

The IPCC (2007) presents new findings on recent changes in precipitation and extreme events:

> Long-term trends from 1900 to 2005 have been observed in precipitation amount over many large regions. Significantly increased precipitation has been observed in eastern parts of North and South America, northern Europe and northern and central Asia. Drying has been observed in the Sahel, the Mediterranean, southern Africa and parts of southern Asia. Precipitation is highly variable spatially and temporally, and data are limited in some regions. Long-term trends have not been observed for the other large regions assessed.
>
> Mid-latitude westerly winds have strengthened in both hemispheres since the 1960s.
>
> More intense and longer droughts have been observed over wider areas since the 1970s, particularly in the tropics and subtropics. Increased drying linked with higher temperatures and decreased precipitation, have contributed to changes in drought. Changes in sea surface temperatures, wind patterns, and decreased snowpack and snow cover have also been linked to droughts.
>
> The frequency of heavy precipitation events has increased over most land areas, consistent with warming and observed increases of atmospheric water vapor.
>
> There is observational evidence for an increase of intense tropical cyclone activity in the North Atlantic since about 1970, correlated with increases of

tropical sea surface temperatures. There are also suggestions of increased intense tropical cyclone activity in some other regions where concerns over data quality are greater. Multi-decadal variability and the quality of the tropical cyclone records prior to routine satellite observations in about 1970 complicate the detection of long-term trends in tropical cyclone activity. There is no clear trend in the annual numbers of tropical cyclones.

5.3.5 Risks for urban settlements

Although concern about the impacts of global climate change in urban systems has arisen, especially related to the magnitude of climate-related natural disasters in the last decade, there are still a few publications that evaluate impacts and vulnerabilities for urban settlements. Some case studies have been published, for example impacts for the cities of London and New York (Rosenzweig and Solecki, 2001) and Mexico City (Magaña, 2006).

The negative impacts of climate change in different regions of the world, such as the decreasing availability of water, the increase in annual river flows, or changes related to glacial melt, sea level rise and extreme events, will affect cities in almost every region; but cities in developing countries will suffer more because of the challenges that are already faced, as well as a lack of resources for infrastructure, and integrated water resources programmes.

Urban areas in developing countries already face enormous problems in terms of poverty, lack of services and segregated human spaces. Thus, the divergent rates and patterns of urbanization and the increase in poverty in urban areas are key elements in the rising vulnerability of urban areas to the negative consequences of global climate change (Sánchez-Rodríguez et al., 2005; Cross 2001; Hamza and Zetter, 1998; Jimenéz-Diaz, 1992).

The IPCC Third assessment report (2001b) addresses impacts for urban settlements as follows:

> Urban settlements feature many of the same impacts of climate change as other settlements—such as air and water pollution, flooding, or consequences of increasingly viable disease vectors. These impacts may take unusual or extremely costly forms in urban areas—for example, flooding that results not from river flooding but from overwhelmed urban storm drains and sewers during extreme rainfall events (which may become more common in the future). Urban settlements also experience the consequences of accommodating migrant populations, the unique aspects of urban heat islands (which affect human health and energy demand), and some of the more severe aspects of air and water pollution.

A recent IHDP report on urbanization and global environmental change by Sánchez-Rodriguez et al. (2005), points to a range of risks to urban settlements related to climate change, some of them are:

> Global environmental change can modify migration settlement patterns between urban and rural areas or within urban areas. Drought, flooding and other consequences associated with global environment can be strong drivers of these demographic changes.

Among the most important consequences of global biophysical processes are impacts on human health. Urban living conditions make urban residents particularly in poor countries, sensitive to problems related to severe deficiency in the supply and operation of public services, infrastructure, sanitation, and health service. Many of theses urban areas already face environmental problems and their inhabitants suffer malnutrition, poor housing condition and other problems associated with poverty and inequity. All this condition plays a role in aggravating the negative consequences of global climate change.

Climate variability and urban water cycle will affect the shape of built environment as they arise amid poorly coordinated administration and planning, the growing influence of an increased globalize economy, growing socio-economic disparities and intensifying environmental burdens.

Urban areas prove to be highly vulnerable in crisis and disasters. Constraints and conflicts may acquire multiple dimensions.

The consequences of climate related impacts also include other climate related problems like the loss of productive activities, and deficiencies in urban functions. These are processes common in a large number of megacities and intermediate urban centers in Latin America, Asia and Africa.

Simon (2007) points out a difference between short-duration disasters and global climate change impacts on urban settlements which are or will be long lasting. Therefore, the policy responses to short-duration extreme events focus on forecasting, evacuation ahead of the event and then resettlement and other post-disaster recovery. Global climate change impacts will not be, for example, for a short period of inundation, but will need to be only the first stage of a comprehensive coping strategy that ultimately re-settled them away from the area likely to be permanently inundated. Thus, it requires a long-term and costly solution.

5.4 ADAPTATION AND INTEGRATION OF CLIMATE CHANGE INTO URBAN WATER RESOURCE MANAGEMENT

5.4.1 Adaptation and sustainable development

Adaptation has the potential to reduce the adverse impacts of climate change and to enhance beneficial impacts, and it is necessary because even with 'best case' mitigation efforts, some climate change cannot be avoided due to the inertia of the global climate system. Thus, adaptation is a necessary strategy at all scales to complement climate change mitigation efforts (IPCC, 2001b). However, adaptation represents costs. And the greater the impacts of climate change, without mitigation efforts, the greater the costs of adaptation and the limits of what it can achieve. As pointed out in the last Stern report (2006), the costs of climate change are likely to accelerate with increasing temperature, while the net benefit of adaptation is likely to fall relative to the cost of climate change.

Therefore, adaptation strategies, especially those related to water, have to integrate into a range of measures compatible with the challenges that all regions, countries and urban settlements are already facing in terms of their sustainable development.

In other words, successful adaptation strategies, will have to be developed in the context of wider strategies for sustainable development, including measures to reduce vulnerability to shocks and stress, this implies that adaptation is highly contextual, and urban and national planning, based on local participation, holds the key to success (UNDP, 2006).

5.4.2 Planning under uncertainties

The problem of integrating climate change impacts into water management is difficult because of the uncertainty associated with estimates of future climate change on hydrological and water resources, as well as extreme events. According to the IPCC (2001b), uncertainties have to be faced when assessing alternatives that include scenario and risk analysis. This requires continuing collection of data and the ability to use scenarios with hydrological models to estimate possible future conditions.

Parry et al. (2005) pointed out that adaptation needs vary across geographical scales (local, national, regional, global), temporal scales and must be addressed in complex and uncertain circumstances. Responding to this process requires interdisciplinary and multiple expertise at the local and international level. Top-down models, based on General Circulation Model scenarios, help to quantify and identify potential impacts of different ecosystems and economic sectors, however, its low resolutions limit the ability to inform on the regional and local impacts of climate change. A new generation of research is now addressing adaptation by taking a bottom-up/vulnerability-driven approach to adaptation that usually involves assessing past and current climate vulnerability, existing coping strategies and how this might be modified to climate change. Both model approaches are necessary in formulating adaptation policy and practice. Box 5.6 presents definitions of sensitivity, adaptive capacity and vulnerability.

BOX 5.6 Sensitivity, adaptive capacity and vulnerability

The following definitions by the IPCC are important to understand the differences between sensitivity, adaptive capacity and vulnerability.

Sensitivity is the degree to which a system is affected, either adversely or beneficially, by climate-related stimuli. Climate-related stimuli encompass all the elements of climate change, including mean climate characteristics, climate variability, and the frequency and magnitude of extremes. The effect may be direct (e.g., a change in crop yield in response to a change in the mean, range, or variability of temperature) or indirect (e.g., damages caused by an increase in the frequency of coastal flooding due to sea-level rise).

Adaptive capacity is the ability of a system to adjust to climate change (including climate variability and extremes) to moderate potential damages, to take advantage of opportunities, or to cope with the consequences.

Vulnerability is the degree to which a system is susceptible to, or unable to cope with, adverse effects of climate change, including climate variability and extremes. Vulnerability is a function of the character, magnitude, and rate of climate change and variation to which a system is exposed, its sensitivity, and its adaptive capacity.

Source: IPCC (2001b).

5.4.3 Supply and demand options

There are two kinds of adaptive measures that can be adopted: the more traditional 'supply-side' techniques, and the 'demand-side' techniques. Table 5.1 presents supply- and demand-side options according to this publication relative to urban water demand and supply (IPCC, 2001b). Supply-side options include infrastructure, such as increasing flood defences, building weirs and locks to manage water levels for navigation, and modifying or extending infrastructure to collect and distribute water to consumers. Demand-side techniques include water demand management, such as encouraging water-efficient irrigation and water pricing initiatives.

Given the scenario results, that flood frequency and intensity are both likely to increase as a result of climate change, special flood management strategies have to be taken in urban areas. From the supply side, new and better infrastructure for flood protection, such as defences, protections, reservoirs and storm drainage systems, have to be built. From the demand side, there are few options but they are important, for example the improvement of flood warning.

5.4.4 Urban water management

To address urban water management, the concept of the urban water cycle has arisen, which provides a unifying concept for addressing climatic, hydrologic, land use, engineering and ecological issues in urban areas. Marsalek et al. (2005) explain in detail the concept. They explain that the basic water management categories encompassed with this approach include:

Reuse of treated wastewater as a basis for disposing potential pollutants or a substitute for other sources of water supply for non potable uses.

Integrated stormwater, groundwater, water supply and wastewater based management, as the basis for: (1) economic and reliable water supply; (2) environmental flow management (deferment of infrastructure expansion, return of water to streams); (3) urban water-scape/landscape provision; (4) substitute non potable uses of water (wastewater and stormwater reuse); and (5) protection of downstream waters form pollution, and

Water conservation (demand management) based approaches including more efficient use of water (water saving devices, irrigation practices), replacing landscape for those having low water demand and use low water consumption industrial processes.

The impacts and adaptation measures for global climate change can be integrated under this approach, taking into account that the ability to adapt to climate variability and climate change is affected by a range of institutional, technological and cultural features at international, national, regional and local levels, in addition to the specific dimensions of the change being experienced. According to the IPCC (2001b), some of these features are: the capacity of water-related institutions; the legal framework for water administration; the wealth of nations in terms of natural resources and ecosystems; human-made capital (especially in the form of water control systems); and human capital (including trained personnel) that determines what nations can 'afford

Table 5.1 Supply-side and demand-side adaptive options: some examples (IPCC, 2001b)

Supply Side		Demand Side	
Option	Comments	Option	Comments
	Municipal water		
– Increase reservoir capacity	Expensive; potential environmental impact	– Incentives to use less water (e.g., through pricing)	Possibly limited opportunity; needs institutional framework
– Extract more from rivers or groundwater	– Potential environmental impact	– Legally enforceable water use standards (e.g., for appliances)	– Potential political impact; usually cost-inefficient
– Alter system operating rules	– Possibly limited opportunity		
– Inter-basin transfer	– Expensive; potential environmental impact	– Increase use of grey water	– Potentially expensive
– Desalinization	– Expensive (high energy use)	– Reduce leakage	– Potentially expensive to reduce to very low levels, especially in old systems
– Seasonal forecasting	– Increasingly feasible		
		– Development of non-water-based sanitation systems	– Possibly too technically advanced for wide application
	Industrial and power station cooling		
– Increase source capacity	– Expensive	– Increase water-use efficiency and water recycling	– Possibly expensive to upgrade
– Use of low-grade water	– Increasingly used		
	Hydropower generation		
– Increase reservoir capacity	– Expensive; potential environmental impact	– Increase efficiency of turbines; encourage energy efficiency	– Possibly expensive to upgrade
	– May not be feasible		
	Pollution control		
– Enhance treatment works	– Potentially expensive	– Reduce volume of effluents to treat (e.g., by charging discharges)	– Requires management of diffuse sources of pollution
		– Catchments management to reduce polluting runoff	
	Flood management		
– Increase flood protection (levees, reservoirs)	– Expensive; potential environmental impact	– Improve flood warning and dissemination	– Technical limitations in flash-flood areas, and unknown effectiveness
– Catchments source control to reduce peak discharges	– Most effective for small floods	– Curb floodplain development	– Potential major political problems

Source: IPCC (2001b)

to commit' to adaptation. This is the major constraint on adaptation to climate change in poorer countries; the state of technology and the framework for the dissemination; mobility of human populations to change residential and work locations in response to severe climate events or climate change; the speed of climate change is crucial in determining the capabilities of societies to adapt and change water management practices; the complexity of management arrangements may also be a factor in response; the ability of water managers to assess current resources and project future resources.

5.4.5 Poverty and equity

Many of the urban settlements in developing countries already face environmental problems and their inhabitants suffer many problems associated with poverty and inequity. These conditions play a role in aggravating the negative consequences of climate change. Lack of resources and infrastructure add to societal characteristics that maximize susceptibility to climate change that include: poverty and low income levels, which prevent long-term planning and provisioning at the household level; lack of water control infrastructures; lack of maintenance and deterioration of existing infrastructure; lack of human capital skills for system planning and management; lack of appropriate, empowered institutions; absence of appropriate land-use planning; high population densities and other factors that inhibit population mobility; increasing demand for water because of rapid population growth; conservative attitudes toward risk (unwillingness to live with some risks as a tradeoff against more goods and services (risk aversion)) and lack of formal links among the various parties involved in water management (IPCC, 2001b).

5.4.6 International aid

International support is a precondition for successful adaptation (UNDP, 2006). But international aid has to be part of an integrated strategy to reduce poverty and adjust adaptation to climate variability, recognizing and supporting poor countries' and communities' own plans and paths. Also, the reduction of international debt has to be taken into account, because, in real terms, it enables economic recovery. International aid has to be understood as direct resources from rich to poor countries rather than debts that have to be paid in the future. As pointed out by a report on poverty and climate change by several agencies, such as the African Development Bank (2003), 'The evidence from past experience suggests that this is best achieved through mainstreaming and integrating climate responses into development and poverty eradication processes, rather than by identifying and treating them separately.' .

5.5 CONCLUSIONS

Half of the world' s population is now classified as living in urban areas, and 60% of the global population is expected to live in cities by 2030. Despite the importance of the urban population, there are very few studies on the impacts of global climate change in urban areas.

Many urban centres already face difficulties in obtain sufficient freshwater to fulfil basic needs for their inhabitants. The situation will be aggravated because, between other factors, population growth in urban centres, and competition of water for different uses, climate change is just an additional factor that puts pressure on freshwater systems.

According to the Stern report and to the IPCC, in future decades, some of the climate change problems related to water will be: 1) increasingly pronounced differences in water availability between regions; 2) serious consequences for people who depend heavily on glacier melt water to maintain supplies during the dry season; 3) several harms related to sea level rise; 4) increasing intensity of storms; 5) *very likely*, hot extremes, heat waves, and heavy precipitation events will continue to become more frequent; and, 6) future tropical cyclones will become more intense.

These changes can affect water quality because of higher water temperatures and changes in the timing, intensity, and duration of precipitation, and water quality is obviously related to health.

The impacts presented are based on studies that evaluate the results of climate models, based on different emission scenarios. These results, although there are uncertainties in the models, predict changes in precipitation for different regions that are presumed to be negative for urban population in main cities.

Adaptation has the potential to reduce adverse impacts of climate change and to enhance beneficial impacts, and it is necessary, because even with 'best case' mitigation efforts, some climate change cannot be avoided due to the inertia of the global climate system. However, adaptation represents costs. And the greater the impacts of climate change, without mitigation efforts, the greater the costs of adaptation and the limits of what it can achieve. To address urban water management, the concept of the urban water cycle has arisen, which provides a unifying concept for addressing climatic, hydrologic, land use, engineering, and ecological issues in urban areas. Under this concept adaptation strategies have to be addressed, integrating climate and water responses into development and poverty eradication. Actual conditions, lack of resources and the regional impacts of climate change show that poor countries and communities will suffer more. Under these circumstances, international support and aid is a condition for successful adaptation for developing countries.

More studies are needed to understand the impacts of global climate change in urban settlements. These have to integrate a multidisciplinary approach, and integrated vision on economic, social, urbanism, hydrological, biological, health, uncertainty and risk, and engineering expertise.

REFERENCES

African Development Bank, Asian Development Bank, Department for International Development, United Kingdom Directorate-General for Development, European Commission Federal Ministry for Economic Cooperation and Development, Germany Ministry of Foreign Affairs – Development Cooperation, The Netherlands Organization for Economic Cooperation and Development, United Nations Development Programme, United Nations Environment Programme and The World Bank. 2003. *Poverty and Climate Change: Reducing the Vulnerability of the Poor through Adaptation. Part 1*. pp. 1–14. Available at: www.ids.ac.uk/ids/pvty/pdf-iles/PovertyAndClimateChangeReportPart12003%5B1%5D.pdf.

Alcamo, J. and Henrichs, T. 2002. Critical regions: A Model Based Estimation of World Water Resources Sensitive to Global Changes, *Aquatic Science*, Vol. 64, pp. 352–362.

Arnell, N.W. 1999. Climate Change and Global Water Resources, *Global Environmental Change*, Vol. 9, pp. S31–S49.

Arnell, N.W. 2004. Climate Change and Global Water Resources: SRES Emissions and Socioeconomic Scenarios, *Global Environmental Change*, Vol. 14, pp. 31–52.

Arnell, N.W. 2006. Climate Change and Water Resources. Schellnhuber, H.J., Cramer, W., Nakicenovic, N., Wigley, T. and Tohe, G. (eds) *Avoiding Dangerous Climate Change*. Cambridge University Press, Cambridge, pp. 167–175.

Burke, E.J., Brown, S.J. and Christidis, N. 2006. Modelling the Recent Evolution of Global Drought and Projections for the Twenty-First Century with the Hadley Centre Climate Model, *Journal of Hydrometeorology*, Vol. 7, pp. 1113–1125.

Collins, M. and the CMIP Modelling Group. 2005. El Nino or La Nina-like Climate Change? *Climate Dynamics*, Vol. 24, pp. 89–104.

Cross, J. 2001. Megacities and Small Towns: Different Perspectives on Hazard Vulnerability, *Environmental Hazards*, Vol. 3, No. 2, pp. 63–80.

Dai, A., Lamb P.J., Trenberth, K.E., Hulme, M., Jones, P.D. and Xie, P. 2004. The Recent Sahel Drought is Real, *International Journal of Climatology*, Vol. 24, pp. 1323–1331.

Fellman, J., Getis, A. and Getis, J. 2003. *Human Geography Landcapes of Human Activities*. 7th edn. Brown Publishers, Dubuque, Iowa.

Hamza, M. and Zetter, R. 1998. Structural Adjustment, Urban Systems and Disaster Vulnerability in Developing Countries, *Cities*, Vol. 15, No. 4, pp. 291–299.

Hungtinton, T.G. 2006. Evidence for Intensificaction of Global Water Cycle: Review and Synthesis, *Journal of Hydrology*, Vol. 319, pp. 1–13.

Intergovernmental Panel on Climate Change. 2001a. *Climate Change 2001, The Scientific Basis*. Cambridge University Press, Cambridge (www.ipcc.ch).

Intergovernmental Panel on Climate Change. 2001b. *Climate Change 2001: Impacts, Adaptation and Vulnerability*. Cambridge University Press, Cambridge (www.ipcc.ch).

Intergovernmental Panel on Climate Change. 2003. *A Multi Millennia Perspective on Drought and Implications for the Future* (www.ipcc.ch).

Intergovernmental Panel on Climate Change. 2007. *Climate change 2007: The Physical Science Basis; Summary for Policy Makers*. Cambridge University Press, Cambridge (www.ipcc.ch).

Jimenéz-Diaz, V. 1992. Landslides in the Squatter Settlements of Caracas; Towards a Better Understanding of Causative Factors, *Environment and Urbanization*, Vol. 4, No. 2, pp. 433–441.

Jiménez, B. 2003. Health Risks in Aquifer Recharge with Recycle Water. Aertgeerts, R. and Angelakis, A. (eds) *State of the Art Report Health Risk in Aquifer Recharge Using Reclaimed Water*. WHO Regional Office for Europe. pp. 54–172.

Karl, T.R. and Trenberth, K.E. 2003. Modern Global Climate Change, *Science*, Vol. 302, No. 5651, pp. 1719–1723.

Kron, W. and Berz, G. 2007. Flood Disasters and Climate Change: Flood Disasters and Climate Change: Trends and Options - (re-)insurer' s View. Lozán, J., Graßl, H., Hupfer, P., Menzel, L. and Schönwiese, C.D. (eds) *Global Change: Enough Water for all?* Hamburg, pp. 268–273.

Lehner, B., Henrichs, T., Döll, P. and Alcamo, J. 2001. *EuroWasser: Model-Based Assessment of European Water Resources and Hydrology in the Face of Global Change*. World Water Series 5, Centre for Environmental Systems Research, Kassel, University of Kassel. Available at: www.usf. unikassel.de/usf/archiv/dokumente/kwws/kwws.5.en.html.

Lenton, R. 2004. Water and Climate Variability: Development Impacts and Coping Strategies, *Water Science and Technology*, Vol. 49, No. 7, pp. 17–24.

Magaña, V. 2006. Vulnerabilidad. Sheinbaum, C. and Vázquez, O. (Coo) *Estrategia Local de Acción Climática del Distrito Federal*. Gobierno del Distrito Federal, Mexico City. Available at www.sma.df.gob.mx/sma/download/archivos/elac/09_vulnerabilidad.pdf.

Marsalek, J., Jiménez-Cisneros, B.E., Malmquist, P.A., Karamouz, M., Goldenfum, J. and Chocat, B. 2005. *Urban Water Cycle Processes and Interactions: UNESCO IHP Program*. Geneva.

Mata, L.J. 2003. Advancing our Understanding of Climate Change and Water Related Problems: An Overview. *Proceeding of the World Climate Change. Conference (WCCC 2003)*. Moscow.

Milly, P.C.D., Wetherald, R.T., Dunne, K.A. and Delworth, T.L. 2002. Increasing Risk of Great Floods in a Changing Climate, *Nature*, Vol. 415, pp. 514–517.

Mirza, M.M.Q. 2003. Three Recent Extreme Floods in Bangladesh, a Hydro-Meteorological Analysis, *Natural Hazards*, Vol. 28, pp. 35–64.

Mirza, M.M.Q., Warrick, R.A. and Ericksen, N.J. 2003. The implications of Climate Change on Floods of the Ganges, Brahmaputra and Meghna Rivers in Bangladesh, *Climatic Change*, Vol. 57, pp. 287–318.

Nicholls, R.J. 2004. Coastal Flooding and Wetland Loss the 21st century: Changes Under the SRES Climate and Socioeconomic Scenarios, *Global Environmental Change*. Vol. 14, pp. 69–86.

Nicholls, R.J. and Tol, R.S.J. 2006. Impacts and Responses to Sea-Level Rise: a Global Analysis of the SRES Scenarios Over 21st Century, *Philosophical Transactions of the Royal Society*, A 364, pp. 1073–1095.

Parry, J.E., Hammile, A. and Drexhage, J. 2005. *Climate Change and Adaptation*. Institutional Institute for Sustainable Development, Winnipeg, Manitoba, Canada. (www.iisd.org).

Rosenzweig, C. and Solecki, W.D. 2001. Global Environmental Change and a Global City: Lessons for New York, *Environment*, Vol. 43, No. 3, pp. 8–18.

Sánchez-Rodríguez, R., Seto, K.C., David, S., Solecki, W.D., Krass, F. and Laumann, G. 2005. *Science Plan: Urbanization and Global Environmental Change IHDP Report 15 International Human Dimmensions Programme on Global Environmental Change*. Bonn (www.ihdp.org).

Simon, D. 2007. Cities and Global Environmental Change: Exploring the Links, *The Geographical Journal*, Vol. 173, No. 1, pp. 75–92.

Stern, N. 2006. *Stern Review on the Economics of Climate Change*. Final Report. Available at: www.hm-treasury.gov.uk/independent_reviews/stern_review_economics_ climate_change/stern_review_report.cfm.

Trenberth, K.E. 2005. Uncertainty in Hurricanes and Global Warming, *Science*, Vol. 308, pp. 1753–1754.

Trenberth, K.E., Dai, A., Rasmussen, R.M. and Parsons, D.B. 2003. The Changing Character of Precipitation, *Bulletin of the American Meteorological Society*, Vol. 84, pp. 1205–1217.

Trenberth, K.E., Overpeck, J. and Solomon, S. 2004. Exploring Drought and its Implications for the Future, *Eos*, Vol. 85, No. 3, pp. 27–29.

United Nations. 2006. *Water: A Shared Responsibility*. The United Nations World Water Development Report 2.

UN Department of Economic and Social Affairs: Population division. 2003. *World Urbanization Prospects: The 2003 Revision*. Available at: www.un.org/esa/population/publications/wup2003/WUP2003Report.pdf.

UN Department of Economic and Social Affairs: Population division. 2007. *World Urbanization prospects: The 2006 Revision*. Available at: www.un.org/esa/population/publications/wpp2006/wpp2006.htm.

United Nations Committee on Economic, Social and Cultural Rights. 2003. (www. un.org).

United Nations Development Programme. 2006. *Human Development Report 2006. Beyond Scarcity: Power, Poverty and the Global Water Crisis*. New York.

United Nations Environment Programme. 2002. *Global Environmental Outlook 3*.

Warren, R., Arnell, N., Nicholls, R., Levy, P. and Price, J. 2006. *Understanding the Regional Impacts of Climate Change*. Research Report Prepared for the Stern Review, Tyndall Centre Working Paper 90 Norwich. Tyndall Centre. Available at: www.tyndall.ac.uk/publications/working_papers/twp90.pdf.

Watson, R.T, Patz, J., Gubler, D.J., Parson, R.A. and Vincent, J.H. 2005. Environmental Health Implications of Global Climate Change, *Journal of Environmental Monitoring*, Vol. 7, pp. 834–843. Available at www.rsc.org/delivery/_ArticleLinking/ DisplayHTMLArticleforfree. cfm? JournalCode=EM&Year=2005&ManuscriptID=b504683a&Iss=9.

WDR. 2003. *World Disaster Report: Focus on Ethics in Aid.* International Federation of Red Cross and Red Crescent Societies.

WDR. 2004. *World Disaster Report: Focus on Community Resilience.* International Federation of Red Cross and Red Crescent Societies.

World Health Organization and United Nations Children's Fund. 2000. *Global Water Supply and Sanitation Assessment 2000 Report.* Geneva.

World Health Organization and United Nations Children's Fund. 2006. *Meeting the MDG Drinking Water and Sanitation Target: The Rural and Urban Challenge of the Decade.* Geneva.

World Health Organization and United Nations Children's Fund. 2000. *Global Water Supply and Sanitation Assessment Report.* Geneva.

Chapter 6

Water source and drinking-water risk management

Francisco Cubillo

Head of Research Development and Innovation, Department, Canal de Isabel II, Madrid, Spain

ABSTRACT: The availability of enough and suitable resources to satisfy the different kinds of water demands in the urban context represents the main basis of supply services. The maintenance of the balance between availability and demand is subject to threats of different nature, and when this balance is broken it has a major impact on and generates costs to citizens, entities and even those responsible for the supply. The break-up of this balance represents one of the main risks which the urban water service must cope with. The problem has not yet been set or solved in a homogeneous way internationally, not even in contexts involving situations of hydric stress. The solutions posited in this paper focus on methodological and operational aspects. From the methodological point of view, a clear difference of setting has been established, with a clear separation between those related to the failure or normal situation and indicators and corresponding conventions to be used in planning, operating and solving contingencies. As for the operational aspects, a division into stages or phases is set with their corresponding effect and cost distribution among the different parties involved. In both approaches, considering risk as a main factor in analysis and decision making is the main pillar for the efficient management of resources and commitments with society and the environment.

6.1 INTRODUCTION

Urban supply systems always cope with the result of a territorial, economic and social development model. They are placed in an environmental and institutional context that dictates many of the peculiarities and conditions under which this activity is developed.

The important temporary evolution that major cities around the world have undergone is in general linked to fast growth, representing a challenge involving opportune adaptation of urban services to the needs and expectations of the citizens and in particular to water. Increases in population, changes in climate conditions, modifications in the availability of resources and increased social demands determine a higher than desirable frequency of episodes in which service conditions do not satisfy citizens' desires and expectations, to the point of even representing a danger to health in some extreme cases.

Within this context, with such important dynamic components, non-desired situations arise with effects and impacts, or simply high probability values of the same occurring.

Societies in the twenty-first century measure the degree of development by the capacity for reaction to these situations, through their prevention, mitigation and effective resolution, among other parameters.

Practices involving the design and operation of supply systems have historically incorporated different principles for facing these types of problems in accordance with the economic capacity of each case and circumstance, and social expectations involving continuity, stability and service quality. Even so, applied technical procedures have adapted to the new legal and cultural frameworks and to the needs that have derived from the changing conditions of each system.

Currently, risk constitutes one of the fundamental components in management of urban supply systems, understood as a combination of the probability of certain threats arising that cause damage or affect one or more social agents.

This chapter posits a methodology for the rigorous incorporation of risk evaluation and management principles in planning and operation tasks pertaining to urban supply systems.

6.2 SECURITY, RELIABILITY AND RISK

One of the main attributes of a supply system is its capacity to satisfy certain service levels under all foreseeable circumstances. Evaluating this capacity is a complex task that comes face-to-face not only with its technical difficulty, but also with a diversity of points of view, measurement parameters and calculation methods.

It is necessary to distinguish the following terms to properly understand this capacity in any supply system:

Security. This is the reflection of a desire or expectation of quality and continuity in the availability of a service or condition. In the case of citizens or users of the water supply service, security would represent expectations of continued compliance with certain availability conditions covering the amount and quality of the water at the points of use and consumption. This desire does not necessarily have to correspond with that established in the regulations, or with the habitual criteria used to plan or design systems, and even less so with the characteristics of a certain system at a given moment in time.

Reliability. This is the capacity of a system to continuously fulfil the 'satisfactory' conditions of a service. The regulatory framework may establish acceptable discontinuity in the compliance of service conditions and, in fact, the entire system is in part subject to some unavoidable punctual interruptions. Reliability quantifies the probable degree of perturbation to resource availability continuity conditions in a range of hydraulic pressure and flow conditions, and compliance with certain water quality conditions. The guarantee is usually measured as the probability of falling within these 'satisfactory' conditions, which do not necessarily coincide with those linked to Security.

Figure 6.1 outlines three types of reliability, depending on whether the availability of resources is evaluated, if capacity for continued compliance with individual users and major consumption areas (municipalities, neighborhoods, etc.) in that pertaining to hydraulic conditions and the quality of supplied water is evaluated, or if mere compliance with the service quality levels in each service connection is evaluated.

Risk. This circumstance corresponds to each system under certain conditions. It involves the probability of events that represent a threat and cause effects arising. In the case of supplies, the effects are always measured in terms of number of users and the time interval during which they do not continuously enjoy the service conditions established as satisfactory or 'standard'. Under the present situation of each system,

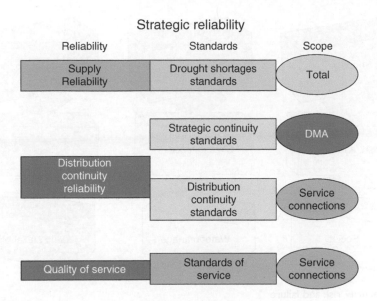

Figure 6.1 **Types of reliability**

there is a risk that corresponds to each one of the possible threats. In turn, a risk will correspond to each threat under any of the scenarios, horizons and future hypotheses considered.

A system may have a risk level that is different to the security level. Risk is complementary and constitutes improved information to that of the guarantee, given that it puts forward a relation with an increased range of effects rather than mere non-compliance with satisfactory conditions.

Failure. This is the circumstance in which service conditions are not complied with. A system may have a certain risk level, even lower than the one considered appropriate for security, and present a failure. Failure may take two forms: 1) an accepted episode, within the ranges of guarantee or probability of occurrence of threat episodes with their related impacts; 2) as a consequence of an event of severity arising that is greater than that foreseen or simply of a nature different to the one considered.

The main consideration in the event of a failure must be minimizing its duration and impact. For this, all measures aimed at resolution must take into account the risk, which always exists, of major impacts taking place in the development of the event that actually generated the failure, or in the resolution process itself.

If failure resolution processes are executed within the established protocols or require non-planned actions, these tend to be respectively called contingencies or crises.

Linked to the concept of failure we find **Recovery**, which is the period that elapses between the time when a failure (of any magnitude) occurs, to the system returning to its normal operating level.

The four terms reflect different concepts and approaches as well as the different parts involved in a supply service. Security is mainly linked to the society receiving the

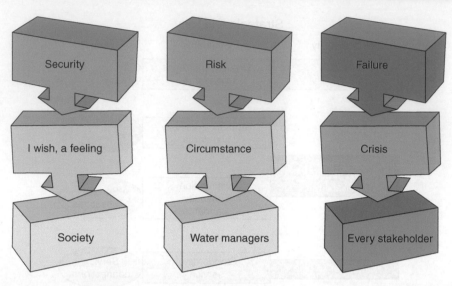

Figure 6.2 Security, risk and failure

service, given that it is a reflection of citizens' culture and expectations. The guarantee is a macro-parameter generally linked to design and planning tasks. The risk, given that it is a circumstance that burdens society as a whole, is a component of management executed from the institutions and agencies responsible for planning and, in particular, for the operation of systems. Risk and its management is, at the least, the implicit mission of those providing resources and handling related infrastructures. In the first place, failure is incumbent to all in which it generates a direct disruption and in the second, to those increasing the probability of repercussion or direct disruption. Finally, it involves those responsible for its resolution and system rehabilitation until it reaches its normal operating situation.

Risk and failure managers are obligated to incorporate a series of precautionary principles. The consideration of how many episodes can be foreseen with a scientific basis is a habitual component in risk evaluation, although also necessary is the inclusion of a set of considerations that will allow incorporation of the eventuality of threats, episodes and circumstances that could be the origin of even more unfavourable scenarios.

6.3 UNCERTAINTY, THREATS AND EFFECTS

Management of the great majority of water supply systems is an exercise in handling uncertainty. Factors that depend on climate conditions are subject to continuous changes, as in general infrastructures are buried and there is a lack of basic knowledge of their useful lifetime and installation conditions. Uses and consumption display a change in basic patterns, which reflect the changes that societies and lifestyles are undergoing, in addition to the fact that, to a large extent, final destinations are neither measured nor controlled, and the quality of water at its natural source, or where it is stored, suffers significant alterations in terms of its content. And there is also a great amount of uncertainty regarding foreseeable responses to the implantation of resolution measures for problems that,

Table 6.1 Frequency of events

High frequency of occurrence events	2–3%	Service connections burst
Medium frequency of occurrence events	0.3–0.5%	Pipe distribution bursts
Medium–Low frequency of occurrence events	0.02–0.05%	Failures in mains or strategic elements
Low frequency of occurrence events	<0.05%	Failures in water treatment process. Lack of records
Very low frequency of occurrence events		Earthquakes, hurricanes, etc. High geographic variation
Events of very variable frequency according to the system and geographic location		Sudden growth of demands or drought or shortage scenarios. Operational failures

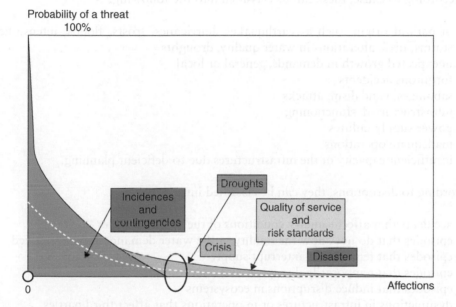

Figure 6.3 Droughts and risk function

fortunately, have not arisen in many supply systems. All these factors comprise uncertainty that must be faced when planning and managing water supply systems.

Risk corresponding to a system is likewise measured in accordance with how uncertainty can be classified, by way of a combination of probability of occurrence of episodes that represent a potential threat or danger to service provisioning and possible disruptions. The quantified combination of these factors is called risk.

All systems have a risk function covering possible threats and consequences. Table 6.1 summarizes some ranges of probability of occurrence of episodes that represent a threat to supply. Figure 6.3 outlines a theoretical risk function in a system as a combination of threats and effects, with segregation in accordance to the magnitude of the impacts. Episodes of shortage with interruptions in the supply conditions have been outlined on the borderline between frequent appearance contingencies and crises with a low probability of occurrence, although with no important disruptions. They contemplate situations that are accepted in the supply standards and are occasionally placed in scenarios of a severity that is greater than that foreseen, in some cases requiring extreme, improvised actions that should in fact be categorized as crises rather than contingencies.

While a hypothetical function of the risks of a system in a certain situation have been outlined in this same figure, what has also been reflected is what the objective risk function itself could be, the same that is defined as a consequence of a normative, regulation, knowledge of citizens' preference or mere establishment from the competent agency.

Potential threats or dangers for a supply system can be enumerated in accordance with their cause and possible disruptions.

According to cause, these can be classified into the following:

- of natural origin, such as earthquakes, hurricanes, frosts, floods, intense heat, storms, fires, alterations in water quality, droughts
- unexpected growth in demands, general or local
- fortuitous accidents
- sabotages, vandalism, attacks
- infrastructure dysfunctioning
- power supply failures
- inadequate operations
- insufficient capacity of the infrastructures due to deficient planning.

According to disruptions, they can be classified into:

- accidents that affect sanitary conditions of the supplied water
- episodes that do not allow the totality of the water demanded to be supplied
- episodes that temporarily interrupt supply
- episodes that temporarily disrupt service conditions
- episodes that induce disruptions in ecosystems
- dysfunctions in infrastructures or in operations that affect third parties
- scenarios that induce foresight of the occurrence of some of the aforementioned episodes and determine alterations caused in the service conditions
- episodes of incapacity to satisfy the amounts and conditions pertaining to essential uses
- episodes of supply shortages.

6.4 PREVENTION, MITIGATION AND RESOLUTION

Policies of supply systems are aimed at maximum user satisfaction, compliance with environmental factors within the regulatory framework and at optimal use of economic resources. Service disruption scenarios, called failures, are events to be avoided, minimized or maintained within the values established as standards or service levels. Two important lines of action are differentiated for this: prevention and mitigation. Each one of them corresponds to temporary different-dimension horizons and they are put forward from different points of view and techniques.

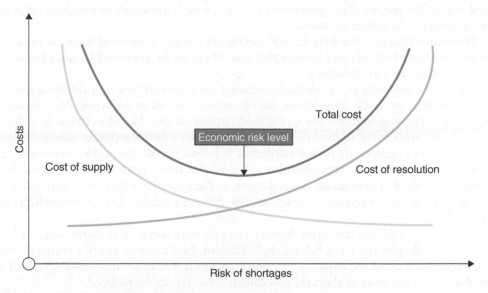

Figure 6.4 Optimum cost for prevention and resolution

The most efficient solution will be the one resulting from a combination of effort and prevention costs, along with the resolution of failures when these are produced. A system with a low reliability level and preventative practices that are not very significant, would give rise to the frequent appearance of failures with the corresponding costs involving their resolution and mitigation. More intense application of preventive measures will translate into a reduction in the frequency and severity of failures, with consequent savings in resolution costs. If diverse options, involving a combination of corresponding preventive practice costs, disruption and resolution costs, are reflected in a scheme, such as the one appearing in Figure 6.4, an optimum value will be obtained. This value would correspond to the economic, environmental and social implications that as a whole must be placed within the reliability and service standards that are established and, when pertaining to security and environmental integration, the value must coincide with the preference of the citizens who are being supplied.

Risk would be the parameter that best reflects each option measured as a combination of the reliability (or probability of incurring failure episodes) and the probability of the intensity of the failure reaching levels of great severity or supply shortages. Associated costs and effects complete the outlook.

Prevention is undertaken at the planning stage, as an analysis of actions required to comply with the standards established in the medium- and long-term horizons. In operational planning, prevention can be described as the anticipated analysis of the most efficient operation (the one closest to the standards with the lowest operating cost) for short-term horizons. Apart from structural and operational actions, availability of procedures, technologies and protocols for anticipation and the correct management of failure episodes must be considered very efficient preventive measures. These management protocols cover failure situations, in their contingency or crisis ranges, and are in themselves a mitigation and resolution procedure, although their mere existence and systematic use are also a preventive practice, which must also be executed prior to the emergence of failure episodes.

Preventive practices for long-term planning are always associated with the implantation of infrastructures and actions that contribute to an improved balance between consumptions and availabilities.

Preventive practices in operational planning are concentrated in a different use of the available resources, intensifying use of certain sources or incorporating strategic reserves to the system under the established circumstances. The objective is to reduce the current probability of incurring in failure levels to a horizon that is nearer.

Mitigation practices are on the borderline between preventive and resolution practices, depending on the interpretation made of the term. Both meanings are valid. What is pursued in prevention is a reduction of the probable effects of a hypothetical failure. In the case of resolution practices, the idea is to reduce the inevitable effects of the failure.

Confronted with the statement 'better to be safe than sorry', it is worthwhile underlining that, despite the truth behind the statement, one must be ready to remedy and that only by way of the capacity to remedy, when it has been impossible to avoid a problem, can the risks of shortage and catastrophic failure be reduced.

6.5 SCARCITY AND DROUGHT, AN OPERATIONAL EXAMPLE

Those situations in which a supply system has a significant probability of not being capable of handling the totality of demands on a short-term horizon, due to a lack of resources in the existing sources, are the ones that tend to be at the origin of particular actions. These actions may involve modification of the normal operating framework for the system and its resource sources or they may determine effects or disruptions for the system's users or other related stakeholders.

The immediate causes giving rise to these situations used to be linked to a precipitation period that is below normal; therefore, these periods tend to be called as droughts. However, it is convenient to keep in mind that low precipitation is not always the main cause, or the only cause. The term drought should be reserved for episodes that are directly linked to meteorological or hydrological phenomena.

Beyond causes linked to natural phenomena are the effects that the same produce. Disruptions affecting rain-fed agriculture due to a period of precipitation with amounts inferior to those required by the crops are to be separately highlighted, as these effects are also deserving of the term drought that is generally applied to them. Other effects depend more on the water system's capacity to store, transport and supply water to different uses and consumptions. In many cases, situations that generate effects or disruptions in the availability of the totality of demanded water after a low precipitation period are also called drought, although they should be called scarcity periods, so disassociating the possible cause from the effect itself, given that, as already mentioned, there is no reason why these are to be the only or main cause of effects.

With all that has been said, one of the main characteristics of scarcity situations should be highlighted: its operational nature, or put another way, its link to decision making by the institution or agency with the capacity and competence to implement measures or disruptions that condition uses and consumptions.

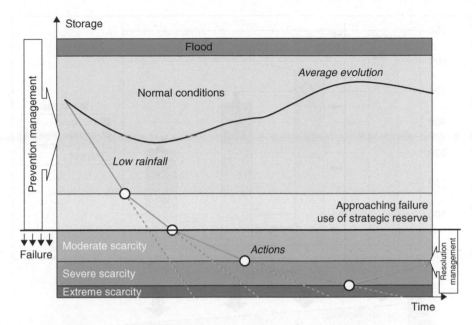

Figure 6.5 Planning and managing storages

Unlike other types of episodes or threats, such as earthquakes or hurricanes, in which consequences and response occur suddenly, in episodes of drought, an evolution in the circumstances and parameters that identify the risk caused by a scarcity of resources for handling the totality of demand on a very short-term horizon takes place gradually. When confronted with this situation, it is those responsible for decision making who actually decide on the time and method for intervention, without waiting until reserves and possibilities have been depleted. This is a decision that transforms the drought situation into a situation of scarcity, given that disruptions are unchained in some of the agents that are linked to water consumption. This is the reason why some authors call these episodes operational drought.

In the opinion of the author, the term scarcity, or operational drought, should be reserved for the existence of effects or limitations on uses and consumptions linked to water supply systems.

Figure 6.5 reflects a resource evolution foreseen in a drought situation. It shows the resulting values from the implementation of preventive practices before the failure with the reinforcement of resources, based on the use of strategic reserves.

In case of a persistent drought situation and going beyond the threshold of a failure, which means the beginning of scarcity or an operational drought situation, the actions to be taken would be the contingency resolution and mitigation. The target is to avoid the reduction of the probability of lack of reserves. The actions would be a combination of measures to reduce demand, with the consequent effects and costs, and integrated management of emergency and complementary resources, strategic reserves and the ordinary ones still available. Figure 6.6 reflects a possible variation of demands and the incorporation of resources in scarcity stages and phases of different severity.

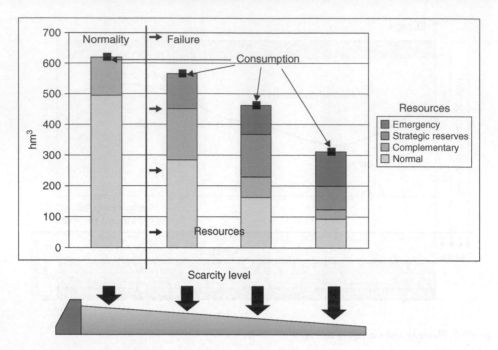

Figure 6.6 **Resources and demand management under scarcity**

The basis for determining the circumstances in which a situation of scarcity should be, should be none other than those linked to the assessment of the probability of a lack of water. In view of these decisions, the probability of a failure situation beginning is no longer important, given that this situation is imminent, and what in fact gains importance is evaluation of the probability of great disruptions actually taking place. Management of scarcity, or of failure, consists of an approach that distributes effects along with a distribution of the probabilities of incurring in more unfavourable situations in one or more areas, activities or users. Altogether, management of scarcity is the management of risks, which in turn is the management, and temporary and spatial distribution of effects.

The approach, pertaining to management of scarcity once the failure has presented itself, involves establishing prevention, mitigation and resolution policies at different coordinate levels. In this case, zero implies lack of water and the objective is to avoid this under all 'foreseeable' circumstances. Gradation of controlled effects must be established as a basic aspect for this planned distribution of effects. Effects will be defined by the intensity of the disruption to be induced in each one of the parts, agents and users, along with their duration.

Duration in each degree, or disruption phase, results from two different considerations: 1) corresponding to maximum established severity, and 2) precaution, constituted in terms of the minimum period prior to incurring in the most severe failure or scarcity as a consequence of the occurrence of a more severe episode than that foreseen for this case. Both temporary intervals are mainly conditioned by the duration of efficient implantation of resolution measures corresponding to this scarcity situation.

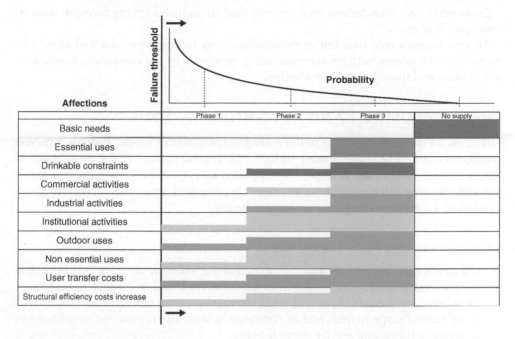

Figure 6.7 **Managing the affected areas**

Both type of measures, increasing available resources as well as reducing demands, require significant temporary intervals which must be taken into account, as is likewise the case with the estimate of values to be obtained in each case and period.

Each case will be established next:

- A hydro-meteorological scenario, which will characterize the severity and probability of occurrence.
- A balance of availabilities and demands as a consequence of actions used to handle the scarcity situation.
- A series of temporary intervals of permanence in the scarcity phases in accordance with said conditions.

What is essentially being defined is a deterioration risk threshold and a jump to the next scarcity severity degree.

Intervals and conditions for each phase are to be calculated once gradation and the management circumstances for each one of the phases has been established, beginning with the most unfavourable one, which must always be the lack of water as a reference to be avoided, thus drawing the line at the initial threshold of the scarcity situation.

The thresholds and action conditions as a whole constitute the scarcity management programme, which is none other than a chained combination of risk scenarios.

Figure 6.7 shows a possible outline of effect distribution in each scarcity phase. Each phase has an occurrence probability dependent on the meteorological regime and the fulfilment of that established in the prevention, mitigation and resolution protocols, mainly the coincidence of the results obtained in each situation with regard to those expected.

Every phase is characterized by a severity and its induced effects foreseen with its corresponding costs.

All this forms a risk function in the resolution of failure scenarios and should be defined in each system with the corresponding approvals by the competent institutions and entities and those potentially affected.

6.6   CONCLUSIONS AND RECOMMENDATIONS

Risk is an inevitable component of the management of water supply systems. Efficient management of these systems must include risk considerations in the compliance of certain service conditions. In order to be applied to episodes of resource shortage, a series of planning and management practices are proposed, structured in their methodological and operational aspects, as detailed below:

6.6.1   Methodological considerations

- Use of adequate parameters and indicators and the will to apply them. It is important to differentiate security indicators from guarantee, risk and emergency indicators.
- Identification or establishment of standard values for the objective service, relative to affection, scope or risk and in consonance with the normative, regulatory or concession frame and service commitments.
- Identification of potential events that represent a danger or threat for the supply system.
- Calculation and assignation of probability of occurrence to potential danger or threat events.
- Establishment and mark-out of procedures and reaction times for the different practices and policies for resolution of situations involving alterations to normal system operation.
- Definition and identification of precautionary values that complement the identified and accepted risk values.
- Establishment of reference thresholds for intervention that correspond to standards, and risk and precaution levels.
- Development and application of procedures for operational prevention and resolution of high-risk situations and resolution of different service failure or disruption episodes.
- Embedding of risk management approaches in planning, dimensioning and operation decision taking.
- Implantation of monitoring and early alarm systems.
- Assurance of real availability pertaining to compatible elasticity and headroom with the policies that are implemented in the components of the supply systems.
- Reducing uncertainties relative to risk causing factors, resolution factors for failure situations and determination of essential uses and basic needs.

6.6.2   Operational considerations

- Identification and definition of horizons, scenarios and conditions for analysis and evaluation of risks.

- Evaluation of the system's risk for 'immediate and short-term' horizons.
- Establishment of 'short-term or immediate' operation guidelines for resources, infrastructures, zoning, monitoring and operation to be applied in each case to assure the objective risk levels, or those that are decided on.
- In the case of foreseeing risk levels in the short or immediate term that are superior to the objective, areas, users and types of risks are to be placed over said values and punctual control, alarm, warning and resolution measures are to be established.
- Calculation of infrastructures or actions necessary to achieving or maintaining the levels of the objective risk for the medium- and long-term horizons. This calculation is to be made for a guideline type of definition pertaining to operational normality.

In the case of foreseeing risk levels that are superior to the objective in the medium or long term, feasible actions are to be identified that will permit the the objective risk to be reached, or which minimize it when compliance with the objective is impossible.

Revision of strategic supply planning, with an integrated focus on social expectations, needs, possibilities and standards. This would consist of calculating the implications involving the need for infrastructures and actions linked to risk levels that are different to those established as an objective, the same for medium- and long-term horizons. This action will be put forward when inevitable risk objectives are not available and when Strategic Planning tasks are being executed that accept the possibility of reconsidering the risk levels. Conclusion of a process of this type would lead to a new establishment of objectives, which in the case of its short and immediate term application would require a revision and updating of all the steps this chapter has detailed.

Chapter 7

Wastewater risks in the urban water cycle

Blanca Jiménez

Instituto de Ingeniería, Universidad Nacional Autónoma de México, Mexico

ABSTRACT: This chapter describes different pollutant sources and analyzes the links between the water cycle and urban activities affecting water quality. Among these, discharges of treated and non-treated wastewater are considered as well as the impacts of landfills, dumping sites, atmospheric pollution, sewers and water main leaks, urban runoff and industries. An ample variety of risks that threaten water sources are discovered, helping to explain the presence of pollutants encountered in water sources. To address this, control measures that go beyond the simplistic approach of building wastewater treatment plants are proposed. Some of them include concepts of cleaner production, the need to avoid producing and using toxic compounds, and above all the need to consider, while planning, that any effluent becomes indirectly a source of water. To put in place the measures proposed, the involvement of all of society and not only the water sector is needed, and thus it is recommended to extend the concept of the integrated water resources management (IWRM) approach to the whole of society, to preserve a clean environment (air, water and soil) as a whole.

7.1 INTRODUCTION

In order to function, cities need enough water and preferably of good quality. Nevertheless, in their functioning cities also reduce water availability in terms of quantity and quality. This is because they concentrate and synergistically combine different human impacts affecting water sources. Impacts are frequently overlooked, because we have a tendency to consider water issues in different professional activities and disciplines and thus, have a fragmented conception of the problems. Water quality impairment by cities represents a growing problem, often aggravated where several cities are in the same region and share water sources as supply or as disposal sites. To control such impacts, first the causes need to be identified, which can only be done by applying the hydrological urban water cycle concept (Figure 7.1). This chapter first describes urban water cycle impacts on water quality. A full description can be found in Marsalek et al., 2006. Due to the book's objective, this chapter will focus on mitigating options; and because much less information is available concerning developing countries a special effort is made to mention what does exist.

7.2 POLLUTANT SOURCES

There are two types of wastewater discharges: point sources, which are located at a specific site and are well identified; and non-point sources, which are produced in a large area and are not well defined.

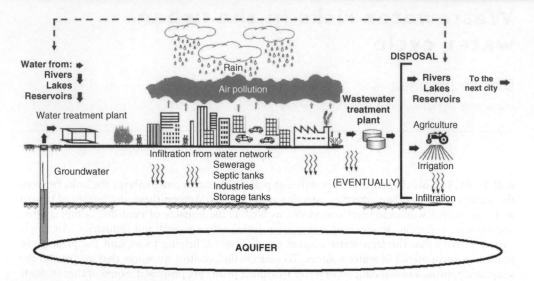

Figure 7.1 Hydrological urban cycle

Source: Adapted from Marsalek et al., 2006

7.2.1 Point sources

Point sources convey three types of wastewater: municipal, industrial and stormwater (when conveyed through a sewer). Municipal and industrial wastewater can be transported alone or combined with stormwater.

7.2.1.1 Municipal wastewater

Municipal discharges contain wastewater produced by households, businesses and industries connected to the municipal network; the relative proportion of each type of residual waste is different for each specific city. Municipal wastewater quality is considered to be similar around the world, which is true for most of the parameters used to design wastewater treatment facilities. A typical composition can be found in Metcalf and Eddy, 2003. In contrast, when the biological content is considered there is a great difference in wastewater quality among developed and developing countries, as result of the notably different health conditions between the two regions (Table 7.1).

Another important municipal wastewater characteristic is its quantity. It depends on: the water use; whether the sewer is or is not combined (i.e. also collects stormwater or not); and – to a much lesser extent – infiltration and exfiltration from sewers.

Total urban water consumption increases at the pace of urban population growth. Urban consumption per capita depends on the level of development. In general, cities in developed countries use more water per capita than cities from developing countries, but the total amount of water consumed depends on the total population, being very big in large cities, particularly megacities. Of the world's 30 biggest cities, 23 are situated in developing countries (Figure 7.2).

Besides water use, the presence of stormwater in municipal sewers increases the volume of wastewater generated to a level that can be very significant during certain periods. The amount of stormwater collected depends on the density of population,

Table 7.1 Comparison of the biological pollutant content in wastewater from developing and developed countries. With information from: Jiménez, 2005

Organism	Developed world	Developing world
Salmonella, MPN/100 mL	$10^3–10^4$	$10^6–10^9$
Enteric viruses, PFU/100 mL	$10^2–10^4$	$10^4–10^6$
Helminth ova, eggs/L	1–9	6–800
Protozoa cysts, organisms/L	10^1	10^3

as urbanization prevents water infiltration to soil and increases the runoff coefficient. Infiltration to sewers (from water mains, foundations drains, springs or underground-water infiltration), as well as exfiltration (sewers leaks), also modifies the amount of wastewater produced.

The volume of wastewater to be disposed of is important because it determines the impact on the environment. A large volume of wastewater discharged into lakes reduces the hydraulic retention time and modifies its biotic and abiotic properties. In rivers, flow velocity is increased and along with it the shore erosion, affecting the amphibious zones (zones in which reeds grow). Also, the water level in rivers can rise, affecting organisms living in the terrestrial zone.

Wastewater is also disposed of into soil, and when the soil is permeable, it results in aquifer recharge, augmenting water availability but also, and depending on the quality of the wastewater and the properties of the soil, polluting groundwater.

7.2.1.2 Industrial wastewater

Industrial wastewater has a very variable quality and volume depending on the production process. Industrial wastewater may be highly biodegradable or not at all, and may or may not contain compounds recalcitrant to treatment. The main concern with industrial wastewater is that it is responsible for the increasing presence (in quantity and quality) of synthetic compounds in the environment (Table 7.2). Aquifer contamination is of great importance due to the stability of the contaminant plume compared to surface waters.

7.2.1.3 Stormwater

Pluvial precipitation varies from region-to-region, and so too does the volume of stormwater. Rainwater in cities dissolves substances and drags particles from the polluted air or from soil to sewers and water bodies. Little is known about this exchange because the same pollutants are rarely measured in the air, pluvial water and water bodies. Urban pluvial water contains sulphurs, nitrogen oxides, acids, particles, hydrocarbons, metals (chromium, cadmium and copper), pathogens and PCBs (polychlorinated biphenyls). The amount of dry matter carried to sewers is a function of the type of urban area, varying from 10 to up to 2300 Kg/ha year of solids in commercial, residential, industrial and roads or freeway areas (OTV, 1994).

In developing countries, in addition to the pollutants mentioned, stormwater conveys a large amount of solid wastes coming from land erosion and garbage collected from dirty streets, due to the lack of proper solid waste management programmes. Unfortunately, there are very little data on this (Box 7.1).

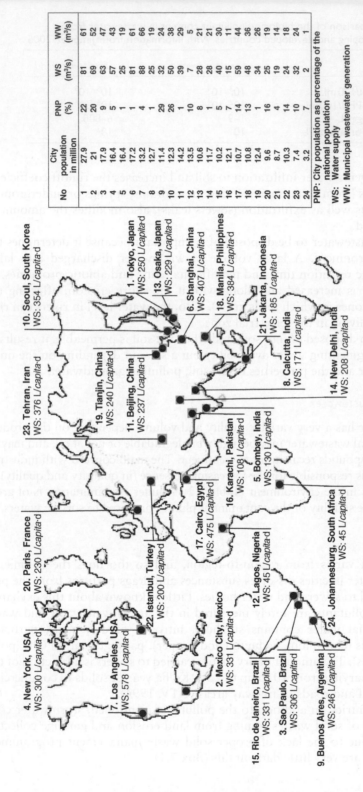

No	City population in million	PNP (%)	WS (m³/s)	WW (m³/s)
1	27.9	22	81	61
2	21	20	69	52
3	17.9	9	63	47
4	16.4	5	57	43
5	16.4	1	25	19
6	17.2	1	81	61
7	13.2	4	88	66
8	12.7	1	25	19
9	11.4	29	32	24
10	12.3	26	50	38
11	14.2	1	39	29
12	13.5	10	7	5
13	10.6	8	28	21
14	11.7	1	28	21
15	12.1	5	40	30
16	12.1	7	15	11
17	10.7	14	59	44
18	10.8	13	48	36
19	12.4	1	34	26
20	9.6	16	25	19
21	8.7	4	19	14
22	10.3	14	24	18
23	7.4	10	32	24
24	3.2	7	2	1

PNP: City population as percentage of the national population
WS: Water supply
WW: Municipal wastewater generation

10. Seoul, South Korea WS: 354 L/capita·d
1. Tokyo, Japan WS: 250 L/capita·d
13. Osaka, Japan WS: 229 L/capita·d
6. Shanghai, China WS: 407 L/capita·d
18. Manila, Philippines WS: 384 L/capita·d
23. Tehran, Iran WS: 376 L/capita·d
21. Jakarta, Indonesia WS: 185 L/capita·d
19. Tianjin, China WS: 240 L/capita·d
11. Beijing, China WS: 237 L/capita·d
8. Calcutta, India WS: 171 L/capita·d
14. New Delhi, India WS: 208 L/capita·d
16. Karachi, Pakistan WS: 108 L/capita·d
5. Bombay, India WS: 130 L/capita·d
20. Paris, France WS: 230 L/capita·d
17. Cairo, Egypt WS: 475 L/capita·d
4. New York, USA WS: 300 L/capita·d
22. Istanbul, Turkey WS: 200 L/capita·d
2. Mexico City, Mexico WS: 331 L/capita·d
12. Lagos, Nigeria WS: 45 L/capita·d
24. Johannesburg, South Africa WS: 45 L/capita·d
7. Los Angeles, USA WS: 578 L/capita·d
15. Rio de Janeiro, Brazil WS: 331 L/capita·d
3. Sao Paulo, Brazil WS: 306 L/capita·d
9. Buenos Aires, Argentina WS: 246 L/capita·d

Figure 7.2 Water supply and municipal wastewater production in megacities. With information from: Aguas Argentinas (2001); Beijing City (2002); Calcutta Municipal Corporation (2000); City of Los Angeles Water Services (1996); Water Supply in Tokyo (2002); Fatma Abdel Rahman Attia (1999); Filibeli (2007); Karamouz (2007); L'eau Paris (2002); Equihua (2007); Lagos State Nigeria (2007); Lagos State Water Corporation (2001); NYC Government (2003); Seoul Metropolitan Government (1999); Shangai Daily (1998); Santos (2003); South Africa Independent Centre (2006); Tutuko (2007); Urban Groundwater Database (2002); Yepes and Augusta (1996).

Table 7.2 Pollutants discharged from different industries, adapted from Foster et al., 2003

Type of industry	Relative use of water	Salinity load	Nutrients load	Organic matter load	Hydrocarbons	Fecal pathogens	Heavy metals	Organic synthetic	Potential index of underground contamination (1–3)
Iron and steel	++	++			++	+	++	++	2
Metal processing	++	+++	++	++	++		+++	+++	3
Mechanical	++	+++	++	++	+++		+++	++	3
Nonferrous metals	+	++	++	++	++		+++	++	2
Non metallic minerals	++	+++	++	++	++		+++	++	1
Petroleum and gas refineries	+	+++	++	+++	+++		++	+++	3
Plastic products	++	++	++	++	++		+	+++	3
Rubber products	+	++	+	++	++		+	+++	2
Organic chemistry	++	++	++	++	++	+	++	++	3
Inorganic chemistry	++	++	++	++	++	+	++		
Pharmaceutics	+++	++	++	+++	++	+	+	+++	3
Wood processing	++	+	++	++	++	+		++	1
Pulp and paper	+++	++	+	+++	++	+		++	2
Detergents and soaps	+++	++	+++	+++	++	+		++	2
Textiles	+++	++	++	++	+	+	+	+++	2
Tanneries	+++	+++	++	+++	++	+++	+++	++	3
Drinks and foods	+++	++	+++	+++		+++		++	1
Pesticides	+++				++			++	3
Fertilizers	++	++	++	+	++		++	++	2
Sugar and alcohol	+++	++	+	++				+++	2
Thermoelectric generation	+++				++		++	++	2
Electric and electronic	+	+	+	+	++		+++	+++	3

Low: +   Medium: ++   High: +++

BOX 7.1 Sediments in Mexico City sewer system. With information from: Jiménez et al., 2004

Mexico City has around 21 million inhabitants. Its combined sewerage system comprises 12,800 km of pipelines 0.3–3 m in diameter, 96 pumping stations with a total capacity of 670 m^3/s, 106 marginal collectors, 12 storm tanks with a total capacity of 130,000 m^3, several inverted siphons to overpass the Metro, 3 rivers, 29 dams, 147 km of open canals and one deep sewer (up to 217 m depth) 155 km long and 3–6.5 m in diameter. This complex system handles in mean conditions 68 m^3/s of wastewater.

The sewer system is cleaned each year. In total, 2.8 millions m^3 of sediments are produced this way. Sediments come from the sludge discharged from 27 wastewater treatment plants, soil erosion from the mountains surrounding the Mexico valley, the uncollected municipal solid wastes and the solids sedimented from wastewater. Sediments contain a large amount of pathogens (fecal coliforms 3–7 log MPN/g TS; *Salmonella* 2–7 log MPN/g TS; viable helminth ova 4–21 eggs/g TS), total petroleum hydrocarbons (TPHs of 89–7955 mg/kg TS), BTEXs (benzene, toluene, ethly-benzene and xylene of 0.006–17 mg/kg TS) and heavy metals. Their organic matter, nitrogen and phosphorus content are also very high. Sediments are disposed of on landfills representing an important loading charge on them. A similar situation has been reported in Brazil and Taiwan. The Mexico City government is looking for new options to deal with this 'solid wastes' problem.

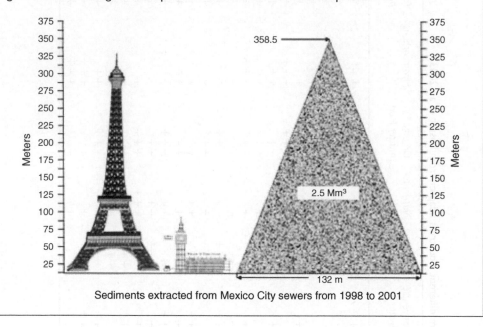

Sediments extracted from Mexico City sewers from 1998 to 2001

If untreated, stormwater disposal has negative effects on water bodies that can be very significant due to the large amount of wastewater discharged in a relatively short period. This is a particular concern in tropical areas with intense periods of rainfall. Stormwater disposal increases turbidity and the solid waste content in rivers and lakes. Effects of runoff discharge become evident only after hours, but can be still perceived weeks following its discharge and several kilometres from the disposal site (OTV, 1994). Stormwater overflows from coastal communities have been linked with recreational beach closings and massive shellfish deaths (Lape and Dwyer, 1994).

7.2.2 Non-point pollutant sources

Water pollutants come not only from urban wastewater discharges. Other urban activities and urban infrastructure are also an important source of pollution. Most of them have been first recognized by groundwater experts who realized that urban soil was an important means of transporting pollution from surface water and soil to groundwater. Table 7.3 contains a list of polluting sources and the pollutants involved.

7.2.2.1 Urban infrastructure

Water network – Due to aging, or bad construction or functioning, water networks leak. In developed countries, urban leaks are around 5–15% of the total water supply while in developing countries values can easily reach percentages of up to 60%. Under permeable soil conditions, water leaks recharge the underlying aquifer, while in other cases leaked water simply enters sewers, unnecessarily increasing the amount of wastewater to dispose of. Even though mains convey substantially clean water, its quality differs from that of the aquifer, modifying its natural composition, particularly with respect to its saline composition. Another difference is the chlorine content.

Sewerage system – Like water networks, sewers also leak. This is a less recognized problem because wastewater is considered a non-valuable source. Infiltration to groundwater from sewers can be as high as 500 mm/yr in highly populated urban areas (Foster et al., 1998). Wastewater introduces viruses, bacteria, organic and inorganic compounds (nitrogen, boron, chlorides and sulphates) and also a wide variety of synthetic products to groundwater. Due to sewage infiltration, the nitrogen content in aquifers underlying cities can be similar or even greater to the one found in aquifers beneath agricultural fields (Wakida and Lerner, 2005). Aquifers recharged with sewage leaks have been reported in Minnesota and Long Island in the United States, Mexico City (around 1 m³/s), Merida and Tijuana all three of them in Mexico, in Santa Cruz, Bolivia, in Bermuda, in Narbonne, France, in Birmingham and Nottingham in UK, in Germany (at least 3 m³/s) and in Madras, India (Jiménez, 2007b). In most of these places, groundwater is also being used as a water source.

Septic tanks and latrines – Septic tanks and latrines are built in peri-urban areas of developing countries, and also worldwide where the soil is hard or has no slope. Unfortunately, septic tanks and latrines frequently leak, discharging partially treated or untreated wastewater into the soil. If the soil is permeable, aquifer recharge occurs, if not, then surface water is negatively affected. Infiltration from septic tanks has been reported in the southern part of Mexico City, in Sana'a, Yemen (representing 80% of the urban recharge, Alderwish and Dottridge, 1999) and in Amman, Jordan (where 250 L/s of the recharge comes from latrines, Salameh et al., 2003). Foster et al. (1998) estimate that recharge from septic tanks can be as high as 500 mm/yr in densely populated areas (>100 person/ha).

Septic tanks are sources of biodegradable organic matter, nitrogen, phosphorus, boron, chlorides, pathogens, suspended solids, trichloroethylene, benzene, ethylene chloride, personal care products and drugs (BGS et al., 1998; US EPA, 1992). Of these, nitrates are the most mobile products and, because they are also persistent, they are frequently detected in aquifers. On-site sanitation systems have also been the proven pathway of pathogen transmission in numerous disease outbreaks in urban settings of developing countries with malfunctioning septic tanks (BGS et al., 1998).

Table 7.3 Sources of urban pollution in surface and groundwater

Origin	Main polluting agents	Relative importance	Concern	
			Developing countries	Developed countries
Urban infrastructure				
Water network	Cl and NMA	+	+++	+
Sewerage system (a)	OM, F, N, T, ED and PCP	+++	+++	++
Septic tanks and latrines	OM, N, ED and PCP	+++	+++	+
Storage or treatment ponds (a)	Variable	++	+++	
Storage tanks (a)	OM, T, HC and DBP	+++	+++	+
Municipal landfills	OM S H T ED PCP	+++	++	+
Hazardous wastes confinement sites (a)	A, OM, S, H, T, ED, PCP, NMA and HC			
Highways drainage soakaways	S and T	+		+
Pipelines (a)	OM, HC and T			
Injection wells	OM, S, H, T, ED and PCP	+++	+++	+
Cemeteries	OM, M, N, NMA and F	++	+++	+
Urban activities				
Industries (a)	Variable	+++	+++	++
Factories and small commerce (a)	Variable	+++	+++	++
Irrigation of amenity areas	N, T, ED* and PCP*	+		++
Application of ice melting substances	T and NMA			++
Transport and transference of material (a)	T and HC	+	+	+
Urban disposal options				
Unsewered sanitation	OM, N, F and T	+++	+++	−
Transportation of polluted water in channels or rivers (a)	OM, N, F, T, H and HC	++	+++	−
Non-treated sewage disposal in soil with impact on water bodies (a)	OM, N, S, F, T, ED and PCP	+++	+++	−
Non treated sewage discharge in river and lakes (a)	OM, N, S, F, T, ED and PCP	+++	+++	−
Treated wastewater disposal	N, NMA, ED, PCP and DBP	++	+	+
Sludge disposal (a)	OM, N, S, F, T, ED and PCP	++	+	++
Uncontrolled dumping sites (a)	OM, S, H, T, ED and PCP	+++	+++	−

(Continued)

Table 7.3 (Continued)

Origin	Main polluting agents	Relative importance	Concern	
			Developing countries	Developed countries
Other urban sources				
Atmospheric pollutants deposition (a)	A, N, M, H and HC	+ +	+ +	+ +
Urban runoff	HC, M, A and B	+ + +	+ + +	+ +
Saline intrusion	NMA	+ +	+ + +	+
Industrial accidental spillage (a)	T and HC	+	+ +	+

(a): Can include industrial compounds	ED: Endocrine disrupters	NMA: Non metal and anions
* Possibly	F: Fecal pathogens	OM: Organic matter
	H: heavy metals	PCP: Personal care products
A: Acids	HC. Hydrocarbons	S: Salinity
Cl: Residual chlorine	N: Nutrients	T: Toxics
DBP: Disinfection by products		+ Magnitude increase

Storage or treatment ponds – Ponds or lagoons are used to treat municipal and industrial wastewater, to evaporate or store industrial liquid wastes or to store stormwater (flood control ponds). Given their low cost, ponds are frequently used in developing countries. Even if some are waterproofed, infiltration to the subsoil occurs depending on the pond's construction and maintenance. In the literature, a common infiltration value of around 10-20 mm/d is cited (Foster et al., 1998).

Storage tanks – Surface or subsurface storage tanks have diverse uses. Subsurface tanks are commonly used to store gasoline and oil. Storage tanks frequently leak due to corrosion, bad connections and aging (Foster et al., 2002). The tank's size does not determine the amount of liquid lost, but does determine how long the leak has been occurring. Because storage tanks are underground, a long time may pass before the leak is detected.

Municipal landfills – Landfills are a potential source of pollution to water bodies, particularly to aquifers. Most of the pollution comes from old (pre-1970s) landfills built without using design criteria to control lixiviation. Leachates contain a wide variety of pollutants in high concentration. When hazardous wastes are mixed with municipal solid wastes, the risks considerably increase. Endocrine disrupters and personal care products have been reported in landfill leachates (Holm et al., 1995). Besides humid climates, pluvial precipitation increases the amount of leachates thus, enlarging the polluted area.

Hazardous waste confinement sites – Although hazardous waste sites are built to not impact on water resources, synthetic compounds have been reported in groundwater located near them.

Highways drainage soakaways – Tyre dust, gasoline and other products used in automobiles accumulate along highways and are removed from asphalt carpets by rain. These compounds are recovered in the stormwater and conveyed with it to disposal sites.

Pipelines – Pipelines are used to transport chemicals (particularly oil products) in cities, and although leaks represent an economic lost, they nevertheless exist. Most of the products pollute soil and can be transported from there to water bodies.

Injection wells – Injection wells are built to reclaim urban stormwater by recharging the aquifer. Because of the quality of stormwater, this can be a source of concern for aquifers.

Cemeteries – Human bodies represent a source of microbiological and organic matter pollutants for water bodies that are relatively important in megacities.

7.2.2.2 Urban activities

Industries – Industries handle liquid and solids products that might be a pollution source to water bodies. Pollution depends on the amount and type of products used. Industries using >100 kg of toxic substances per day (hydrocarbons, organic synthetic solvents, heavy metals, etc.) represent the higher risk (Foster et al., 2002). The types of pollutants involved depend on the type of industry.

Factories and small commerce – In cities, small factories and services (mechanical and dry cleaning services, etc.) use toxic substances, such as chlorinated solvents, aromatic hydrocarbons and pesticides, among others. Although they are supposed to properly manage their wastes, in practice this does not always happen. Given the widespread presence of this type of business, globally they introduce a significant amount of noxious compounds to water.

Irrigation of amenity areas – Green areas (parks, gardens, landscaped areas, tree-lined streets and avenues) are frequently over-irrigated in cities (BGS et al., 1998). Although these areas normally represent a small portion of the urban area, the volume of used water is often large. Water applied in excess is recovered in either aquifers, surface water bodies or sewers. Irrigation drainage can contain fertilizers and pesticides applied to green areas. Other compounds may be found if irrigation is performed with reused water.

Application of ice melting substances – In cities with cold climates, during winter, different substances (such as calcium chloride, sand or salt) and even treated wastewater (Funamizu et al., 2001) are applied to streets to melt snow. Viklander (1999) showed that melted ice contains, besides ice melting substances, suspended solids, heavy metals and phosphorus dragged from streets. Water from melted snow, which has a variable pH, is captured in sewers and disposed of as storm runoff.

Transport and transference of material – In cities different kinds of material are always being moved from one deposit to another. Transport takes place via pipelines, trucks and even trains. Leaks from these operations can bring pollutants into the environment with localized effects. These materials may either remain fixed to the soil or be transported to water bodies.

7.2.2.3 Disposal practices

Unsewered sanitation – In peri-urban areas of the developing world, frequently there is no sanitation service. This constitutes a continuous source of pollution to water bodies and soil.

Transportation of polluted water in channels or rivers – Rivers and open channels are commonly used to convey wastewater in developing countries. Infiltration from

---

BOX 7.2 Urban agriculture

Urban agriculture is practised in cities regardless of their climate, due to a combination of: wastewater generation, demand for fresh food products and people living in a marginal situation with no job options. It is practised in very small parcels irrigated with waste or polluted water that flows in open channels in urban and peri-urban areas to grow trees, fodder and any other produce that can be introduced to the market in small quantities (flowers and vegetables) or be used as part of the family diet. Wastewater channels can also be used to wash clothes, for cattle drinking, aquaculture, personal cleaning and even as a water supply if the water's appearance is sufficiently acceptable (Cockram and Feldman, 1996; Ensink et al., 2004b). Urban agriculture contributes to food security and thus, improves nutrition and living standards, paradoxically improving health (van der Hoek et al., 2001).

Urban agriculture is being practised over several million hectares, in 0.5–2 ha parcels. It is estimated that 10–70% of the population living in cities of the developing world rely on it (Cornish and Lawrence, 2001; IMWI, 2003). The social situation around it is very complex, making it difficult to implement traditional norms that require constant intervention and monitoring from authorities, therefore urban agriculture is a practice generally tolerated (Cornish and Lawrence, 2001; Ensink et al., 2004a).

Urban agriculture has developed its own methods to somehow mitigate its negative effects at an accessible cost while preserving the advantages of using wastewater. The lack of scientific studies demonstrating their effectiveness as well as a different conception of how reuse should be performed in the developing world makes their dissemination difficult, although they have proven to be useful in practice (Ensink et al., 2004a and b; Jiménez and Garduño, 2001). By the year 2025, there will be 292 cities with more than one million inhabitants mainly in poor countries, a situation that risks increasing urban agriculture.

---

unlined rivers or channels affects water quality as well as the final disposal of the polluted water.

Non-treated municipal wastewater disposal on soil – In the developing world, wastewater treatment is not common practice. Asian countries treat an average of 35% of the total wastewater they produce, while in Latin America-Caribbean and Africa the numbers drop to 22% and close to 0%, respectively (WHO/UNICEF, 2000). Non-treated wastewater is often used to irrigate by disposing it of directly on soil or indirectly by discharging it into rivers or channels from where water is taken to irrigate (IMWI, 2003). This situation occurs not only due to the lack of wastewater treatment, but also because three-quarters of the world's irrigated area is located in the developing world (United Nations, 2003) and there is a high dependence on water for food production. It is considered that at least 20 million hectares of 50 countries (around 10% of irrigated land) are irrigated with raw or partially treated wastewater (United Nations, 2003). Approximately one-tenth of the world's population consumes crops irrigated with wastewater (Smit and Nasr, 1992). Agricultural reuse of non-treated wastewater exists even inside cities and is called urban agriculture (Box 7.2). One risk when using wastewater to irrigate is the possibility of recharging an aquifer that is being used for other purposes (Box 7.3), such cases have been reported in some parts of the world (Jiménez, 2007b).

Non-treated wastewater disposal – In the same way that non-treated wastewater is disposed of in soil, it is also discharged to rivers and lakes.

BOX 7.3 Use of an aquifer recharged with non-treated wastewater as a drinking source. With information from: Jiménez and Chavez, 2004 and Jiménez, 2005

Mexico City produces around 68 m³/s of wastewater. Of the total wastewater produced, around 10% is treated and reused within the same city. The rest has been sent to a valley located north of the city, since the end of the nineteenth century. This valley is called the Tula valley but is colloquially known as ' "El Mezquital Valley'. Since the arrival of wastewater it has been used for agricultural irrigation. This is because the valley is a semi-arid region (pluvial precipitation of 550 mm while evaporation is 1,700 mm). Thanks to the use of wastewater, the agricultural yield has increased 150% for maize, 67% for wheat and 100% for barley, producing three crops per year instead of only one. Because wastewater plays an important role in the local economy, farmers asked the President to grant it to them. This was done in 1955, when the irrigation district was officially recognized.

The irrigation system is very complex. It comprises nine dams (three containing freshwater and six containing wastewater), three rivers and 858 km of unlined canals. One of the dams (Endho Dam) with a capacity of 202,250 hm³ is probably one of the world's biggest wastewater dams. Nevertheless, the use of wastewater to irrigate has also caused a 16-fold increase in helminthiasis (worms diseases) among children under five. Besides this effect, the use of large amounts of wastewater to irrigate (1.5–2.2 m³/m².yr) combined with the transport and storage of raw wastewater in unlined infrastructure has provoked the artificial recharge of the Tula Valley aquifer. This recharge is of at least 25 m³/s, i.e., 13 times the natural recharge. As a result, the Tula River flow (partially fed from the aquifer) increased from 1.6 m³/s to more than 12.7 m³/s between 1945 and 1995, and the water table rose from being 50 m below the ground level in 1940 to the surface creating artesian wells with flows varying from 40 to 600 L/s. The groundwater is being used as a water supply for 500,000 inhabitants.

Treated wastewater disposal – Like non-treated wastewater, treated water needs also to be disposed of into the environment. It was believe that treated wastewater could be assimilated by the environment, but unfortunately, there is increasing evidence that it might not be the case. Pollutants in very low concentrations with unexpected effects are now being found in natural water sources (Box 7.4). And, instead of finding 'first-use water' (i.e. water that because it has not been used has no pollutants) we are finding 'used water' (i.e., water containing anthropogenic pollutants). This happens simply because water sources have been required to play a contradictory role: on the one hand to supply good quality water, and on the other, to convey, receive, de-pollute, dilute or store used water (and even other kinds of discharges, as presented in this chapter). Because the urban population is increasing, the amount of treated wastewater (mostly in developed countries) that is being discharged in a relatively small geographical area is increasing, surpassing the local environment's capacity to 'assimilate' pollution.

Sludge – Water and wastewater treatment processes end with two products: treated water and sludge. This latter concentrates the pollutants removed plus chemicals added during treatment, and as treated water, sludge needs a disposal option. A detailed analysis of the sludge impact is presented in Chapter 8.

Dumping sites – In cities in developing countries, waste solids management is an ongoing process. Just for main Latin American cities, wastes collection varies from 65–100%, but only 33 % of landfills meet environmental criteria (Acurio et al., 1997). In Africa, solids wastes collection in cities varies from 8–100%. For this reason, it is not

---

BOX 7.4 Indirect reuse of treated wastewater in Berlin. With information from: Heberer and Reddersen, 2001

The Berlin area is a densely populated region, with a population of around four million people, where emerging pollutants have been found in the water used as supply. This is because waterways are used to receive effluents (treated wastewater) but also as source of drinking-water after bank filtration. Infiltrated water represents up to 75% of the groundwater used in drinking water production. In 1990, clofibric acid was detected in Berlin groundwater samples. Between 1992 and 1995, clofibric acid was measured in concentrations at the ng/L-level in groundwater and in Berlin tap-water samples (Heberer and Stan, 1997). The presence of these residues was associated with the infiltration of sewage effluents into the soil. Later, other polar PhACs were also identified as well as diclofenac, ibuprofen, propyphenazone, gemfibrozil and primidone. Although the concentrations found are very low, the presence of these compounds shows that wastewater is being indirectly reused as a drinking source.

Use of some of the emergent pollutants found:
Clofibric acid: Drug used as lipid regulator.
Diclofenac: Anti-inflammatory
Ibuprofen: Drug used to reduce or dissipate inflammation
Gemfribrozil: Lipid regulator

---

uncommon to find uncontrolled dumping sites that are a source of pollution to surface and underground water.

7.2.2.4 Other sources

Atmospheric deposition – Air pollutants may end up in water courses either by direct deposition or after settling in soil and transference from there to water. Actually, air pollution is considered as one of the world's most extensive pollution problems. The sources of air pollutants are varied (Table 7.4), as is the type and amount of pollutants produced. Pollutants in waterbodies that may originate from atmospheric sources (at least partly) include nitrogen compounds, sulphur compounds, mercury, pesticides and other toxics (US EPA, 2007a). Nevertheless, only some air pollutants are routinely measured during air monitoring campaigns, the most common ones being sulphur dioxide, nitrogen oxides, particles, carbon monoxide and ozone. Sulphur, nitrogen oxides, CO and CO_2 are related to acid rain. Nitrogen oxides and other nitrogen compounds contribute to eutrophication in reservoirs, as well as to increasing the nitrate content in underground water. Particles are easily transported to soil and water, and are a source of concern because they adsorb metals and toxic organic compounds. As an example, fine particles produced by diesel vehicles contain 1,3 benzene and butadiene, two well-known carcinogenic compounds (Fernandez-Bremauntz, 2001). Carbon dioxide and methane contribute to climate change producing, indirectly, different impacts on watercourses (Kundzewicz et al., 2007). Little information is available on how air pollutants reach the water and soil. In some cases, air pollutants deposition can be very important, deserving special control programmes to stop the impairment of water. For instance, in the United States, the EPA has set one, based on the Clean Water Act, to reduce the

Table 7.4 Sources of air pollutants

Point sources
Electricity generation, food industry, apparel industry, chemical industry, wood industry, metallic and non-metallic industry, and printing products

Mobile sources
Cars, taxis, buses, gasoline trucks, diesel vehicles, gas trucks and motorcycles

Area sources
Solvents consumption; surfaces cleaning; building impermeabilization; covering of industrial surfaces; dry cleaning; graphic arts; bakeries; automotive painting; street and road painting; PL gas distribution and gas tank leaks; unburned hydrocarbons PL gases; gasoline distribution, storage and sale; airplane re-filling; trains; landfills; asphalt application; wastewater treatment plants; sterilization and combustion in hospitals; domestic combustion; commercial-institutional combustion; forest fires; fires; and unpaved roads and streets.

impairment of watercourses by mercury coming from industrial atmospheric discharges (US EPA, 2007b).

The possibility of air pollutants in water depend on their Henry's Law constant (via atmospheric deposition), their persistence and their solubility. Formic, acetic and propionic acids, which are organic acids reported in polluted air, display high solubility in water and have toxic effects on living beings. Hydrocarbons have proven to be transferred from the air to the soil and water. Capella and Pegueros in 1998 reported the systematic presence of gasoline and oil derived compounds in Mexico City's wastewater, for instance. Later, in 2002, Molina and Molina and Bravo et al. reported the presence of toluene, benzene and formaldehydes in the polluted air of Mexico (Box 7.5).

Acid rain – Rain pH under normal conditions is of 5.6 due to the dissolution of the CO_2 naturally contained in the atmosphere and forming carbonic acid in water. But, when SOx and NOx pollutants are present in the air they produce H_2SO_4 and HNO_3 in atmospheric water, lowering the pH below 5.6.[1] Acid water falls as acid rain, snow, dew, fog, water snow or hail. Acid precipitations corrode metallic surfaces, damage urban infrastructure, cars, paintings and historical monuments. Acid fog forms fine particles that reduce visibility. Acid deposition causes damage to plants and trees. In rivers, lakes and reservoirs with low buffer capacity (which occurs in areas with non-carbonated detritic rocks – sandstones – or crystalline rocks – granites or gneisses), acid rain modifies the pH. Watercourse acidification diminishes water biota and reduces its diversity. Acidification also provokes metal solubilization, making it available for biota producing toxic effects and limiting water uses (Chapman, 1992). Acidification is one of the biggest problems observed in lakes and reservoirs[2] of temperate areas as a result of acid deposition from polluted air. The acidification of lakes and reservoirs leaches sediments found at their bottom and hydrolizes iron oxides, manganese and aluminum as well as other toxic metals. Aluminum dissolution is produced at a pH of 4.5 and cause asphyxia to fish because it is deposited as aluminum oxide in gills.

---

[1]The contribution of organic acids to acid rain is marginal.
[2]Water velocity in lakes and reservoirs is slow, of around 0.001 to 0.01 m/s (Chapman, 1992); as a result, hydraulic retention time is high (months to years) and pollutants are retained for longer periods than in rivers.

BOX 7.5 Pollutants found in Mexico City's air (See also colour plate 4)

Mexico City has nearly 3 million vehicles. It is estimated that due to its intense human activity each year 19,889 tons of <10 micron particles, 22,466 ton of SO_2, 1,768,836 ton of CO, 205,885 ton of NOx and 465,021 ton of hydrocarbons are discharged to the urban air. The content of formic acid in the air is 2–24 µg/m³, that of acetic acid is 0.5–7 µg/m³ and that of propionic is 0–18 µg/m³ (Ruiz, 2001). Additionally, close to 200 volatile organic compounds with at least 2–13 carbons have been identified in the urban air. The most common ones reported were alkenes (52–60%), followed by aromatic compounds (14–19%), olefins (9–11%) and oxygenated compounds (1–2%). Although these latter ones were found in low concentrations, they are responsible for the ozone's presence, until now considered the city's main air problem. The average concentration of hydrocarbons is 8.8 ppm, a value which is much bigger than the value of 2 ppm for Los Angeles city in the 1980s (Molina and Molina, 2002).

Hydrocarbons come from automobiles (82,000 ton/yr) and from solvents used as cleaning products (77,000 ton/yr), as well as from the evaporation and leaks of unburned hydrocarbons (27,000 tons/yr). Other sources of air pollutants in Mexico City are architectonic surface covering, domestic gas leaks, dry clean services, landfills, the chemical industry and graphical arts.

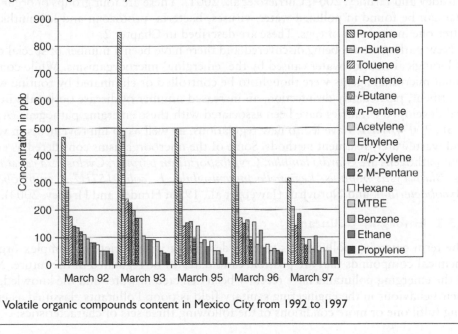

Volatile organic compounds content in Mexico City from 1992 to 1997

Acid rain can be formed and deposited locally, but can also be transported hundreds to thousands of kilometres by winds. The area affected can be as high as 10^6 km. In soil, acid rain changes the nutrient availability for plants and modifies its structure and permeability by extracting the aluminum content.

7.3 POLLUTANTS

In this section, the meaning of some pollutants is discussed. Given that there is a lot of literature dealing with conventional ones, the main focus is on two groups. The first one deals with biological pollutants, highlighting their difference and importance for developing countries and the relevance of emerging pathogens in developed ones. The second one relates to emerging pollutants, due to the recent concerns they are causing worldwide and the implication they have for water management.

7.3.1 Conventional parameters

Table 7.5 summarizes the main conventional parameters used to describe pollution in wastewater. More information can be found in most of the relevant literature on wastewater treatment.

7.3.2 Biological pollutants

Cities are highly sensitive to biological pollutants owing to their high population density and the particular infectivity of pathogens. This is true not only for cities in developing countries but also cities in developed ones, as was unfortunately demonstrated in Milwaukee, US when 403,000 people became sick and more than 50 died after a *Cryptosporidium* infection was transmitted through the drinking-water supply (Hrudey and Hrudey, 2004; Curriero et al., 2001). There are four groups of organisms that can be found in polluted water: viruses, bacteria, protozoan and helminths, the latter one in the form of eggs. These are described in Chapter 2.

New pathogens are being discovered and there have been a number of special cases of biological risk for water caused by the 'emerging' microorganisms. While conventional microbial agents were thought to be controlled or eliminated by routine water treatment, particularly disinfection, an increased number of disease outbreaks in several developed countries have been associated with these emerging pathogens (Asano et al., 2007). These have led to new regulations, as well as to improvements in water and wastewater treatment methods. Some of the microorganisms considered as emergent pathogens are *Giardia lamblia, Cryptosporidium parvum, Cyclospora cayetanensis, Blastocystis hominis, Legionella pnuemophila, E. coli 0157H7, Campylobacter Mycobacterium*, and Norovirus (Jawetz et al., 1996; Hrudey and Hrudey, 2004).

7.3.3 Emerging pollutants

The term emerging pollutants is used to describe a wide variety of complex organic chemical compounds that are possible candidates to be regulated in the future. Most of the emerging pollutants have recently been detected in water and the knowledge of their behaviour in the engineering sanitary field is poor. Pollutants classified as emerging fulfil one or more conditions of the following three sets of characteristics:

a. Quantity
    ● are present in a very low content in water (in the order of μ or nanog/L)
    ● are toxic at very low concentrations.
b. Detection
    ● are not detectable using traditional parameters to measure pollution, such as BOD, COT and even with toxicity tests
    ● require complex analytical equipment to be detected, such as GC-MS (gas chromatography coupled with mass spectrophotometer)
    ● have been detected in wastewater, sewage effluents and as well as in water sources.

Table 7.5 Selected parameters used to evaluate conventional pollution

Parameter	Significance
Acids	They modify water's pH, modifying by these different biochemical reactions necessary to depollute water and sustain life.
Dissolved oxygen	Dissolved oxygen is needed for the normal development of biota in water, but also to oxidize organic pollution. Therefore, when water is polluted, biota competes with organic matter for the dissolved oxygen. And, if there is not enough, even for very short periods, there is a rapid decrease in aquatic communities, particularly fish.
Heavy metals	Most of metals are adsorbed into particles at normal pH and redox potential conditions but can be released in acid and reduced ones. Bacteria can solubilize metals (Hg, As and Pb) from sediments transforming them into volatile organometallic compounds. The health and environmental effects depend on the type of metal involved and its concentration, the most dangerous being arsenic, beryllium, cadmium, chromium, lead and mercury.
Nutrients (N and P)	In surface water bodies, nutrients can cause eutrophication. Nutrients cause green-blue algae to bloom; algae produce toxins, increase chlorine demand and affect potable water taste and odour. In groundwater, an increase in nitrogen (usually as nitrates) can impair its use for human consumption, although the link between nitrates and health has recently been questioned by Fewtrell, 2004.
Organic matter	Organic matter refers to different kinds of organic compounds. Biodegradable ones (measured as BOD or biochemical oxygen demand) are those that can be degraded by microorganisms in water. Not biodegradable may be toxic to life.
Salts	Water salinization is a natural process that occurs to a greater degree in rivers and lakes with high evaporation rates. Saline discharges accelerate this process. The higher the salt content in water the higher the treatment cost and after a certain limit (1,500–2,000 mg/L) the use of water is economically prohibited. The ocean level rise as result of climate change is expected to increase the saline content of coastal aquifers. Salt water is harmful to agriculture.
Surfactants	They tend to accumulate in the air–water interface, reducing water oxygenation. They are also toxic and affect the surface properties of compounds
Suspended matter	Suspended solids can clog urban infrastructure and form sediments in water bodies. Suspended matter has a very varied chemical composition because it includes several types of compounds. The particular effects depend on its composition.
Oil and greases	They accumulate on water surfaces limiting the oxygen exchange between water and the atmosphere. They clog fish gills and adhere to aquatic animals causing severe damage.
Temperature	Modifies the natural depollution process in water bodies and affects all the reactions taking place in water. High temperatures increase the amount of pollutants dissolved in water but reduce the amount of dissolved gases, especially oxygen and with it reduces the self-depollution capacity. High temperatures can kill aquatic life. In lakes and dams, temperature determines stratification, and if changed, ecological cycles can be completely altered.

c. Effects
- display negative effects on animals raising concerns about the effects produced in humans
- have no negative effect but due to their function are suspected to affect the environment or health
- have toxic effects on humans.

The sources of emerging pollutants are diverse. They come from urban runoff, polluted soil, atmospheric deposition (such as polycyclic aromatic hydrocarbons or PAHs), sanitary landfills or dumping site leachates, industrial spills, disinfection by-products, naturally occurring hormones and wastewater (treated or non-treated). Wastewater treatment processes have not been designed to remove them, thus, they are randomly removed during conventional treatment (Halling-Sorensen et al., 1998; Andreozzi et al., 2003). Two groups of compounds that are part of emerging pollutants are: endocrine disrupter compounds (EDC) and personal care and pharmaceutical products (PCPP). These compounds and their characteristics are discussed in Chapter 2.

7.3.3.1 Content in water

Non-treated and treated wastewater – Emerging pollutants are rarely measured in untreated and treated wastewater due to the cost of the analysis and the difficulty of obtaining 'clean samples' for detecting very low doses of compounds with very varied chemical characteristics, the presence of which is unknown. And, when present, these compounds have been found to be highly variable in content during the day, from sample-to-sample and from site-to-site (Gray and Sedlak, 2003). Table 7.6 presents data concerning the content reported of some emerging pollutants in non-treated and treated wastewater, but due to the limited available data, the wide content variation and the lack of follow-up in wastewater treatment plants, reported concentrations should be considered only as indicative. The presence of these substances in treated wastewater indicates that wastewater treatment methods do not remove or partially remove these constituents.

7.3.3.2 Content in surface and groundwater

As a consequence of discharging emerging pollutants in treated and non-treated wastewater to the environment, they have also been detected in surface water (Table 7.7), groundwater (Table 7.8) and in some cases even in drinking water (Table 7.8). Finally, a wide variety and high concentration of emerging pollutants have been detected in landfill leachates (Fent et al., 2006).

7.4 MANAGEMENT

Traditionally the management of wastewater has consisted of collecting it in sewers to transport it and treat it in wastewater treatment plants. This has been performed successfully in developed countries but not often in developing ones. Until now, it has been expected that, by applying this practice, wastewater treatment plants combined with the environment's depolluting capability would control all noxious substances

Table 7.6 Content of emerging compounds measured in untreated and treated wastewater in μg/L unless otherwise indicated

Compounds	Untreated wastewater	Country	Treated wastewater	Country
Acetylsalicylic acid	3.2	Germany	0.05–1.51	Germany and Canada
Bezafibrate	0.42–5	Austria, Canada, Finland, Germany, Mexico and Spain	ND–4.6	Austria, Canada, Finland, France, Greece, Italy, Sweden (7WWTP) Germany, Spain and Switzerland
Carbamazepine	0.7–1.5	Austria and Canada	ND–6.3	France, Greece, Germany, Italy, Sweden and Switzerland
Clofibric acid	0.15–1	Germany, Switzerland and UK	ND–2 μg/L	France, Germany, Greece, Italy, Sweden, Switzerland, UK and USA
Clotrimazole	0.031	UK	0.14	UK
Diazepam	0.59–1.18	The Netherlands	0.1–0.66	The Netherlands
Ethinylestradiol	0.003	Italy	0.0004	Italy
Oestrogen type compounds	0.50–0.15	Japan, Mexico	0.001–0.013	Japan
Fenofibric acid	0.44	Germany	0.05–1.19	Germany
Gemfibrozil	0.01–0.7	Canada, Mexico and Germany	0.06–4.76	Canada, France, Greece, Germany, Italy, Sweden and Switzerland
Ibuprofen	0.2–38	Austria, Canada, Finland, Mexico, Spain, Switzerland, USA and UK	0.02–7.11	Austria, Canada, Finland, France, Germany Greece, Italy, Mexico, Spain, Sweden, Germany, Switzerland, UK and USA
Ifosfamide	0.007–0.029	Germany	0.010–0.043	Germany
Ketoprofen	0.25–5.7	Canada, Finland, Germany, Mexico, Spain, Switzerland and USA	ND–1.62	Finland, France, Greece, Italy, Sweden, Germany, Spain, Switzerland and USA
Mefenamic acid	1.6–3.2	Switzerland and UK	0.8–2.3	Switzerland and UK
Metoprolol	0.21–2.60	Mexico	0.01–2.2	France, Greece, Germany, Italy, Sweden and Switzerland
Naproxen	0.5–40.7	Canada, Finland, Mexico, Germany Spain and USA	ND–12.5	Canada, Finland, France, Germany, Greece, Italy, Sweden, Spain, Switzerland and USA
N-Nitrosodimethylamine	1.5	USA	0.1	USA
Paracetamol	6.9	UK	0*	UK
Propranolol	70	UK	0.01–0.30	France, Greece, Germany, Italy, Sweden, Switzerland and UK
Salycilic acid	0.550–330	Germany, Canada, Mexico and Spain	0.05–3.6	Canada and Spain
Trimethoprim	0.11–0.32	Mexico	0.02–0.56	France, Greece, Germany, Italy, Sweden, Switzerland and UK
Tamoxifen	0.15	UK	0.20	UK
X-ray contrast media	0.18–7.5	Germany	0.14–8.1	Germany

References: Garrison et al., 1976; Rogers et al., 1986; Stumpf et al., 1996 and 1999; Buser et al. 1998; Ternes et al., 1998; Ternes et al., 1999; Kummerer et al., 1997; Heberer et al., 1998; Hirsch et al., 1999; Baronti et al., 2000; Matsui et al., 2000; Sedlak et al., 2000 and 2003; Golet et al., 2001; Ollers et al., 2001; Heberer, 2002; Andreozzi et al., 2003; Kolodziej et al., 2003; Metcalfe et al., 2003; Ashton et al., 2004; Kreuzinger et al., 2004a, Carballa et al., 2004; van der Hoeven, 2004; Srenn e: al., 2004, Thomas and Foster, 2004; Lindquist et al., 2005; Suter et al., 2005; Quintana et al., 2005; Tauxe-Wuersch et al., 2005; Roberts and Thomas, 2005; Asano et al., 2007 and Gibson et al., 2008.

Table 7.7 Selected observations of emerging pollutants in surface water

Observation	Place	Reference
Pharmaceutical presence in rivers up to 1 µg/L	UK	Richardson and Bowron, 1985
Acetylsalic acid (< 0.05 µg/L), bezafibrate (0.005–0.38 µg/L), clofibric acid (0.005–0.30 µg/L), diclofenac (0.005–0.49 µg/L), fenofibric acid (0.005–0.17 µg/L) and Ibuprofen (0.05–0.28 µg/L)	Germany	Heberer et al., 1998 and Stumpf et al., 1996
Twenty different drugs and four corresponding metabolites were found to be in streams and river water. Some of them were: bezafibrate, carbamazepine clofibric acid (at 0.55 µg/L), diclofenac, fenofibric acid, gemfibrozil, ibuprofen, indometacine, metoprolol, naproxen, phenazone, propranolol, and salicylic acid	Germany	Ternes et al., 1998
1,4 dioxane was frequently found in the lower part of the Mississippi River	USA	Johns et al., 1998
In an extended monitoring study, 95 micro-pollutants in 139 streams downstream of urban areas and livestock production were detected. In some sites as many as 38 out of the 95 compounds were found in a single water sample (the average number of compounds in a sample was seven). Among the most frequently detected compounds were steroids, an insect repellent (N, diethyltoluamide), caffeine, triclosan (an anti-microbial compound), antibiotics, a fire retardant, 4-nonylphenol and some pharmaceuticals. Ibuprofen and its metabolites have been detected in surface water of up to 1 µg/L. The frequency found was as follows: (a) > 75% for caffeine, coprostanol, cholesterol, Tris (2-chlroethyl phosphate) N-N-diethyltoluamide, (b) >50% and < 75% for Ethanol, 2-butoxy-phosphate, 4-Nonylphenol, 4-Nonylphenol monoethoxylate, triclosan; (c) >25% but < 50% for acetaminophen, 5-methyl-1H-benzotriazole, 1,4-dichlorobenzene, cotinine, fluoranthene, bisphenol A, 4-octylphenol monoethoxylate, 1,7,-dimethylxanthine, pyrene, trimethoprim and (d) < 25% but > 15% for carbaryl, tetrachloroethylene, erythromycin-H_2O, estriol, lincomycin, 4-octylpheno diethoxylate, phthalic anhydride and sulfamethoxazole	USA	Kolpin et al., 2002
The anti-epileptic carbamazepine and clofibric acid, a metabolite of the lipid lowering agents clofibrate, etofibrate and etofyllin clofibrate were detected.	Switzerland and North Sea	Buser et al., 1998 and Weigel et al., 2002
In the stretch of the River Elbe between Dresden and Magdeburg where big population centres are located, endocrine disrupter and the associated effects in the resident fish have been detected.	Germany and Czech Republic	Hecker et al., 2002

Description	Location	Reference
In all sampling sites atenolol, bezafibrate and furosemide were found and ranitidine, clofibric acid, diazepam were detected in some.	At the river Po and Lambro, Italy	Calamari et al., 2003
In water samples upstream and downstream of selected towns and cities during high-normal and low-flow conditions pharmaceuticals and other organic wastewater contaminants were found. Prescription drugs were only frequently detected during low-flow conditions.	Iowa, USA	Kolpin et al., 2004
Diclofenac, ibuprofen, carbamazepine, and other various antibiotics and lipid regulators were detected in the Elbe River and its tributaries between the source and the city of Hamburg	Germany	Wiegel et al., 2004
In surface water, carbamazepine was found with maximum concentrations of 1.2μg/L	Germany	Wiegel et al., 2004
Ibuprofen was found in estuaries and in seawater	UK and North sea	Thomas and Hilton, 2004 and Wiegel et al., 2004
Carbamazepine was found in 44 rivers with a mean concentration of 0.60 μg/L in water and 4.2 μg/mg in the sediment.	USA	Thacker, 2005
Oestrogen was detected.	River Zenne, Brussels; River Seine, Paris; River Manzanares, Madrid and River Lambro, Milan	Suter et al., 2005
Nonyl phenol presence.	River Zenne, Brussels; River Seine, Paris; River Manzanares, Madrid and River Lambro, Milan	Suter et al., 2005
Different pharmaceuticals were detected in concentrations ranging from 0.04 to 2.4 μg/L.	Tyne estuary in the UK	Roberts and Thomas, 2005
From a detailed literature review, it was established that the pharmaceuticals most frequently present in surface water were: acetyl salicylic acid, albuterol, bezafribrat, caffeine, carbamazepine, cimetidine, cloflbric acid, codeine, cotinine, diclofenac, diazepam, fenoprofen, fluoxetine, gemfibrazol, ibuprofen, indometacin, ketoprofen, mefenamic acid, metformin HCl, metoprolol, naproxen, paracetamol, phenazone, propanolol, ranitidine, tamoxifen and salicylic acid.	Several countries	Fent et al., 2006

Table 7.8 Examples of emerging pollutants found in groundwater and drinking water

Observation	Country	Reference
Groundwater		
X-ray contrast in groundwater and in drinking water after treatment	Germany	Putschew et al., 2000
Ibuprofen, naproxen, Salicylic acid, Triclosan, carbamazepine salicylic acid, carbamazepine, triclosan at traces level in an aquifer recharged with wastewater.	Mexico	Gibson et al., 2008
Drinking water		
1,4 dioxane	USA	Kraybill, 1977
In Atlanta, at trace concentrations (ng/L), the following compounds were detected: acetaminophen, 2-butoxy-phosphate, 2,6-d-t-butylphenol, fluoranthene, caffeine, cotinine, tri(2-Chloroethyl)phosphate, triclosan, phthalic anhydride, pyrene, thanol- and tributyl phosphate.	USA	Henderson et al., 2001
Phenazone, propiphenazone and clofibric acid in potable water collected in the vicinity of Berlin, Germany	Germany	Heberer and Stan, 1997 and Reddersen et al., 2002
Clofibric acid at concentrations as high as 270 ng/L in drinking water from Berlin.	Germany	Heberer and Dumnbier, 2000 cited by Snyder et al., 2003

contained in wastewater. However, a look at the urban water cycle begs the question: is this approach enough to control all risks? This next section discusses management options from a different perspective; wastewater treatment will be considered further.

7.4.1 Changing the concept of pollution sources

As discussed, pollutants in water come not only from wastewater discharges but also from other urban sources. So, a first step is to recognize all such sources, to consider that through the (urban) water cycle, water is being reused rather than being depolluted, contrary to what was previously thought. That way, it should be easier to harmonize the different and incompatible 'services' water bodies are expected to perform including: 1) extractive or in-stream uses (for production activities or social well-being); 2) environmental conservation; and 3) conveying and treating effluents (treated or non-treated wastewater).

7.4.2 Gathering useful information

It is not usual to have professionals, research centres or government offices gathering information about pollutants transported and exchanged through the urban water cycle with the rest of the environment (soil and air). Moreover, although some of this pollution information is already collected, it is done in a fragmented fashion by different disciplinary fields (air, water and soil experts) and different government offices (water works, environmental ministries, solid waste management offices, etc.) thus, rendering its analysis impossible in practice. It would be a good idea for one institution to gather such information and process it into a common framework.

7.4.3 Monitoring campaigns

Additionally, it would be useful to develop monitoring programmes to measure the pollutant exchange occurring within the urban water cycle. This does not necessarily mean implementing new costly monitoring programmes but establishing links between current environmental programmes and choosing parameters that could be used as appropriate indicators. This information would be a key part of an Integrated Water Resources Management programme.

7.4.4 Water sources management

Once the interaction between the different components of the urban environment and water sources has been properly established, options for controlling negative effects should be defined. It is particularly important to acknowledge the indirect reuse of water (Jiménez, 2007b), to create awareness of the importance of properly closing the urban hydrological cycle. This will mean more complete and complex activities than simply installing waste-water treatment plants and building more complex water treatment plants.

7.4.4.1 Groundwater

Groundwater plays a major role in providing the water supply for domestic purposes in several cities around the world (BGS et al., 1998). But, at the same time, it is the receptor of much urban and industrial wastewater, along with leachate percolation from diverse solid and liquid wastes. Moreover, even though incidental recharge is recognized as one of the three recharging groundwater processes (Tuinhof and Heederik, 2002), its impact on water quality is seldom studied. Thus, at the present time, urban impacts on groundwater are still hard to understand, have not been properly quantified, and much less adequately quantified. Under these circumstances, it is hard to establish proper management strategies for addressing risks. Thus, it is important to know the capacity of the soil in the vadose zone to retain pollutants. In particular, it is important to determine the soil's capacity to retain recalcitrant organic compounds, as in many places this would be the only economically feasible way of dealing with them. For some regions this is an urgent task, since reports on aquifer recharge with non-treated wastewater have been documented in several part of the world, for example, Miraflores, in Lima, Peru; Wagi Dhuleil in Jordan; the Mezquital Valley and the Leon Valley, both in Mexico, and Hat Yai, in Thailand (Foster et al., 2003). In the Mezquital Valley alone, there are more than 500,000 people consuming water from this aquifer (Jiménez and Chávez, 2004).

7.4.4.2 Surface water

Because, for society, surface water is a visible source, more care is often taken over its protection. Therefore, it normally receives discharges consisting of treated water (at least in developed countries). But, as discussed in the section referring to emerging pollutants, even this treated wastewater contains emerging pollutants. This may be a concern because, for instance, in the Thames River in England, seven out of ten parts of the water extracted for drinking purposes are treated effluent, as is the case in the Santa Ana River in California where drinking water largely consists of wastewater effluent (Sedlak et al., 2003). Endocrine disrupters have been detected in several rivers that are used as a water source in Europe (Suter et al., 2005) as result of the discharge of treated wastewater. Because this is a concern, it is important to recognize such indirect reuse of water for drinking purposes, to begin controlling possible problems.

7.4.5 Pollutant management

Targeted pollutants vary across regions, depending on their needs and economical development. In this section, the management of biological and chemical pollutants is analyzed.

7.4.5.1 Biological pollutants

Controlling health risks produced through wastewater is the most important task water manager's face. And for that, a reliable monitoring system is needed. Unfortunately, because of all the numerous pathogens in wastewater (and water) and the complexity of the laboratory analyses needed, this is a costly task. Therefore, currently only indicators are measured and, although there is evidence that they cannot represent all kinds of pathogens, universally the fecal coliform or even the *E. Coli* group is used. But, under certain conditions these indicators should be accompanied with complementary ones, considering the local diseases transmitted through water as well as the wastewater disposal methods.

Developing countries – Due to the high infection rates in the population, a wide variety of pathogens are present in wastewater, a situation that combined with a lack of sanitation reflects the lower quality of water sources. If to this is added the frequently poorly managed water systems, the cost of having safe water substantially increases (Box 7.6).

BOX 7.6 What is the cost of access to safe water in Mexico City

In Mexico City water sources have been deteriorating with time, while potabilization systems are using the same technology as in the 1940s. Additionally, the quality of the supplied water is further deteriorated during its distribution. The water network operates intermittently and at low pressure. To increase the reliability on the water supply, people install storage tanks and pumping systems at home. As result of all these problems, tap water has an unreliable quality. To have water safe to drink, people need to buy bottled water or potabilize tap water at home. For a 4-member family living with a minimum wage this represents 6–10% of its income. Most common in-house potabilization methods are by boiling, addition of chemical disinfectants (chlorine, colloidal silver or iodine) or small potabilization systems using ozone, UV-light or Silver colloids[3]).

Cost per cubic metre of in house potabilization methods

Item	Cost per cubic metre in US$ to supply a 4-member family
Tap water	0.6
In house storage tanks and pumping system*	0.4
Cost of bottle water (lowest commercial price)	2.2
Cost of bottle water (lowest commercial price)	3.2
Cost to produce safe water by boiling it	1.8

*Considering a 20-year depreciation

[3]Individual disinfection systems double the price to disinfect water

BOX 7.7 Different treatment process efficiencies to remove microorganisms from wastewater (See also colour plate 5)

Figure shows the removal of microorganisms during different treatment stages. Helminth ova and protozoans are best removed by methods used to remove solids (coagulation–flocculation and filtration). Bacteria are best controlled through conventional disinfection methods (chlorination, UV-light disinfection and ozonation).

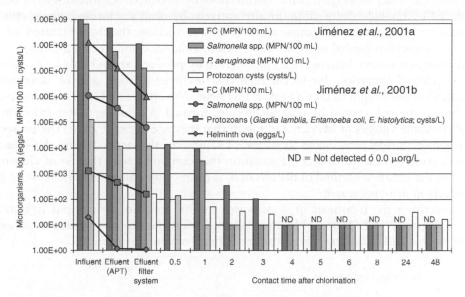

Removal of microorganisms from wastewater using different treatment processes

Source: Jiménez et al., 2001a and b.

Table 7.9 In-house disinfection methods for different kinds of pathogens

Organisms	In-house disinfection methods
Giardia	Boiling water
	Iodine is a better disinfectant than chlorine
Cryptosporidium	Highly resistant to chlorine, removal must be performed through filtration
	Heating water at 45°C for 5–20 minutes or above 65°C for 1 minute
Entamoeba histolytica	Resistant to chlorine and iodine at doses normally used to kill bacteria
	Filtration is efficient

Considering this situation, the main target of water works is to efficiently disinfect water and wastewater, and for that the efficacy of different treatment methods to reduce pathogens should be considered (Box 7.7). The selection of a treatment should be based on the maximum removal of all kinds of organisms. In addition, developing countries' city managers should promote education and public campaigns in which in-house potabilization methods are efficient for safe water depending on the pathogens involved (Table 7.9).

For monitoring programmes in cities where wastewater is reused for agricultural irrigation, two types of indicators (fecal coliforms and helminth ova) should be used, according to WHO (2006).

Developed countries – There are several examples in developed countries of wastewater and water treatment schemes that are being modified to control emerging pathogens. Filtration is the treatment step most frequently added to remove protozoa, such as *Cryptosporidium*. In addition, water treatment plants are being modified to control the risks associated with disinfection by-products. Chlorination is being replaced by UV-light disinfection, or alternatively, by-products formed during chlorination are being removed. Because all these options increase the cost of treated water, further research is needed to reduce costs and to enhance disinfection efficiency.

Although it is often believed that chlorination by-products should only be a concern for developed countries, due to the high cost involved in their control, developing countries should begin to take a look at this issue. Navarro et al. (2007), for instance, evaluated the risk caused by organochlorides formed in drinking water from small cities in some villages of Mexico and found that they are not negligible. The presence of organochlorides was linked to a lack of sanitation that is increasing the organic matter content in water sources, a situation that combined with the use of chlorination, as the cheapest method of disinfection, is promoting a significant increase of by-products in drinking water.

Concerning the treatment of wastewater for its reuse in agricultural irrigation, according to WHO (2006), risks in developed countries are mainly due to viruses, thus, norms and technology should address this problem in particular.

7.4.5.2 Chemical compounds

In principle, because chemicals are man-made there are two ways to control them: 1) through the implementation of 'integrated water' policies; and 2) through wastewater treatment. Integrated water policies certainly need to go beyond the water sector, because they imply not only establishing criteria for wastewater discharges but also promoting norms and incentives for industries and the social sector, in order to not produce, use and indiscriminately discard chemicals. This can be a difficult task for water managers if they are not able to proactively collaborate with other sectors.

Also, to control chemical compound pollution in water, a better management of urban soil, solid wastes, industries, atmospheric pollution and other polluting urban activities is needed and, for that, it is important to raise awareness in society about the impact cities have on water through different activities.

7.4.6 Urban infrastructure and urban activities

As discussed in Section 7.2.2.1, urban infrastructure and urban activities are a source of water pollutants. Tables 7.10 and 7.11 present different options for managing their impact.

7.4.7 Climate change

Climate change is a new aspect affecting water quality and quantity. According to Kundzewicz et al. (2007), three effects will be observed in water: fresh water salinization,

Table 7.10 Management options for controlling water pollution from urban infrastructure; those considered as the main targets for developing countries are highlighted with *

Source	Management option
Water network leaks	• Monitoring and surveillance • Proper design, construction of new pipes* • Proper maintenance of the network • Network sectorization* • Repair and replacement of pipes particularly in the oldest areas • Maintenance of a proper level of pressure in the network*
Sewerage leaks	• Proper design and construction* • Monitoring and surveillance • Replacement of sewers to fit new capacity demand. • Balancing sewer leaks control with installing sewers where they are not available* • Repair and maintenance of the sewer system • Exfiltration control • Use of separated sewers for wastewater and stormwater in new urban areas
Septic tanks and other individual sanitation systems	• Proper design and maintenance* • Proper selection of sites where this technology is safe and suitable* • Monitoring of underground water impact* • Economic and political support to perform training programmes for their proper construction and maintenance*
Storage or treatment ponds	• Planning to define sites where they should be allowed* • Proper design and construction • Maintenance and surveillance considering a suitable budget to perform this activity*
Storage tanks	• Development of norms for proper design and location • Proper construction and maintenance* • Monitoring • Surveillance in sensitive areas*
Municipal landfills	• The replacement of dumping sites with landfills* • Proper site selection* • Design criteria to control leachates infiltration • Proper maintenance • Monitoring programmes
Hazardous waste confinement sites	• Segregation of hazardous wastes from municipal wastes* • Planning to define possible location* • Installation of hazardous waste confinement sites • Proper maintenance and surveillance • Reduction of hazardous waste generation* • Public involvement to proper handle hazardous wastes*
Transportation of polluted water in channels or rivers	• Wastewater treatment* • Lining
Highway drainages or soakaways	• Collection and disposal in soil* • Collection and treatment
Pipelines	• Surveillance • Selection of alternative methods for transporting high-risk substances • Surveillance covering sensitive areas only*

(Continued)

Table 7.10 (Continued)

Source	Management option
Injections wells	• Setting norms for their construction* • Proper operation • Monitoring programmes • Monitoring programmes only in sensitive areas*
Cemeteries	• Planning for location, proper design and surveillance* • Impervious tombs and corrosion resistant coffins
Wastewater treatment plants	• For developed countries mostly, aging wastewater collection infrastructure and upgrading treatment plants is an issue. • Promoting safe water reuse* • Infrastructure to treat stormwater.

Table 7.11 Options for managing urban activities polluting water; those considered as the main targets for developing countries highlighted with *

Activity	Management options
Industries	• Planning location considering environmental criteria • Development of an appropriate scheme of norms and incentives to protect the environment • Setting achievable norms: modest but enforceable* • Surveillance and monitoring • Limiting the use of recalcitrant compounds*
Irrigation of amenity areas	• Use of low water consumption irrigation methods • Use of gardens needing low watering
Application of ice melting substances	• Evaluating risks to determining its relative magnitude • Limiting the amount of salts to be used
Transport and transference of material	• Promote clean production programmes in industries

especially in coastal areas due to a rise in the sea level; changes in water availability in terms of quantity and quality due to more frequent extreme events and temperature increases; and a rise in the CO_2 content of water. Although these effects are expected to happen with a high degree of reliability, changes at a local level are not yet known. Some water utilities have already begun to implement programmes to face climate change effects, the main activities focus on the efficient use of water, the reuse of wastewater and the improvement of water treatment plants to face lower influent quality during extreme events, particularly higher pluvial precipitation (Kundzewicz et al., 2007). Chapter 5 deals in more detail with climate change effects.

7.4.8 Education and research

As seen, several pollutants in water have their origin in a wide variety of urban activities, and controlling those means getting society involved. For proactive public participation there needs to be a general awareness of the problems that are occurring and their cause, as well as how society can participate in the solution. To do this, educational and information programmes tailored to local needs and concerns must be developed.

Finally, because little is still known about many aspects of the urban water cycle and its interaction with the rest of the environment, research must be done in this field with a holistic approach by different disciplines.

7.5 TREATMENT

The treatment of wastewater, while not a caveat for all the negative factors affecting water quality, is still a very important option. This section presents specific aspects of wastewater treatment with special emphasis on conventional treatment methods for biological pollutants (pathogens) and emerging pollutants.

7.5.1 Biological pollutants

Table 7.12 presents the organism's removal or inactivation achieved under different wastewater treatment processes. This table is a guide for selecting a process, but to design a complete treatment scheme the pre-treatment to each process need to be considered.

Developing countries – In developing countries the main issue is still the proper disposal of feces, particularly in low-income peri-urban areas. This combined with a high content of pathogens in municipal wastewater, implies a proper selection of a treatment process to control disease dissemination. In general, coupling any kind of wastewater treatment process (biological or physico-chemical) with a filtration step before disinfection will notably reduce the content of protozoan cysts and oocysts, and that of helminth eggs. However, this is not always feasible for economic reasons and, therefore, alternative methods using less technology must be employed.

As shown in Table 7.1, wastewater from different countries contains different amounts of helminth ova, which is why they are not considered in all countries' norms, as is the case for organic matter measured as BOD or fecal coliforms. Of all the organisms that can be present in wastewater, protozoans and helminth eggs are the most difficult to remove. Consequently, it follows that technology to treat wastewater should also differ among countries. When wastewater is used for irrigation, WHO (2006) recommends a content of ≤1 eggs/L to irrigate crops that are eaten uncooked. For fish culture, trematode eggs (*Schistosoma* spp., *Clonorchis sinensis* and *Fasciolopsis buski*) must be zero, as these worms multiply by the tens of thousands in their first intermediate aquatic host (an aquatic snail). Helminth eggs are resistant to chlorination, UV-light and ozonation but they are inactivated if the temperature rises above 40°C for several days or moisture is reduced to less than 5% (Jiménez, 2007a). These conditions are not normally encountered in wastewater treatment processes. Thus, helminth ova are removed from wastewater to be inactivated in sludge. To remove helminth eggs from wastewater, processes to remove suspended solids, such as sedimentation or filtration, are used because they are particles measuring between 20–80 μm in size.

Developed countries – As mentioned for developed countries, emerging organisms are the main target. Studies performed with *Cryptosporidium* oocysts, which are highly resistant to chlorination, led to the addition of a filtration step to water and wastewater treatment plants. Also, because the main health concern stems from the presence of viruses, turbidity in effluents for some states and programmes, has been reduced to <0.1 NTU when they are to be reused (Asano et al., 2007).

Table 7.12 Reduction or inactivation different biological pollutants in wastewater. Adapted from: WHO, 2006

Treatment process	Log unit microorganisms removal[a]			Removal, %
	Viruses	Bacteria	Protozoan (oo)cysts	Helminth eggs
Natural systems				
Waste stabilization ponds	1–4	1–6	1–4	90–99[(1)(5)]
Wastewater storage and treatment reservoirs	1–4	1–6	1–4	90–99[(1)(4)]
Constructed wetlands	1–2	0.5–3	0.5–2	90[(1)(5)]
Primary treatment				
Primary sedimentation	0–1	0–1	0–1	0–<1[(1)]
Chemically enhanced primary treatment	1–2	1–2	1–2	90–99[(1)(5)]
Anaerobic upflow sludge blanket reactors	0–1	1–2	0–1	96[(1)]
Secondary treatment				
Activated sludge + secondary sedimentation	0–2	1–2	0–1	1–<2[(1)]
Trickling filters + secondary sedimentation	0–2	1–2	0–1	1–2[(3)]
Aerated lagoon or oxidation ditch + settling pond	1–2	1–2	0–1	1–3[(3)]
Tertiary treatment				
Coagulation/flocculation	1–3	0–1	1–3	2[(1)]
High-rate granular or slow-rate sand filtration	1–3	0–3	0–3	>99[(1)]
Dual-media filtration	1–3	0–1	1–3	2–3[(3)]
Membrane bioreactors	2.5–>6	3.5–>6	>6	>3[(3)]
Disinfection				
Chlorination (free chlorine)	1–3	2–6	0–1.5	0–<1[(1)]
Ozonation	3–6	2–6	1–2	0–2[(2)]
UV irradiation	1–>3	2–>4	>3	0[(2)]

[a] The log unit reductions are log_{10} unit reductions defined as log_{10}(initial pathogen concentration/final pathogen concentration). Thus, a 1-log unit reduction = 90% reduction; a 2-log unit reduction = 99% reduction; a 3-log unit reduction = 99.9% reduction; and so on.
(a) Tested with up to 2 log initial content. Might have greater efficiencies than reported
(1) Have been tested on a full scale
(2) From laboratory data
(3) Theoretical efficiency based on removal mechanisms
(4) Total helminth egg removal is only achieved when wetlands are coupled with a filtration step
(5) Tested with high helminth egg content

7.5.2 Emerging pollutants

The content of some emerging pollutants could be reduced by optimizing wastewater treatment processes for this purpose. Table 7.13 contains the theoretical performance expected in different unit processes after emergent pollutants' properties (molecular size, hydrophobicity, functional group composition). Most compounds of concern are relatively polar (log K_{OW} < 3) and as a result, only a few (e.g., nonylphenol, fluroanthene, pyrene) are expected to be removed through coagulation–flocculation or precipitation. In contrast, they are expected to adsorb into particles (organic or clay, Snyder et al., 2003). In the following section, the results of the application of some treatment processes are presented.

Table 7.13 Theoretical removal of EDC and PPCP in treatment processes. From: Snyder et al., 2003

Classification	AC	BAC	O₃/ AOPs	UV	Cl₂/Cl O₂	COA/ FLOC	SOF/ MO	NF	RO	B/P/AS
Endocrine Disrupter Compounds (EDC)										
Pesticides	E	E	L-E	E	P-E	P	G	G	E	E {P}
Industrial chemicals	E	E	F-G	E	P	P-L	P-L	E	E	G-E {B}
Steroids	E	E	E	E	E	P	P-L	G	E	L-E {B}
Metals	G	G	P	P	P	F-G	F-G	G	E	P {B}, E {AS}
Inorganics	P-L	F	P	P	P	P	G	G	E	P-L
Organometallics	G-E	G-E	L-E	F-G	P-F	P-L	P-L	G-E	E	L-E
Pharmaceuticals and personal care products										
Antibiotics	F-G	E	L-E	F-G	P-G	P-L	P-L	E	E	E {B} G-E {P}
Antidepressants	G-E	G-E	L-E	F-G	P-F	P-L	P-L	G-E	E	G-E
Anti-inflammatory	E	G-E	E	E	P-F	P	P-L	G-E	E	E {B}
Lipid regulators	E	E	E	F-G	P-F	P	P-L	G-E	E	P {B}
X-ray contrast media	G-E	G-E	L-E	F-G	P-F	P-L	P-L	G-E	E	E {B and P}
Psychiatric control	G-E	G-E	L-E	F-G	P-F	P-L	P-L	G-E	E	G-E
Synthetic musks	G-E	G-E	L-E	E	P-F	P-L	P-L	G-E	E	E {B}
Sunscreens	G-E	G-E	L-E	F-G	P-F	P-L	P-L	G-E	E	G-E
Antimicrobials	G-E	G-E	L-E	F-G	P-F	P-L	P-L	G-E	E	F {P}
Surfactans/detergents	E	E	F-G	F-G	P	P-L	P-L	E	E	L-E{B}

AC: Activated carbon
BAC: Biodegradation I Activated Carbon
O₃: Ozonation
AOPs: Advanced oxidation process
UV-light decomposition
E: excellent (90%)
G: good (70–90%)
F: fair 0–70
L: low 0–40
P: poor <20

COA-FLOC: Coagulation-flocculation
Cl/ClO₂: Chlorine dioxide oxidation
SOF/MA: Softening metal oxide precipitation
NF: Nanofiltration
RO: Reverse osmoses
B/P/AS: Biodegradation/Photodegradation
(solar)/Activated sludge

Biological treatment – Table 7.14 shows pharmaceutical removal rates in wastewater treatment facilities, most of them in biological secondary treatment plants. As observed, removal varies widely, which is not surprising given that pharmaceuticals are compounds belonging to different chemical groups and displaying very dissimilar properties. According to Fent et al. (2006) removal takes place mainly at the secondary treatment stage, whereas in the primary one, removal occurs only partially. Nevertheless, efficiencies reported might reflect only the transformation of one compound to another product. Removal in biological treatment plants may occur through either biodegradation or adsorption, which are believed to be the two most important processes in wastewater treatment for removing emerging pollutants. The removal mechanism is not yet well understood in detail, but for those that are biodegraded, degradation increases as the hydraulic and the cellular retention time increases in activated sludge systems (Kreuzinger et al., 2004b and Suter et al., 2005).

Coagulation flocculation – Flocculation has proven ineffective at removing traces of organic pollutants (Zhang and Emary, 1999; Ternes et al., 2002; Adams et al.,

Table 7.14 Removal efficiency in sewage treatment plants. With information from: Fent et al., 2006

Compound	Maximum or range removal in %
Acetylsalicylic acid	81
Atenolol	0–10
Bezafibrate	4–100
Carbamazepine	0–53
Clofibric acid	0–91
Clotrimazole	55
Diazepam	93
Dextropropoxyphene	0
Diclofenac	0–71
Ethinylestradiol	85
Fenofibric acid	6–64
Gemfibrozil	10–75
Ibuprofen	4–100
Ifosfamide	0
Ketoprofen	8–100
Mefenamic acid	0–50
Metoprolol	0–83
Naproxen	15–100
Paracetamol	100
Propranolol	0–96
Salicylic acid	99
Tamoxifen	0
X-ray contrast media	0

2002) and the poor removal observed has been linked to their previous adsorption in suspended solids.

Adsorption – Absorption in activated carbon has been used to remove EDCs and PPCPs (Gillogly et al., 1998; Zhang and Emary, 1999; Fuerhacker et al., 2001; Tanghe and Verstraete, 2001; Ternes et al., 2002; Bruce et al., 2002 and Snyder et al., 2003). Performance depends on the activated carbon properties (surface area, pore size distribution, surface charge and oxygen content) but will also be different for each pollutant according to its shape, size, charge and hydrophobicity. Hydrophobic interactions are the dominant removal mechanisms for most compounds. As a result, activated carbon will, in principle, efficiently remove most non-polar organic compounds (i.e. those with log $K_{OW} > 2$). Because actual performance depends on the strength of the polar interactions between the compounds properties and the quality of water, it is difficult to predict removal, and experimental tests need to be performed (Snyder et al., 2003). Additionally, in wastewater treatment, organic matter competes for adsorption sites, thus, decreasing the activated carbon capacity for removing targeted pollutants (Adams et al., 2002).

Oxidation – Oxidation is used to decompose emerging pollutants. Ozone combined, or not, with hydrogen peroxide has produced relatively good results (Zwiener and Frimmel, 2000; Ternes et al., 2002) using concentrations generally applied to drinking water. During the process, the ozone consumption by organic matter needs to be taken into account. For instance, to degrade pharmaceuticals (>90%) the quantity of ozone applied needs to be at least equal to the DOC (dissolved organic value). During oxidation, care must be taken with regard to the formation of new products that might have greater toxicity than the pollutants initially present.

Laboratory reactivity constants with ozone have been reported for diclofenac, ethynylestradiol, carbamazepine, sulfamethoxazole, bezafibrate, ibuprofen, diazepam and iopramide, with values ranging from 1 to $10^6 M^{-1}s^{-1}$ (Huber et al., 2003). Oestrogen steroids and nonylphenols react with ozone under conditions similar to those applied in drinking-water treatment systems. At the bench scale, the removal of diclofenac, carbamazepine and bezafibrate has been reported but not for clofibric acid (Sacher et al., 2000). Instead, clofibric acid along with ibuprofen, and diclofenac, have been removed with ozone combined with hydrogen peroxide (0.4 to 0.7 mg H_2O_2/mg ozone dosed, Carlson et al., 2000 cited in Snyder et al., 2003).

Chlorination – Chlorine dioxide (ClO_2) oxidizes herbicides, pesticides and PAHs at doses of 1–2 mg/L in some minutes to hours (Ravacha and Blits, 1985; Lopez et al., 1997). The transformation of several amine-containing antibiotics (Adams and Kuzhikannil, 2000; Huang and Sedlak, 2001; Sedlak et al., 2003) and caffeine (Gould and Richards, 1984) has been observed in laboratory experiments with chlorine. But, besides having a less oxidant effect than ozone, chlorine holds the risk of forming noxious by-products.

Ultraviolet (UV) irradiation – UV alone or as part of an AOP (using ozone or hydrogen peroxide) system has been used to oxidize pesticides, micro-pollutants, antibiotics and fragrances (such as nitromusks) according to Chiron et al., 2000 and Snyder et al., 2003. Because several EDCs and PPCPs have chromophores that lead to the adsorption of light at UV wavelengths, many may be amenable to transformation during UV treatment. However, typical UV doses required for disinfection (i.e. 5 to 30 mJ/cm^2) are several orders of magnitude lower than those needed to treat micro-pollutants. Therefore, UV is not considered an economic option.

Membranes – Apart from micro-filtration, membrane filtration has been reported as an efficient method for removing EDC/PPCP. Reverse osmoses is considered the best membranes process for removing emerging compounds, although it is not very efficient at treating low molecular weight uncharged compounds (Adams et al., 2002; Sedlak et al., 2003). Actual removal through the membrane processes (NF and UF) needs to be assessed through laboratory tests because the presence of cations and organic matter can notably affect efficiency (Snyder et al., 2003).

Photolysis – Natural attenuation through photolysis using sunlight seems to be an important degradation process for some compounds in surface water (Fent et al., 2006). Photolysis has proved useful at removing diclofenac in full-scale installations (Buser et al., 1998). At laboratories, some pharmaceuticals (sulfamethoxazole, ofloxacin and propranolol) have been removed by direct and indirect photolysis (Andreozzi et al., 2003). Besides the pollutants' properties, the efficiency of photodegradation depends on the strength of solar irradiation and, therefore, as a natural removal mechanism it will be different for different latitudes, water bodies and seasons of the year. Water constituents may also act as photosensitizers generating hydroxyl radicals and single oxygen enhancing degradation (i.e. nitrates, humic acids) or as interferences. Estradiol and 17α-Ethinyl Estradiol removal in wetlands is said to occur by photolysis with half-lives of approximately two weeks, using the dissolved organic matter content as a photoreactive transient (Gray and Sedlak, 2003).

SAT – Soil-aquifer treatment (SAT) and river bank filtration have proved effective at removing several EDCs and PPCPs through natural processes (Snyder et al., 2003). However, little is known about the mechanisms involved or the new products formed. Data from sites where wastewater effluent is used to recharge aquifers indicate that many of the PPCPs are removed during the first centimetres of passage through soil (Drewes et al., 2001 cited in Snyder et al., 2003). However, certain recalcitrant compounds, including the anti-epileptics, carbamazepine, primidone and iodinated X-ray contrast media are not retained and reach the saturated soil zone (Drewes et al., 2001 cited in Snyder et al., 2003).

Sludge – Depending on their hydrophobic and electrostatic properties, emerging pollutants can or cannot be absorbed by particulates and microorganisms during wastewater treatment. Once in sludge, they concentrate and tend to remain there, since they are not degraded. During sewage treatment, it is likely that many organic compounds, particularly hydrophobic compounds, will be absorbed and concentrated in the sludge (Fent et al., 2006; Ternes et al., 2002).

7.5.3 Criteria for selecting wastewater treatment processes

Guidance for properly selecting a wastewater treatment process is presented in Table 7.15. Due to the notable difference between the level of treatment in developed and developing countries, less experience is available in the latter ones and therefore process selection is a more delicate issue.

Table 7.15 Criteria for selecting wastewater treatment operation and processes

Process applicability
- Must be evaluated based on past experience, data from full-scale plants, published data and from pilot and plant studies.
- If few or unusual conditions are encountered (atypical wastewater characteristics) pilot plant studies are essential.

For developing countries:
- Since much less experience is available, a good wastewater characterization is needed as well as a request during bids that the applicability of the processes should be demonstrated under local conditions.
- Bids need to limit accepted technology based on experience in similar conditions.

Performance
- Performance needs to be expressed in terms of the effluent quality but as well on its VARIABILITY, and must be consistent with the effluent discharge requirements as well as with the future use of treated wastewater.
- Performance needs also to be considered in terms of its RELIABILITY, as it may vary according to the process type.

Influent wastewater variability
- Consider wastewater characteristic variations in probabilistic terms for rain and dry seasons.

For developing countries:
- It is important to have a statistically representative wastewater characterization considering parameters not only defined in norms but also those that might interfere with the treatment processes.
- Design data should not be based on bibliography data especially that coming from other countries.
- Since segregation and pre-treatment of industrial discharge is not common, wastewater might often contain inhibiting constituents. An evaluation of these is important although in not as intensive way as the targeted pollutants.

(Continued)

Table 7.15 (Continued)

Reliability
● Achievable performance needs to be expressed in statistical terms and in short and long terms, taking into account water flow and wastewater quality variations.

For developing countries:
− 'Unusual' situations and 'emergencies' are common. Selecting robust albeit more expensive processes might be cheaper long term, in economic terms as well as in terms of the negative effects that malfunctioning can produce in environment and health.

Process sizing
● Reactor sizing is based on the governing reaction, kinetics and kinetic coefficients. If kinetic expressions are not available, process loading criteria are used, but not always with good results, even in developed countries.

For developing countries:
− Most of the available information used in the design process comes from the developed world, where wastewater and climatic conditions, among others, are different, and so bibliographic kinetic data and load criteria use should be avoided as much as possible.
− If experimental data is not available, the adjustment of published data to local conditions, such as pressure and temperature, should always be checked in bids.

Applicable flow range and flow variations
● The process should be matched to the expected ranges of flow rates. And, whenever possible, considering the presence of stormwater.

For developing countries:
− For those located in regions with high pluvial precipitation concentrated in short periods, treatment processes must be able to deal with major flow and wastewater quality variations.
− Alternatively, the use of flow equalization tanks and their cost should be considered.

Residual treatment and disposal
● The types and amounts of solids, liquid and gaseous residuals produced must be estimated.
● Use pilot plant studies to identify and quantify residuals.

For developing countries:
− By-products and wastewater treatment residues are often disregarded in proposals, in order to offer a lower operating cost. To avoid this, it is important to clearly state in bids that any residues must be quantified and the management options considered.
− The management of hazardous chemicals in wastewater treatment plants is also an item frequently overlooked.

Sludge processing
● A wastewater treatment system is a factory that produces water and sludge, where sludge treatment can represent 50% of the total cost. Therefore, the proper design, operation and maintenance of the sludge treatment system must be considered with the same level of importance as that of the wastewater treatment.

For developing countries:
− Sludge treatment and disposal will increase at the pace of wastewater treatment. Revalorization of sludge as biosolids (treated sludge) can be a very interesting option in countries with poor soils or where the cost of fertilizers is high for farmers. Biosolids can also be used as cover material in landfills.

Climatic constraints
● Temperature affects the reaction rate of most chemicals and biological processes; therefore local water temperature should be taken into account when selecting a process.

For developing countries:
− In most developing countries, temperature is relative high, so problems arise mostly due to high temperatures not low ones. Warm temperature may accelerate odour generation and also limit the amount of oxygen that can be dissolved.

(Continued)

Table 7.15 (Continued)

Environmental constraints
- Environmental factors, such as prevailing winds and wind directions and proximity to residential areas, may restrict or affect the use of certain processes, especially where odour may be produced.
- A wastewater treatment plant can cause a negative impact if not properly designed.
- The disposal site restrictions of the treated wastewater need to be considered regardless of the norms to be met.

Water and sludge reuse
- Water reuse can be a way of making wastewater treatment more interesting in economic terms
- For countries located in water-stressed areas, besides being a disposal option reuse serves to alleviate water scarcity.

For developing countries:
- Land degradation is costing 5–10% of their agricultural production (Young, 1998) and fertilizers have often a prohibited cost for farmers; in both cases biosolids can be used to remedy these problems. Nutrients contained in sewage are as well interesting to reclaim.

Ancillary processes
- Wastewater treatment plants are often accompanied by ancillary (complementary) processes that are very varied and beyond the scope of wastewater treatment operators. It is important therefore to know before selecting a process what its external needs are as well as the capacity of the local area to supply them.

Chemical requirements
- The type and amount of chemicals to be used need to be considered as well as their cost both now and in the future.
- When treated wastewater or sludge are to be reused, chemical selection needs to consider disposal option requirements. Chemicals can either limit or benefit future use depending on how well selected they are.

For developing countries:
- Although the use of chemicals is often prohibited, an economical comparison is worth making, especially if chemicals are available locally.

Energy requirements
- The present and future cost of the energy used needs to be considered.
- Efficient use of energy and the possibility of recovering/producing energy for in-plant use must form part of the selection criteria that in the long term will contribute to properly closing the urban water cycle. For developing countries trading carbon bonds should not be neglected.

Personnel requirements
- The number of people as well as their skill level needs to be well defined.

For developing countries:
- The most common situation is a high availability of low skilled personnel working for low salaries. Thus, selected processes may have a high labour demand but cannot be very sophisticated. Alternatively, intense training programmes should be considered.

Complexity and compatibility
- Define operations needs under routine an emergency conditions.
- Define the type and need to make repairs.
- It is important that the items selected be compatible for a good operation.

For developing countries:
- Consider that in some cases cheap equipment is very costly if frequent repair is needed.
- Equipment and the spare parts must be available within a logical period. Obsolete equipment is very difficult to repair.
- Normally, few items are produced or available locally, therefore overall equipment selection needs to consider compatibility between different equipment trades.

(Continued)

Table 7.15 (Continued)

Adaptability
- Many treatment plants will need to adapt to future conditions and not all systems have the same capability to be adapted.

Economic life-cycle analysis
- Cost evaluation must consider initial capital cost and long-term operating and maintenance costs. The plant with lowest initial capital cost may not be the most effective with respect to operating and maintenance costs.
- The nature of the available funding will affect the choice of the process.

Land availability
- It is important to consider the size of the selected treatment process with respect to available land, including buffering zones for future expansions.

For developing countries:
- Land is not always available and when available it is not necessarily cheap.
- Considering the fast growth of cities in the developing world and the possibility of building plants by modules, it is very useful to consider buffering zones either to increase treatment capacity, complete the treatment process or even to avoid building human settlements near to treatment facilities.

7.6 WASTEWATER DISPOSAL

In the middle of the nineteenth century Edwin Chadwick proposed that 'stormwater should go to the rivers while wastewater should go to soil'. Unfortunately, his words were not heeded and both stormwater and wastewater were sent together into rivers, as the cheapest way to get rid of wastewater. However, at the present time, and in view of the consequences, using separate sewers is increasingly being considered, as well as disposing of treated wastewater in the soil rather than in surface water.

7.6.1 Soil disposal

7.6.1.1 Soil disposal and aquifer storage

Soil has proven to have a greater capacity to depollute water than surface water bodies (Bouwer et al., 1980; Rice and Bouwer, 1984; Bouwer 1989). Soil can eliminate suspended solids, nutrients, metals, recalcitrant organic compounds, endocrine disrupters, pharmaceuticals and pathogens through several processes, such as filtration, biological degradation, ion exchange and adsorption (Fujita et al., 1996; Quanrud et al., 1996, Wilson et al., 1995, Asano et al., 2007; Heberer, 2002). This principle is used for a wastewater treatment method known as SAT (Soil-aquifer Treatment). A SAT system operates by combining flood and dry cycles to maintain aerobic-anoxic conditions in soil with percolating rates depending on the type of soil. For good functioning, hydraulic (29 to $111 \, m^3/m^2yr$) and mass loads should be limited. To avoid aquifer pollution, application of wastewater (preferentially partially treated) is restricted to sites where groundwater is at a minimum of 3 m depth. SAT systems are also used as recharge methods using aquifers as storing systems (Lance and Whisler, 1976 and Lance et al., 1980). SAT-ASR (Aquifer storage recover system) has several advantages, see Table 7.16. Aquifers can be an economic disposal option as they are also useful to store water, preventing its evaporation in arid and semi-arid countries.

Table 7.16 Objectives of Aquifer Storage Recovery Systems (ASR). From Jiménez, 2003

Objectives

- Water storage for use in different seasons of the year
- Long-term storage
- Storage for emergencies
- Strategic reserve
- Daily storage
- Reduction of disinfection by-products
- Restoration of the phreatic level
- Pressure and flow maintenance in the distribution network
- Improvement of the water quality
- Saline intrusion prevention
- Agricultural supply

- Nutrient leaching control
- Increase in well production
- Delay of new water supply systems
- Storage of reclaimed water
- Soil treatment
- Refinement of water quality
- Stabilization of aggressive water
- Hydraulic control of pollution plumes.
- Water temperature maintenance for fish production
- Reduction of environmental effects caused by spills
- Salinity lixiviation from soils

ASR systems have been used for a long time in several parts of the world. The first system to operate in the United States was built in 1968. By 1994, 7.2 m^3/s of fresh and saline water was being stored in aquifers (Pyne, 1995).

7.6.1.2 Soil disposal and agriculture

Cities are facing increasing food demands. Coupling wastewater disposal with agricultural reuse is an important way of freeing up first use water to supply cities and to reclaim used water for food production (Box 7.8). Reuse of wastewater to grow crops can contribute to a secure food supply in developing countries (IMWI, 2003). The economic and even health benefits of reusing non-treated wastewater have been documented by Raschid-Sally et al., 2005. Agricultural reuse allows: 1) reclaiming the nutrients contained in wastewater reuse for boosting crop yields; 2) a year-round crop production; and 3) a wider variety of the crops, especially in arid and semi-arid areas (IMWI, 2003). Wastewater reuse for agriculture is an interesting option, provided negative health risks are controlled (WHO, 2006). Actually, the reuse of wastewater for this purpose is already a common practice in several countries that is almost always linked to cities and involves both treated and non-treated wastewater (Jiménez, 2006).

7.6.2 Disposal in water bodies

7.6.2.1 Eutrophication

Eutrophication is the natural process of water bodies 'aging'. Contamination with nutrients accelerates this phenomenon, which is commonly observed in lakes and dams. Unfortunately, treated wastewater commonly accelerates this phenomenon because secondary effluents contain large amounts of nutrients. Nutrients (nitrogen and phosphorus) promote algae and aquatic plants growth, which covers the surface

BOX 7.8 From wastewater to milk: the Andhra Pradesh case

For many of people that live in arid and semi-arid regions, wastewater produces their food. It also represents the only option to have an income. This is the case of Andhra Pradesh, India, where wastewaters from the Hyderabad and Secunderabad conurbation are used to irrigate, directly and indirectly, 40,600 ha and to produce fish (Buechler and Devi, 2003). The population in both cities is around 6 million people living in a region with 750 mm of precipitation. Different societies have developed in the urban, peri-urban or rural areas around the wastewater (Buechler and Devi, 2003).

Urban zone – In a 5 km strip along the Musi river, wastewater is used directly by around 250 families to irrigate 100 ha. The parcels as a whole form a green belt area around the city. Main crops are fodder grass (65%), banana (20%), coconut palms (10%), green leafy vegetables (4%), and fruits and flowers (crossandra and jasmine to adorn women's long braids). Relying on the use of wastewater a complex social organization has developed comprising 1) *Landowner Farmers* (who rent their land for grass production; 2) *Dairy Producers* (who save 67% of the production cost by growing fodder with wastewater); 3) *Casual Labourers* (1,260 per day to handle fodder grass and 40,000 in related activities performed mainly by women, as it is their only possibility of work); 4) *Permanent labourers* (men who work on one plot of land year-round and are from drought-prone states), and 5) *Caretakers of parcels* (who live for free with their families on the land belonging to other people in small huts or tents).

Peri-urban zone – In recent decades, in 0.6 to 12 ha parcels, an intense culture of grass occupying 95% of 40,000 ha has developed, the rest being used for crops that can be commercialized or used for household consumption. In the peri-urban area, ambulatory street vendors and women with market stands are very common.

Rural area – 1,000 ha are sown, mainly with fodder grass. In addition, livestock, toddy tapping and aquaculture are also performed with wastewater. As in the other cases, crops for family consumption and commercialization are also grown.

What is the fodder for? India is number one in world milk production. The annual growth rate for milk production is 6%, compared to a world rate of only 1%. Andhra Pradesh contributes 6% of the national milk production. Cattle are mainly fed with fodder grown in wastewater-irrigated fields in the urban, peri-urban and rural areas of the Musi River. Cattle drink wastewater or bore well water that is also used for domestic purposes. 25% of the milk production in Andhra Pradesh is also used for household consumption. The FAO projects that by 2010, 70% of undernourished people will live in south Asia and Sub-Saharan Africa. Will wastewater be a way of alleviating hunger?

and prevents sunlight and oxygen from entering the water. Common affectations observed in eutrophicated reservoirs are:

- increased algae and plant production
- consumption of the oxygen in the hypolimnion
- release of Fe, Mn, NH_4 and metals from the hypolimnion
- presence of plagues, such as schistosomas (flukes) and mosquitoes
- loss of biodiversity (in special higher trophic levels) and displacement of native species
- obstruction of channels and drains in irrigation zones and hydroelectric plants
- restrictions on tourist, recreational and fishing activities.

Although eutrophication is thought to be a problem only in lakes and reservoirs, it has been observed in rivers since the 1950–1960s in areas with slow flow (Chapman, 1992). The distinctive eutrophication symptom is the proliferation of aquatic weeds, such as water lilies or hyacinths (*Eichhornia crassipes*), hydrilla (*Hydrilla verticillata*), cattail (*Thypa* sp) and duckweed (*Lemna* sp). Water lilies grow in a great variety of habitats – from continental pools, marshes, drains, channels, lakes dams to low flow rivers – and adapt to a wide variety of environmental conditions. They can even survive for long periods in oligotrophic waters, although optimal growth takes place in eutrophic conditions. *Hydrilla verticillata* has it origins in Central Africa and has proliferated in warm regions of the world. *Thypa* sp. is an herbaceous plant found throughout North America, Europe and Asia, mainly in the temperate, subtropical and tropical zones. It lives near the shore of dams, channels, pools and marshes and grows densely in humid habitats or in fresh or brackish waters in sizes of up to 1 m depth. *Lemna* sp. is an aquatic plant that floats on the surface of lakes, pools and bogs. Its fast vegetative propagation in aquatic environment causes high evapotranspiration rates and therefore important water losses that can have tragic effects in shallow water bodies. It is considered cosmopolitan due to its location in tropical zones of America, Asia and Africa.

To reduce aquatic weed density (plants/m^2), rather than eliminate them, which is almost impossible, the following five methods are commonly used:

1. Biological control – Living organisms are used to control weeds. In theory, this is a cheap option because no maintenance, equipment or personnel cost is involved. However, this is only partially true because besides needing an ideal biological controlling agent, surveillance is also required. An ideal controlling agent meets the following requirements: it is selective to controlling weeds; it is able to survive in a new environment where it has never lived before; and its reproduction rate is equal to or slightly higher than the weed reproduction rate. To control water lilies, 70 arthropod species, 32 fungi and 6 bacteria species have been reported in 12 countries with limited results (Jiménez, 2001).

2. Mechanical control – These methods destroy or remove weeds using tools that can be operated manually or with equipment. The simplest procedure consists of harvesting weeds from water using winnowing forks or manual rakes. This is an expensive method due to the labour involved, but can be useful for controlling infested areas at their initial phase or to maintain areas where weeds have already been reduced to a low density. It is estimated that one man can harvest 393 kg/h (2.5 tons/d) of weed. Equipment consists of floating harvesting machines that cut plants, store them and transport them to the shore.

3. Chemical control – Pesticides are also used to control weeds. The chemicals added modify the plant's metabolism preventing growth and causing death. Some substances that have been used are Clarosan 500, Diquat, 2,4-D, Glyphosphate, Paraquat and Simazine. Due to their toxicity, these substances can only be applied under controlled conditions and if a certain amount of time is left before the water is used.

4. Habitat manipulation – By controlling the water level in reservoirs it is possible to control weed growth. When the water level decreases, weeds remain on the shore. If weeds are kept in the dry area for certain periods (21 days for water lilies) they die. The usefulness of this method is limited to places where the water level can be controlled.

5. Nutrient control – Weed growth is caused by a high N or P content in water, and so lowering their concentration through wastewater treatment is another alternative. Unfortunately, the cost remains high.

Due to their low efficiency or cost, in practice two or more of the methods described are often combined to control weeds. A common problem to all of them is what to do with weeds, since no revalorization method is economically convenient.

7.6.2.2 Coupling wastewater disposal with water reuse

The rapid growth of the urban population is increasing the amount of treated wastewater (in developed countries) and non-treated wastewater (in developing ones) that is being discharged into relatively small geographical areas, surpassing the local environment's capacity to 'assimilate' pollution. Therefore, the idea of setting wastewater discharge criteria assuming that the environment will restore the water's original quality is not enough in many cases. Actually, the lack of natural restoration is the reason why drinking-water standards are becoming more stringent and comprehensive, and more sophisticated technologies are needed to treat water. It is also the reason why chlorination is now generating by-products (organochlorides)[4] which require an expensive additional treatment step.

Natural reuse is occurring in developed countries, but cases are not documented as such. As mentioned, in the Thames River in England, during dry periods 70% of the water used as supply downstream comes from treated effluent. In California's Santa Ana River, a large part of the supply consists of treated wastewater. In Tucson Arizona, supply comes from an aquifer having a natural recharge of one-tenth that of the one with treated wastewater. In Berlin, 17–35% of the city's water supply comes from an advanced treated effluent (Ziegler, 2001). But there are also cases where non-treated wastewater is being reused after recharging aquifers; these include Hanoi in Vietnam, Leon Guanajuato, the Tula (or Mezquital) Valley, Mexico City and Merida, all four of them in Mexico, Hat Yai in Thailand, Ica Valley in Peru, Santa Cruz in Bolivia, Sana'a in Yemen, and in Hungary along the rivers Danube, Drava and Tisza (Jiménez, 2007b).

Therefore, whichever way you look at it, water contained in surface water bodies can no longer be classified as 'first use water', on the contrary, it must be accepted that water is being reused. This unintentional reuse makes it important to find a better way to close the urban water cycle. Indirectly it also means developing new regulations, treatment processes, monitoring programmes, and research and educational programmes that keep water reuse in mind.

7.7 CONCLUSIONS

Water pollution analyzed through the urban water cycle reveals new pollutant sources and unexpected risks. It also highlights the natural reuse of water. Recognizing

---

[4]When raw water contains organic matter, during chlorination organochlorides (chloroform CH_3, dibromochloromethane $CHClBr_2$, bromodichloromethane $CHCl_2Br$ and tribromomethane $CHBr_3$, mainly) that are compounds with cancer are formed.

material and energy exchange between water and urban soil and air creates awareness about the importance of properly closing the urban water cycle. To achieve this, it is necessary to perform several activities involving society as a whole, not only water professionals. For example, new concepts of cleaner production need to be introduced, the definition of quality of life needs to be adjusted to avoid the use of toxic recalcitrant compounds, and – from the water-side sector – a truly integrated water resources management (IWRM) approach needs to be put in place. To date, the management of water quality has mostly been done by implementing wastewater treatment facilities, but to have clean water it is also necessary to control air and soil pollution, as well as any pollutant discharge to the environment. This is a major challenge requiring different strategies to be implemented, depending on whether we are dealing with the developing or the developed world. Unfortunately, most of those strategies have yet to be developed. Some activities that could be put in place are:

- Control leakage from water mains and sewer systems.
- Promote cleaner industries.
- Improve the location and quality of wastewater discharges.
- Promote sanitation programmes by ensuring good construction and operating procedures.
- Design wastewater treatment processes as a part of a reuse project, i.e. not only for ecological reasons.
- Plan cities' growth taking into account water resources (availability and planned reuse).
- Develop national policies (and eventually even international ones) that help establish an agreement between localities to protect water sources.
- Increase water source monitoring.
- Use an irrigation 'hydraulic barrier' to protect water supplies from diffuse pollution
- Set up appropriate sewers and sewage disposal methods in areas of high groundwater vulnerability and/or source protection areas.
- Restrict the disposal of industrial effluents or other residues to the ground in vulnerable areas.
- Promote education on the water cycle that considers pollutant flow and water reuse.
- Recognize that incremental changes in the intensification of agricultural cultivation increase surface runoff and/or aquifer recharge.

REFERENCES

Acurio, G., Rossin, A., Teixeira, P.T. and Zepeda, F. 1997. *Diagnosis of the Situation of the Municipal Solid Residues Management in America Latina and the Caribbean.* Publication of the Inter-American Development Bank and the Pan-American Organization, Washington, DC. Julio de 1997, No. ENV.97-107, pp. 130. [In Spanish].

Adams, C., Wang, Y., Loftin, K. and Meyer, M. 2002. Removal of Antibiotics from Surface and Distilled Water in Conventional Water Treatment Processes, *Journal of Environmental Engineering*, Vol. 128, No. 3, pp. 253–260.

Adams, C.D. and Kuzhikannil, J.J. 2000. Effects of UV/H_2O_2 Peroxidation on the Aerobic Biodegradability of Quaternary Amine Surfactants, *Water Research*, Vol. 34, pp. 668–672.

Aguas Argentinas. 2001. Historia de obras sanitarias. Available at: www. fcapital,com,ar/fcapital/odisea/OdiseaAguas/Odisea/trabajo%20ET%20N17.

Alderwish, A. and Dottridge, J. 1999. Urban Recharge and its Influence on Groundwater Quality in Sana'a, Yemen. Chilton, J. (ed) *Groundwater in the Urban Environment, Selected City Profile*. Rotterdam, Balkema, The Netherlands, pp. 85–90.

Andreozzi, R., Raffaele, M. and Nicklas, P. 2003. Pharmaceuticals in STP Effluents and their Solar Photodegradation in Aquatic Environment, *Chemosphere*, Vol. 50, No. 10, pp. 1319–1330.

Asano, T., Burton, F., Leverenz, H., Tsuchihashi, R. and Tchobanoglous, G. 2007. *Water Reuse: Issues, Technologies and Applications*, McGraw Hill, New York, pp. 1570.

Ashton, D., Hilton, M. and Thomas, K.V. 2004. Investigating the Environmental Transport of Human Pharmaceuticals to Streams in the United Kingdom, *Science of the Total Environment*, Vol. 333, No. 1–3, pp. 167–184.

Baronti, C., Curini, R., D'Ascenzo, G., Di Corcia, A., Gentili, A. and Samperi, R. 2000. Monitoring Natural and Synthetic Estrogens at Activated Sludge Sewage Treatment Plants and in a Receiving River Water, *Environmental Science and Technology*, Vol. 34, No. 24, pp. 5059–5066.

Beijing City. 2002. Available at: http://host3.iges.or.jp/kitakyushu/Meetings/Thematic%20Seminar/PPP/ Beijing/2%20Beijing.pdf.

BGS, CNA, SAPAL, WAJ, DMR and PSU. 1998. *Protecting Groundwater Beneath Wastewater Recharge Sites*. Technical Report WC/98/39, Ed. British Geological Survey, Wallingford, UK, pp. 150.

Bouwer, H. 1989. Groundwater Recharge with Sewage Effluent, *Water Science and Technology*, Vol. 23, pp. 2099–2108.

Bouwer, H., Rice, R., Lance, J. and Gilbert, R. 1980. Rapid Infiltration Research at Flushing Meadows Project Arizona, *Journal of the Water Pollution Control Federation*, Vol. 52, pp. 2457–2470.

Bravo, H., Sosa, R., Sánchez, P., Bueno, E. and González, L. 2002. Concentrations of Benzene and Toluene in the Atmosphere of the Southwestern area at the Mexico City Metropolitan Zone, *Atmospheric Environment*, Vol. 36, No. 23, pp. 3843–3849.

Bruce, D., Westerhoff, P. and Brawleychesworth, A. 2002. Removal of 2-methylisoborneol and geosmin in surface water treatment plants in Arizona, *Journal of Water Supply Research and Technology AQUA*, Vol. 51, pp. 183–198.

Buechler, S. and Devi, G. 2003. *Household Food Security and Wastewater-Dependent Livelihood Activities along the Musi River in Andhra Pradesh, India*. Report submitted to the World Health Organization (WHO), Geneva, Switzerland and background paper for 2006 World Health Organization's publication Health Guidelines for the Use of Wastewater in Agriculture and Aquaculture. Available at: www.who.int/water_sanitation_health/ wastewater/gsuww/en/index.html and www.who.int/water_sanitation_health/wastewater/ gwwufoodsecurity.pdf.

Buser, H.R., Poiger, T. and Müller, M.D. 1998. Occurrence and Fate of the Pharmaceutical Drug Diclofenac in Surface Waters: rapid photodegradation in a lake, *Environmental Science and Technology*, Vol. 32, No. 22, pp. 3449–3456.

Calamari, D., Zuccato, E., Castiglioni, S., Bagnati, R. and Fanelli, R. 2003. Strategic Survey of Therapeutic Drugs in the Rivers Po and Lambro in northern Italy, *Environmental Science and Technology*, Vol. 37, No. 7, pp. 1241–1248.

Calcutta Municipal Corporation. 2000. *Calmanac*. Available at: www.calmanac.org/cmcnew/cmc/water/00.htm?showmenu=no.

Capella, S. and Pegueros, A. 1998. Analysis by Solid Phase Micro-extraction (_-SPE) and Capillary Gas Chromatography-mass Spectrometric Detection (CGC-MSD) of Chlorinated Benzenes and Phenols in Treated Wastewater from Mexico's City Municipal Effluent. *Proceedings of the 20th International Symposium on Capillary Chromatography*, Edition en CD, p. 4.

Carballa, M., Omil, F., Lema, J.M., Llompart, M., Garcia-Jares, C., Rodriguez, I., Gomez, M. and Ternes, T. 2004. Behavior of Pharmaceuticals, Cosmetics and Hormones in a Sewage Treatment Plant, *Water Research*, Vol. 38, No. 12, pp. 2918–2926.

Chapman, D. (ed.) 1992. *Water Quality Assessment.* WHO/UNESCO/UNEP. Chapman & Hall, London, UK.

Chiron, S., Fernandez-Alba, A., Rodriguez, A. and Garcia-Calvo, E. 2000. Pesticide Chemical Oxidation: State-of-the-art, *Water Research,* Vol. 34, No. 2, pp. 366–377.

City of Los Angeles Water Services. 1996. *Water Supply Fact Sheet, Sources of Water of Los Angeles.* Available at: www.ladwp.com/water/supply/facts/index.htm

Cockram, M. and Feldman, S. 1996. The Beautiful City: Gardens in Third World Cities, *African Urban Quarterly, Vol.* 11, No. 2–3, pp. 202–208.

Cornish, G.A. and Lawrence, P. 2001. *Informal Irrigation in Peri-urban Areas: A Summary of Findings and Recommendations.* Report OD 144 HR Wallingford/DFID.

Curriero, F., Patz, J., Rose, J. and Lele, S. 2001. The Association Between Extreme Precipitation and Waterborne Disease Outbreaks in the United States, 1948–1994, *American Journal of Public Health,* Vol. 91, pp. 1194–1199.

Ensink, J., Mahmood, T., van der Hoek, W., Raschid-Sally, L. and Amerasinghe, F. 2004a. A Nationwide Assessment of Wastewater use in Pakistan: An obscure activity or a vitally important one? *Water Policy,* Vol. 6, pp. 197–206.

Ensink, J., Simmons, J. and van der Hoek, W. 2004b. Wastewater use in Pakistan: The cases of Haroonabad and Faisalabad. Scott, C., Faruqui, N. and Raschid-Sally, L. *Wastewater use in irrigated agriculture.* CAB International, Wallingford, pp. 91–102.

Equihua, L. 2007. Personnel Communication.

Fatma Abdel Rahman Attia. 1999. *Cidob d'afers Internacionals. Water and Development in Greater Cairo (Egypt).* Available at: www.cidob.org/Ingles/Publicaciones/Afers/45-46abdel.html.

Fent, K., Weston, A.A. and Caminada, D. 2006. Ecotoxicology of human pharmaceuticals, *Aquatic Toxicology,* Vol. 76, pp. 122–159.

Fernandez-Bremauntz, A. 2001. Contingencies and Quality of the air in the City of Mexico, *Bulletin of development and investigation into the quality of the air in great cities,* Year I, No. 01, July December, pp 2–3. [in Spanish].

Fewtrell, L. 2004. Drinking-water Nitrate and Methemoglobinemia. Global Burden of Disease: A discussion, *Environmental Health Perspectives,* Vol. 112, No. 14, pp. 1371–1374.

Filibeli, A. 2007. Personal communication.

Foster, S., Garduño, H., Tuinhof, A., Kemper, K. and Nanni, M. 2003. *Urban Wastewater as Groundwater Recharge: Evaluating and Managing the Risks and Benefits.* GWMate Briefing Note Series No. 12, Ed. The World Bank, Oxford, UK.

Foster, S., Hirata, R., Gomes, D., D'Elia, M. and Paris, M. 2002. *Groundwater Quality Protection: a Guide for Water Service Companies, Municipal Authorities and Environment Agencies.* World Bank Group and Global Water Partnership (eds). The World Bank, Washington, DC.

Foster, S., Lawrence, A. and Morris, B. 1998. *Groundwater in Urban Development Assessing Management Needs and Formulating Policy Strategies.* World Bank, Technical paper, No. 390, The World Bank, Washington, DC.

Fuerhacker, M., Dürauer, A. and Jungbauer, A. 2001. Adsorption Isotherms of 17β-estradiol on Granular Activated Carbon (GAC), *Chemosphere,* Vol. 44, No. 7, pp. 1573–1579.

Fujita, Y., Ding, W.H. and Reinhard, M. 1996. Identification of Wastewater Dissolved Organic Carbon Characteristics in Reclaimed Wastewater and Recharged Groundwater, *Water Environment Research,* Vol. 68, No. 5, pp. 867–876.

Funamizu, N., Iida, M., Sakakura, Y. and Takakuwa, T. 2001. Reuse of Heat Energy in Wastewater: Implementation Examples in Japan, *Water Science and Technology,* Vol. 43, No. 10, pp. 277–286.

Garrison, A.W., Pope, J.D. and Allen, F.R. 1976. Analysis of Organic Compounds in Domestic Wastewater. Keith, C.H. (ed.) *Identification and analysis of organic pollutants in water.* Ann Arbor Science, Michigan, US, pp. 517–566.

Gibson, R., Becerril-Bravo, E., Silva-Castro, V. and Jiménez B. (2007) Determination of Acidic Pharmaceuticals and Potential Endocrine Disrupting Compounds in Wastewaters and Spring

Waters by Selective Elution and Analysis by Gas Chromatography – Mass Spectrometry, *Journal of Chromatography A*. 1169(1–2): 31–39.

Gillogly, T.E.T., Snoeyink, V.L., Elarde, J.R., Wilson, C.M., and Royal, E.P. 1998. ^{14}C-MIB Adsorption on PAC in Natural Water, *Journal American Water Works Association*, Vol. 90, No. 1, pp. 98–108.

Golet, E.M., Alder, A.C., Hartmann, A., Ternes, T. and Giger, W. 2001. Trace Determination of Fluoroquinolone Antibacterial Agents in Urban Wastewater by Solid-phase Extraction and Liquid Chromatography with Fluorescence Detection, *Analytical Chemistry*, Vol. 73, No. 15, pp. 3632–3638.

Gould, J.P. and Richards, J.T. 1984. The Kinetics and Products of the Chlorination of Caffeine in Aqueous-solution, *Water Research*, Vol. 18, pp. 1001–1009.

Gray, J.L. and Sedlak, D.L. 2003. Removal of 17-B-estradiol and 17-a-ethinyl Estradiol in Engineered Treatment Wetlands. *National Ground Water Association, International Conference on Pharmaceuticals and Endocrine Disrupters*, Minneapolis, MN, March 19–21.

Halling-Sorensen, B., Nielsen, S.N., Lanzky, P.F., Ingerslev, F., Lutzhoft, H.C. and Jorgensen, S.E. 1998. Occurrence, Fate and Effects of Pharmaceutical Substances in the Environment, *Chemosphere*, Vol. 36, pp. 357–393.

Heberer, T. 2002. Occurrence, Fate, and Removal of Pharmaceutical Residues in the Aquatic Environment: A review of recent research data, *Toxicology Letters*, Vol. 131, No. 1–2, pp. 5–17.

Heberer, T. and Stan, H.J. 1997. Determination of Clofibric Acid and N-(Phenylsulfonyl)-Sarcosine in Sewage, River and Drinking Water, *International Journal of Environment Analytical Chemistry*, Vol. 67, No. 1–4, pp. 113–123.

Heberer, T., Schmidt-Baumler, K. and Stan, H.J. 1998. Occurrence and Distribution of Organic Contaminants in Aquatic System in Berlin. Part I: Drug residues and other polar contaminants in Berlin surface and groundwater, *Acta Hydrochimica et Hydrobiologica*, Vol. 26, No. 5, pp. 272–278.

Heberer, Th. and Reddersen, K. 2001. Occurrence and Fate of Pharmaceutical Residues in the Aquatic System of Berlin as an Example for Urban Ecosystems. *2nd International Conference on Pharmaceuticals and Endocrine Disrupting Chemicals in Water*, October 9–11, 2001, Hyatt on Nicollet Mall, Minneapolis, Minnesota pp. 12–25. Available at: www.epa.gov/esd/chemistry/ppcp/images/ngwa_abs.pdf.

Hecker, M., Tyler, C.R., Hoffmann, M., Maddix, S. and Karbe, L. 2002. Plasma Biomarkers in Fish Provide Evidence for Endocrine Modulation in the Elbe River, Germany, *Environmental Science and Technology*, Vol. 36, No. 11, pp. 2311–2321.

Henderson, A.K., Moll, D.M., Frick, E.A. and Zaugg, S.D. 2001. *Presence of Wastewater Tracers and Endocrine Disrupting Chemicals in Treated Wastewater Effluent and in Municipal Drinking Water*. National Groundwater Association, Atlanta, GA.

Hirsch, R., Ternes, T., Haberer, K. and Kratz, K.L. 1999. Occurrence of Antibiotics in the Aquatic Environment, *Science Total Environment*, Vol. 225, No. 1–2, pp. 108–118.

Holm, J.V., Rugge, K., Bjerg, P.L. and Christensen, T.H. 1995. Occurrence and Distribution of Pharmaceutical Organic-compounds in the Groundwater Downgradient of a Landfill (Grindsted, Denmark), *Environmental Science and Technology*, Vol. 29, No. 5, pp. 1415–1420.

Hrudey, S.E. and Hrudey, E.J. (eds) 2004. *Safe Drinking Water, Lessons from recent Outbreaks in Affluent Nations*. IWA publishing, London.

Huang, C.H. and Sedlak, D.L. 2001. Analysis of Estrogenic Hormones in Municipal Wastewater Effluent and Surface Water Using ELISA and GC/MS/MS, *Environmental Toxicology and Chemistry*, Vol. 20, pp. 133–139.

Huber, M.M., Canonica, S., Park, G.Y. and von Gunten, U. 2003. Oxidation of Pharmaceuticals During Ozonation and Advanced Oxidation Processes (AOPs), *Environmental Science and Technology*, Vol. 37, No. 5, pp. 1016–1024.

IMWI, International Water Management Institute. 2003. *Water Policy Briefing Issue 9. Putting research knowledge into action.* Colombo, Sri Lanka, International Water Management Institute. Available at: www.iwmi.cgiar.org/waterpolicybriefing/files/wpb09.pdf.

Jawetz, E., Melnick, J. and Adelberg, E. 1996. *Medical Microbiology*, 23rd edn. Manual Moderno, Mexico. [In Spanish].

Jiménez, B. 2001. *Environmental Pollution, Cause, Effects and Appropriate Technology.* Limusa, 925 pp. Mexico [In Spanish].

Jiménez, B. 2003. Health Risks in Aquifers Recharged Using Reclaimed Water. Chapter 3 In: Aertgeerts, R. and Angelakis, A. (eds) *Health risks in aquifers recharged using recycled water in the cutting-edge report.* WHO Regional Office for Europe, pp. 54–172.

Jiménez, B. 2005. Treatment Technology and Standards for Agricultural Wastewater Reuse: A case study in Mexico, *Irrigation and Drainage*, Vol. 54, No. 1, pp. S23–S33.

Jiménez B. 2006. Irrigation in Developing Countries Using Wastewater, *International Review for Environmental Strategies* (IRES), Vol. 6, No. 2, pp. 229–250.

Jiménez, B. 2007a. Helminth ova Removal from Wastewater for Agriculture and Aquaculture Reuse, *Water Science and Technology*, Vol. 55, No. 1–2, pp. 485–493.

Jiménez, B. 2007b. Coming to Terms With Nature. Water Reuse New Paradigm Towards Integrated Water Resource Management. Chapter 2.20A.6.3 In: *The Encyclopedia of Life Support System.* EOLSS Publishers Co Ltd. UNESCO, Water and Health theme. Available at: www.eolss.net/ E2-20A-toc.aspx.

Jiménez, B. and Chávez, A. 2004. Quality Assessment of an Aquifer Recharged with Wastewater for its Potential use as Drinking Dource: 'El Mezquital Valley' case, *Water Science and Technology*, Vol. 50, No. 2, pp. 269–273.

Jiménez, B. and Garduño, H. 2001. *Social, Political and Scientific Dilemmas for Massive Wastewater Reuse in the World in Navigating Through Waters: Ethical issues in the water industry.* Davis and McGin (eds). Edited by AWWA.

Jiménez, B., Chávez, A., Maya, C. and Jardines, L. 2001a. The Removal of a Diversity of Microorganisms in Different Stages of Wastewater Treatment, *Water Science and Technology*, Vol. 43, No. 10, pp. 155–162.

Jiménez, B., Mendez, J., Barrios, J., Salgado, G. and Sheinbaum, C. 2004. Characterization and Evaluation of Potential Reuse Options for Wastewater Sludge and Combined Sewer System in Mexico, *Water Science and Technology*, Vol. 49, No. 10, pp. 171–178.

Jiménez, B.E., Maya, C. and Salgado, G. 2001b. The Elimination of Helminth ova, Fecal Coliforms, Salmonella and Protozoan Cysts by Various Physicochemical Processes in Wastewater and Sludge, *Water Science and Technology*, Vol. 43, No. 12, pp. 179–182.

Johns, M., Marshal, W. and Toles, C. 1998. Agricultural By-Products as Granular Activated Carbons for Absorbing Dissolved Metals and Organic, *Journal of Chemical Technology and Biotechnology*, Vol. 71, pp. 131–140.

Karamouz, M. 2007. Personal Communication.

Kolodziej, E.P., Gray, J.L. and Sedlak, D.L. 2003. Quantification of Steroid Hormones with Pheromonal Properties in Municipal Wastewater Effluent, *Environmental Toxicology and Chemistry*, Vol. 22, No. 11, pp. 2622–2629.

Kolpin, D.W., Furlong, E.T., Meyer, M.T., Thurman, E.M., Zaugg, S.D., Barber, L.B. and Buxton, H.T. 2002. Pharmaceuticals, Hormones and Other Organic Waste Contaminants in U.S. streams, 1999–2000: A national reconnaissance, *Environmental Science and Technology*, Vol. 36, pp. 1202–1211.

Kolpin, D.W., Skopec, M., Meyer, M.T., Furlong, E.T. and Zaugg, S.D. 2004. Urban Contribution of Pharmaceuticals and other Organic Wastewater Contaminants to Streams during Differing Flow Conditions, *Science of the Total Environment*, Vol. 328, No. 1–3, pp. 119–130.

Kraybill, H. 1977. Global Distribution of Carcinogenic Pollutants in Water, *Ann New York Academy of Science*, Vol. 298, pp. 80–89.

Kreuzinger, N., Clara, M. and Strenn, B. 2004a. Carbamazepine as a Possible Anthropogenic Marker in the Aquatic Environment: Investigations on the Behaviour of Carbamazepine in Wastewater Treatment and During Groundwater Infiltration, *Water Research*, Vol. 38, No. 4, pp. 947–954.

Kreuzinger, N., Clara, M., Strenn, B. and Kroiss, H. 2004b. Relevance of the Sludge Retention Time (SRT) as Design Criteria for Wastewater Treatment Plants for the Removal of Endocrine Disruptors and Pharmaceuticals from Wastewater, *Water Science and Technology*, Vol. 50, No. 5, pp. 149–156.

Kummerer, K., Steger-Hartmann, T. and Meyer, M. 1997. Biodegradability of the Anti-tumour Agent Ifosfamide and its Occurrence in Hospital Effluents and Communal Sewage, *Water Research*, Vol. 31, No. 11, pp. 2705–2710.

Kundzewicz, Z., Mata, L.J., Arnell, N., Döll, P., Kabat, P., Jiménez, B., Miller, K., Oki, T., Sen, Z. and Shiklomanov, I. 2007. *Freshwater Resources and Their Management*. In IV Intergovernmental Panel on Climate Change, Fourth assessment, WMO.

L'eau de Paris. 2002. Available at: www.paris.fr/fr/environnement/eau/.

Lagos State Nigeria. 2007. *Power & Water Supply*. Lagos State Government. Available at: www. lagosstate.gov.ng/Agenda/Power&Water.htm.

Lagos State Water Corporation. 2001. *Agenda Consulting Nigeria*. Available at: www.lagoswater. com/psp/psp%20tit%20bits.htm.

Lance, J.C. and Whisler, F.D. 1976. Stimulation of Denitrification in Soil Columns by Adding Organic Carbon to Wastewater, *Journal Water Pollution Control Federation*, Vol. 48, pp. 346.

Lance, J.C., Rice, R. and Gilbert, R. 1980. Renovation of Wastewater by Soil Columns Flooded with Primary Effluent, *Journal Water Pollution Control Federation*, Vol. 52, No. 2, pp. 381–388.

Lape, J. and Dwyer, T. 1994. A new policy on CSOs, *Water Environment and Technology*, Vol. 6, No. 66, pp. 54–58.

Lindquist, N., Tuhkanen, T. and Kronberg, L. 2005. Occurrence of Acidic Pharmaceuticals in raw and Treated Sewages and in Receiving Waters, *Water Research*, Vol. 39, pp. 2219–2228.

Lopez, A., Mascolo, G., Tiravanti, G. and Passino, R. 1997. Degradation of Herbicides (Ametryn and Isoproturon) During Water Disinfection by Means of two Oxidants (Hypochlorite and Chlorine Dioxide), *Water Science and Technology*, Vol. 35, pp. 129–136.

Marsalek, J., Jiménez-Cisneros, B., Malmquist, P.A., Karamouz, M., Goldenfum, J. and Chocat, B. 2006. *Urban Water Cycle Processes and Interactions*. International Hydrological Programme IHP-VI, Technical Documents in Hydrology, No. 78, UNESCO, Paris, pp. 92.

Matsui, S., Takigami, H., Taniguchi, N., Adachi, J., Kawami, H. and Shimizu, Y. 2000. Estrogen and Estrogen Mimic Contamination in Water and the role of Sewage Treatment, *Water Science and Technology*, Vol. 42, No. 12, pp. 173–179.

Metcalf and Eddy, Inc. 2003. *Wastewater Engineering, Treatment and Reuse*. 4th edn., McGraw-Hill, New York.

Metcalfe, C.D., Koenig, B.G., Bennie, D.T., Servos, M., Ternes, T.A. and Hirsch, R. 2003. Occurrence of Neutral and Acidic Drugs in the Effluents of Canadian Sewage Treatment Plants, *Environmental Toxicology and Chemistry*, Vol. 22, No. 12, pp. 2872–2880.

Molina, L. and Molina, M. (eds). 2002. *Air Quality in the Mexico MegaCity: An Integrated Assessment*. Kluwer Academic Publishers, February, ISBN 1-4020-0452-4.

Navarro, I., Jiménez, B. and Lucario, S. 2007. Assessment of Potential Cancers Risks From Thrihalomethanes in Water Supply in Mexican rural communities. En: UNESCO-IHP Symposium, Paris, 12–14 September 2007, *New directions in Urban Water Management*.

NYC Government. 2003. Available at: www.nyc.gov/portal/site/nycgov/menuitem.9e96a73ffb 670207a62fa24601c789a0/.

Ollers, S., Singer, H.P., Fassler, P. and Muller, S.R. 2001. Simultaneous Quantification of Neutral and Acidic Pharmaceuticals and Pesticides in low-ng/L Level in Surface and Wastewater, *Journal of Chromatography A*, Vol. 911, No. 2, pp. 225–234.

OTV. 1994. *Depolluer les eaux pluviales*. Ed. Lavoisier TEC & DOC. Paris, France [In French].

Putschew, A., Wischnack, S. and Jekel, M. 2000. Occurrence of Triiodinated X-ray Contrast Agents in the Aquatic Environment, *The Science of the Total Environment*, Vol. 255, No. 1, pp. 129–134.

Pyne, R. 1995. *Groundwater Recharge and Wells: A Guide to Aquifer Storage Recovery*. Lewis Publishers, CRC Press, New York, US.

Quanrud, D.M., Arnold, R.G., Wilson, L.G., Gordon, H., Graham, D. and Amy, G. 1996. Soil fate of Organics During Column Studies of Aquifer Treatment, *Journal of Environmental Engineering*, Vol. 122, No. 4, pp. 314–321.

Quintana, J.B., Weiss, S. and Reemtsma, T. 2005. Pathways and Metabolites of Microbial Degradation of Selected Acidic Pharmaceutical and their Occurrence in Municipal Wastewater Treated by a Membrane Bioreactor, *Water Research*, Vol. 39, pp. 2654–2664.

Raschid-Sally, L., Carr, R. and Buechler, S. 2005. Livelihoods and Sanitation Solutions for poor Countries Through Wastewater Agriculture, *Special Issue on Wastewater Irrigation of the International Commission on Irrigation and Drainage (ICID) Journal Irrigation and Drainage*, Vol. 54, July.

Ravacha, C. and Blits, R. 1985. The Different Reaction Mechanisms by Which Chlorine and Chlorine Dioxide React with Polycyclic Aromatic-hydrocarbons (PAH) in water, *Water Research*, Vol. 19, pp. 1273–1281.

Reddersen, K., Heberer, T. and Dunnbier, U. 2002. Identification and Significance of Phenazone Drugs and their Metabolites in Ground- and Drinking Water, *Chemosphere*, Vol. 49, No. 6, pp. 539–544.

Rice, R. and Bouwer, H. 1984. Soil Aquifer Treatment Using Primary Effluent, *Journal Water Pollution Control Federation*, Vol. 56, No. 1, pp. 84–88.

Richardson, M.L. and Bowron, J.M. 1985. The Fate of Pharmaceutical Chemicals in the Aquatic Environment, *Journal of Pharmacy and Pharmacology*, Vol. 37, No. 1, pp. 1–12.

Roberts, P.H. and Thomas, K.V. 2005. The Occurrence of Selected Pharmaceuticals in Wastewater Effluent and Surface Waters of the Lower Tyne Catchment, *Science of the Total Environment*, Vol. 356, pp. 143–153.

Rogers, I.H., Birtwell, I.K. and Kruznyski, G.M. 1986. Organic Extractables in Municipal Wastewater of Vancouver, British Columbia, *Canadian Journal Water Pollution Research*, Vol. 21, pp. 187–204.

Ruiz, S.L. 2001. Analyses of Gases and Particles Suspended in the Atmosphere of the city of Mexico, *Bulletin of development and investigation into the quality of the air in great cities*, Year I, No. 1 July, pp. 4–5.

Sacher, F., Haist-Gulde, B., Brauch, H.J., Preuß, G., Wilme, U., Zullei-Seibert, N., Meisenheimer, M., Welsch, H. and Ternes, T.A. 2000. Behaviour of Selected Pharmaceuticals During Drinking Water Treatment. *219th ACS National Meeting*, San Francisco, CA, pp. 116–118.

Salameh, E., Alawi, M., Batarseh, M. and Jiries, A. 2003. Determination of Trihalomethanes and the Ionic Composition of Groundwater at Amman City, Jordan, *Hydrogeological Journal*, Vol. 10, pp. 332–339.

Santos, E. 2003. *Manila Water Supply Regulation, Metropolitan Waterworks and Sewerage System*. Regulatory Office, Manila Philippines, 3rd World Water Forum, Osaka, Japan. Available at: www.adb.org/Documents/Events/2003/3WWF/Cities_Santos.pdf.

Sedlak, D., Gray, J. and Pinkston, K. 2000. Understanding Microcontaminants in Recycled Water, *Environmental Science and Technology*, Vol. 34, No. 23, pp. 508–515A.

Sedlak, D., Pinkston, K., Gray, J. and Kolodziej, E. 2003. Approaches for Quantifying the Attenuation of Wastewater-derived Contaminants in the Aquatic Environment, *Chimia International Journal of Chemistry*, Vol. 57, No. 9, pp. 567–569.

Seoul Metropolitan Government, 1999. Available at: http://english.metro.seoul.kr/government/ policies/ statistics/water/.

Shanghai Daily. 1998. Available at: www.shanghai-window.com/shanghai/shdaily/data/8m/ body 980805.html.

Smit, J. and Nasr, J. 1992. Urban Agriculture for Sustainable Cities: Using Waste and Idle Land and Water Bodies as Resources, *Environment and Urbanization*, Vol. 4, No. 2, pp. 141–152.

Snyder, S.A., Westerhoff, P., Yoon, Y. and Sedlak, D.L. 2003. Pharmaceuticals, Personal Care Products, and Endocrine Disruptors in Water: Implications for the Water Industry, *Environmental Engineering Science*, Vol. 20, No. 5, pp. 449–469.

South Africa, Independent Centre. 2006. Available at: http://southafrica.indymedia.org/news/ 2006/ 07/10745.php.

Strenn, B., Clara, M., Gans, O., Kreuzinger, N. 2004. Carbamazepine, Diclofenac, Ibuprofen and Bezafibrate—Investigations on the Behaviour of Selected Pharmaceuticals During Wastewater Treatment, *Water Science and Technology*, Vol. 50, No. 5, pp. 269–276.

Stumpf, M., Ternes, T.A., Haberer, K., Seel, P. and Baumann, P.W. 1996. Determination of Drugs in Sewage Treatment Plants and River Water, *Vom Wasser*, Vol. 86, pp. 291–303.

Stumpf, M., Ternes, T.A., Wilken, R.D., Rodrigues, S.V. and Baumann, W. 1999. Polar Drug Residues in Sewage and Natural Waters in the State of Rio de Janeiro, Brazil, *Science of the Total Environment*, Vol. 225, No. 1/2, pp. 135–141.

Suter, M.J.F., Giger, W., Aerni, H.R., Wettstein, F.E., Svenson, A., Hylland, K., Nakari, T., Johnson, A.C., Gerritsen, A., Gibert, M., Jürgens, M. and Pickering, A. 2005. Comparing Steroid Estrogen, and Nonylphenol Content Across a Range of European Sewage Plants with Different Treatment and Management Practices, *Water Research*, Vol. 39, No. 1, pp. 47–58.

Tanghe, T. and Verstraete, W. 2001. Adsorption of Nonylphenol onto Granular Activated Carbon, *Water Air and Soil Pollution*, Vol. 131, pp. 61–72.

Tauxe-Wuersch, A., de Alencastro, L.F., Grandjean, D. and Tarradellas, J. 2005. Occurrence of Several Acidic Drugs in Sewage Treatment Plants in Switzerland and Risk Assessment, *Water Research*, Vol. 39, pp. 1761–1772.

Ternes, T., Meisenheimer, M., McDowell, D., Sacher, F., Brauch, H.J., Haist-Glude, B., Preuss, G., Wilme, U. and Zulei-Seibert, N. 2002. Removal of Pharmaceuticals During Drinking Water Treatment, *Environmental Science and Technology*, Vol. 36, pp. 3855–3863.

Ternes, T., Stumpf, M., Schuppert, B. and Haberer, K. 1998. Simultaneous Determination of Antiseptics and Acidic Drugs in Sewage and River Water, *Vom Wasser*, Vol. 90, pp. 295–309.

Ternes, T.A., Kreckel, P. and Meuller, J. 1999. Behaviour and Occurrence of Estrogens in Municipal Sewage Treatment Plants—II. Aerobic Batch Experiments with Activated Sludge, *Science of the Total Environment*, Vol. 225, pp. 91–99.

Thacker, P.D. 2005. Pharmaceutical Data Elude Researchers, *Environmental Science and Technology*, Vol. 39, No. 9, pp. 193A–194A.

Thomas, K.V. and Hilton, M.J. 2004. The Occurrence of Selected Human Pharmaceutical Compounds in UK Estuaries, *Marine Pollution Bulletin*, Vol. 49, No. 5/6, pp. 436–444.

Thomas, P.M. and Foster, G.D. 2004. Determination of Nonsteroidal Anti-inflammatory Drugs, Caffeine, and Triclosan in Wastewater by gas Chromatography–mass Spectrometry, *Journal of Environmental Science and Health*, Vol. 39, No. 8, pp. 1969–1978.

Tuinhof, A. and Heederik, J. (eds). 2002. *Management of Aquifer Recharge and Subsurface Storage: Making Better use of our Largest Reservoir*. Netherlands National Committee for the IAH Publication, No. 4. Wageningen, The Netherlands. pp. 106.

Tutuko, K. 2007. Jakarta, Indonesia. Jakarta water supply. In *Jakarta*. pp. 19–69. Available at: www.pecc.org/community/papers/SCTFReports/HongKong/jakarta.pdf.

United Nations. 2003. *Water for People, Water for Life*. The United Nations world water development report. Barcelona: UNESCO.

Urban Groundwater Database. 2002. Available at: www.clw.csiro.au/UGD/DB/Rio-de-Janeiro/ Rio-de-Janeiro.html#water.

US EPA. 1992. *Guidelines for Water Reuse.* EPA/625/R-92/004. Environmental Protection Agency, Center for Environmental Research Information, Cincinnati, US, pp. 264.

US EPA. 2007a. *Air Pollution and Water Quality.* Available at: www.epa.gov/owow/airdeposition/ [consulted on July 18, 2007].

US EPA. 2007b. *Listing Waters Impaired by Atmospheric Mercury Under Clean Water Act Section 303(d).* Available at: www.epa.gov/owow/tmdl/mercury5m/ [consulted on July 18, 2007].

van der Hoek, W., Sakthivadivel, R., Renshaw, M., Silver, J.B., Birley, M.H. and Konradsen, F. 2001. *Alternate Wet/dry Irrigation in Rice Cultivation: A Practical Way to Save Water and Control Malaria and Japanese Encephalitis?* Research Report 47. Colombo, Sri Lanka, International Water Management Institute, pp. 30.

van der Hoeven, N. 2004. Current Issues in Statistics and Models for Ecotoxicological Risk Assessment, *Acta Biotheorica*, Vol. 52, No. 3, pp. 201–217.

Viklander, M. 1999. Substances in Urban Snow. A comparison of the contamination of snow in different parts of the city of Luleå, Sweden, *Water, Air and Soil Pollution*, Vol. 114, No. 3–4, pp 377–394.

Wakida, F. and Lerner, D. 2005. Non-Agricultural Sources of Groundwater Nitrate: A Review and Case Study, *Water Research*, Vol. 39, pp. 3–16.

Water Supply in Tokyo. 2002. *Outline of Tokyo Water Works.* Available at: www.waterworks. metro.tokyo.jp/eng/supply/supply_02.htm.

Weigel, S., Kuhlmann, J. and Huhnerfuss, H. 2002. Drugs and Personal Care Products as Ubiquitous Pollutants: Occurrence and Distribution of Clofibric Acid, Caffeine and DEET in the North Sea, *Science of the Total Environmental*, Vol. 295, No. 1–3, pp. 131–141.

World Health Organization (WHO). 2007. Weekly Epidemiological Record. 82(32):285–296. Available at: www.who.int/wer.

WHO, World Health Organization. 2006. *Guidelines for the Safe Use of Wastewater, Excreta and Greywater. Vol. 2 Wastewater use in agriculture.* WHO Library Cataloguing-in-Publication Data, Geneva, p. 213.

WHO/UNICEF, World Health Organization/United Nations Children's Fund. 2000. *Global Water Supply and Sanitation Assessment Report.* Joint Monitoring Program for Water Supply and Sanitation. Geneva, WHO.

Wiegel, S., Aulinger, A., Brockmeyer, R., Harms, H., Löffler, J., Reincke, H., Schmidt, R., Stachel, B., Von Tumpling, W. and Wanke, A. 2004. Pharmaceuticals in the River Elbe and its Tributaries, *Chemosphere*, Vol. 57, No. 2, pp. 107–126.

Wilson, L., Amy, L., Gerba, C., Gordon, H., Johnson, B. and Miller, J. 1995. Water Soil Quality Changes During Aquifer Treatment of Tertiary Effluent, *Water Environment Research*, Vol. 67, No. 3, pp. 371–376.

Yepes, G. and Augusta D. 1996. *Water and Wastewater Utilities: Indicators.* 2nd edn. The World Bank, Washington, DC, Water and Sanitation Division. Available at: www.worldbank. org/html/fpd/water/pdf/indicators.pdf.

Young, A. 1998. *Land resources: Now and for the Future Land degradation.* Cambridge University Press, Cambridge UK, Vol. 7, pp. 319.

Zhang, T.C. and Emary, S.C. 1999. Jar tests for Evaluation of Atrazine Removal at Drinking water Treatment Plants, *Environmental Engineering Science*, Vol. 16, No. 6, pp. 417–432.

Ziegler, D. 2001. *Untersuchungen Zur Nachhaltige Wirkung des Uferfiltration im Wasserkreislauf Berlins.* PhD. Thesis at the Technical University of Berlin [In German].

Zwiener, C. and Frimmel, F.H. 2000. Oxidative Treatment of Pharmaceuticals in Water, *Water Research*, Vol. 34, No. 6, pp. 1881–1885.

Chapter 8

Risks associated with biosolids reuse in agriculture

Stephen R. Smith

Department of Civil and Environmental Engineering, Imperial College, London, United Kingdom

ABSTRACT: Sewage sludge is a proven fertilizer and soil conditioner and can be reused beneficially in agriculture without detriment to the environment or human health. The technical database on the fate of sewage sludge in soil is extensive, enabling management practices to be defined that protect the environment to ensure safe and secure reuse of sludge for crop production. Hygiene is the principal concern, particularly where agricultural practices are labour intensive and farm workers may come into direct contact with sludge. Under these circumstances, biosolids are supplied to the farm gate after treatment to eliminate pathogens. A dual barrier approach, employing treatment and land use restrictions, is effective in controlling the potential transmission of enteric disease when sludge is applied in mechanized farming systems. Potentially toxic elements (PTEs) slowly accumulate in sludge-treated soil, but this can be controlled to avoid deleterious effects on crops, soil microbial processes or the food chain. Organic contaminants sorb strongly to soil, volatilize or readily biodegrade, providing barriers to their transmission to the food chain; the consensus of scientific data is that they are not hazardous to human health or the environment in sludge. Effective trade effluent and source controls on contaminants are critical to sustainable recycling programmes.

8.1 INTRODUCTION

Sewage sludge is the necessary and inevitable by-product of urban wastewater treatment. It is a consequence of the effective removal of the polluting load from urban wastewater so that the treated effluent can be safely discharged to the environment to improve surface water quality. As environmental improvements involving the expansion of wastewater treatment are introduced, so the production of sewage sludge will also increase and require a considered strategy for its management.

Beneficial reuse as an alternative fertilizer and soil conditioner in agriculture is by far the most important approach to managing the sludge generated by urban wastewater treatment. In the UK, EU and US, more than 50% of sludge production is spread beneficially on farmland. There are a number of reasons for this:

- Recycling resources is intuitively a common-sense approach; it is preferred to disposal in the EU waste management hierarchy and agricultural use is the only outlet where significant benefit is gained from the important nutrient and organic matter resources contained in sludge.
- It is thoroughly researched (Smith, 1996), and guidance to protect the environment is well defined (US EPA 1992a, 1992b, 1993, 1999; UK DoE, 1996).

- The alternatives are: expensive and unpopular with the general public (e.g., incineration); have long-term potential environmental consequences and are considered to be unsustainable (e.g., landfill); may be important and allow recovery of value from sludge, but are only available locally (e.g., land restoration); or are at early stages of development (e.g., aggregates, activated carbon production, fuel replacements) and are therefore unlikely to be a significant outlet for sludge in the short to medium term.

Despite the positive aspects of agricultural recycling, some countries have banned the use of sludge in agriculture directly, through primary legislation (e.g., Switzerland), or have introduced environmental controls that are so stringent that sludge cannot meet the quality requirements (e.g., the Netherlands). Under these circumstances, the main option proposed for the management of sludge is usually incineration. However, this must be seen as a disposal route for sludge. In contrast to incineration of municipal solid waste, for example, there is essentially no net recovery of energy, as all the calorific value contained in sludge is utilized due to the heat demand for drying and the electricity demand of operating the incinerator itself (Mininni et al., 1997; Thierbach and Hanssen, 2002).

The majority of the potentially polluting substances present in urban wastewater are transferred to the sewage sludge during wastewater treatment. The organic component is putrescible and potentially malodorous and the solids are also a repository for pathogenic and parasitic microorganisms and chemical contaminants that are discharged to the sewer from connected domestic and industrial sources. However, all of these potentially negative environmental aspects can be dealt with from a technical standpoint to facilitate the safe reuse of sludge on agricultural land.

This chapter addresses the priority environmental and agronomic issues that impinge on the long-term and sustainable agricultural use of sewage sludge. These include:

- nutrient and agronomic value
- microbiological quality
- potentially toxic elements (PTEs)
- organic contaminants.

8.2 NUTRIENT AND AGRONOMIC VALUE

Fertilizer replacement is the primary benefit associated with recycling sewage sludge on farmland. Sludge has significant nitrogen (N) and phosphorus (P) fertilizer value, but potassium (K) is eluted during the wastewater treatment process and concentrations in sludge are small and not agronomically important. Sludge may have significant liming value (e.g., if treated by a lime conditioning process); it also contains organic matter, sulphur (S), magnesium (Mg) and trace elements. Guidance is available on the optimum use of biosolids to maximize the agronomic value, whilst avoiding unnecessary losses of the applied nutrients to the environment (e.g., MAFF, 2000).

Sludge contains organic and inorganic N, and the concentrations or stability of these forms depend on the type of sludge treatment process and determine the N fertilizer value. Nitrogen mineralization during anaerobic digestion of sludge, for example, increases the ammonium (NH_4^+) content, providing a soluble and plant available source of N equivalent to inorganic fertilizer. Microbial stabilization processes degrade the organic N fraction during sludge treatment (e.g., anaerobic digestion and composting) and a proportion of this residue is relatively resistant to further mineralization in the soil, providing a slower release N source. Recent experimental investigations in the UK (Smith et al., 2002a; Morris et al., 2003) have updated the agronomic data on new sludge products that have been introduced with the development of enhanced treatment processes for sludge (ADAS, 2001). N fertilizer values of a range of sludge products are shown in Table 8.1. The availability of P in sludge is generally assumed to be 50% of the total P content (MAFF, 2000), although this may be reduced by dewatering and drying sludge, and by the increase in sludge Fe content from chemical removal of P during wastewater treatment, or it may be higher for sludge from biological P removal (Smith et al., 2002b; O' Connor et al., 2004). A single application of sludge may supply most of the P requirements of a crop rotation.

In warm climates, the apparent nutrient value of sludge is increased, presumably because the rate and extent of mineralization of the organic N fraction is greater, compared to temperate climates. For example, extensive fertilizer value field trials conducted in Egypt in the Nile Delta by WRc (1999), as part of the Cairo Sludge Disposal Study, showed the N equivalency value of digested sludge in the first season of application was 50% relative to mineral N (Figure 8.1).

Nitrogen equivalency, based on linear regression functions in Figure 8.1:

$y = a + bx$; where a is the yield with no N addition and b is the increase in yield per kg of N in fertilizer or sludge
$bS/bN \times 100 =$ N fertilizer equivalency
$0.008/0.017 \times 100 = 50\%$

Table 8.1 Mean N contents and first year N fertilizer values of different biosolids products applied to three field experiments in 2000–2003 with perennial ryegrass relative to ammonium nitrate fertilizer (Smith et al., 2002a; Morris et al., 2003)

Biosolids type	TN (% DS)	Min-N (% DS)	Min-N (% TN)	Org-N (% DS)	Available N (%TN)	Mineralizable N (% Org-N)
Dewatered digested	5.6	0.89	16	4.7	30	17
Dewatered, prepasteurized digested	4.1	0.56	14	3.5	38	28
Dewatered, thermally hydrolysed digested	5.1	0.86	17	4.3	30	16
Thermally dried digested	4.6	0.11	2.4	4.5	37	36
Thermally dried raw	4.4	0.03	0.7	4.4	30	29
Lime stabilized raw	1.7	0.07	4.3	1.7	37	34
Composted with greenwaste	1.5	0.02	4.9	1.4	21	17

Abbreviations: Total nitrogen – TN
                Mineral nitrogen – Min-N
                Organic nitrogen – Org-N
                Dry solids – DS

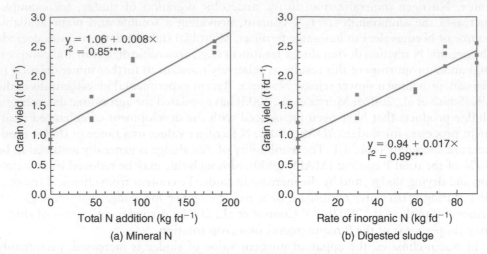

Figure 8.1 Yield response of winter wheat to (a) mineral N fertilizer and (b) anaerobically digested sewage sludge applied to clay soil, Ain Shams, N. Cairo, Egypt

Note: 1 feddan (fd) = 0.42 ha

Source: WRc, 1999

Table 8.2 Some significant water-borne pathogens in sewage sludge

Bacteria	Protozoa	Viruses	Helminths
Salmonella spp	Entamoeba histolytica	Enteroviruses	Taenia saginata
Escherichia coli	Giardia lamblia	Rotavirus	Ascaris lumbricoides
Campylobacter	Cryptosporidium parvum		

8.3 MICROBIOLOGICAL QUALITY

Sewage sludge may contain a wide range of water-borne microbial pathogens that are discharged to the sewer in the feces of infected individuals. These agents potentially cause enteric disease and infections arise through fecal–oral transmission. Some of the principal pathogen types are listed in Table 8.2.

There are two approaches to controlling the spread of infectious disease from land application of sewage sludge. These are:

1. Providing barriers to disease transmission (WHO, 1981) by treating sludge to significantly reduce its pathogen content, coupled with land use restrictions, particularly avoiding applications to crops that come in direct contact with the soil and may be consumed uncooked (Figure 8.2).
2. Treating sludge to reduce its pathogen content to background or undetectable levels so that it can be used without further restriction.

The multiple barrier approach has been effective in preventing the spread of disease when sewage sludge is used in mechanized farming systems. It is the basis of the

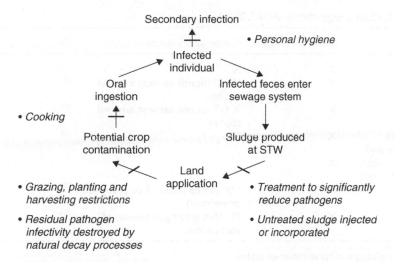

Figure 8.2 Barriers to the transmission of enteric disease when sewage sludge is used in agriculture

guidance adopted for the agricultural use of sewage sludge in the UK (UK DoE, 1996) and it is the principle behind the Class B pathogen requirements of US EPA's *Part 503 Rule on the Standards for Use or Disposal of Sewage Sludge* (US EPA, 1993, 1999). Assurance about the effectiveness of these measures is provided by the absence of any epidemiological evidence linking disease in the population to the agricultural use of sludge where the multiple barrier approach has been followed (RCEP, 1996).

In the UK, the main pathogens of economic concern, which were used to define effective treatment processes to significantly reduce the pathogen content of sludge (UK DoE, 1996), included *Salmonella* and the human-beef tapeworm, *Taenia saginata* (Carrington et al., 1998). The survival of a range of other organisms exposed to these conditions have also been recently evaluated (UKWIR, 2002; Horan et al., 2004). These requirements were further strengthened by the Safe Sludge Matrix (ADAS, 2001), which ended the practice of spreading untreated sludge on agricultural land in the UK (Table 8.3). The Matrix adopts a dual level approach to the microbiological controls for the agricultural use of sludge, depending on the extent of pathogen destruction during sludge treatment, synonymous with the US EPA Class A and Class B pathogen reduction requirements (US EPA, 1993). Thus, sludges that have undergone a treatment process to significantly reduce the residual numbers of enteric organisms (e.g., mesophilic anaerobic digestion) are defined as 'conventionally treated' and sludges treated so as to effectively eliminate the pathogenic content are described as 'enhanced treated'. Numerical microbiological standards also apply to these different treatment classes (Table 8.4) and more detailed land use restrictions (Table 8.3) are also stipulated (ADAS, 2001), which were intended to become compulsory in the UK in a revised Statutory Instrument for the agricultural use of sludge (UK SI, 1989; Sweet et al., 2001). The differences in *E. coli* requirements between the two classes of treated sludge indicate the much lower risk to health associated with using sludge in agriculture compared to its unrestricted use in domestic situations or for producing ready-to-eat crops.

Table 8.3 The Safe Sludge Matrix (ADAS, 2001)

Crop group	Untreated sludge	Conventionally treated sludges	Enhanced treated sludges
Fruit	X	X	[2] √
Salads	X	X (30 month harvest interval applies)	[2] √
Vegetables	X	X (12 month harvest interval applies)	[2] √
Horticulture	X	X	[2] √
Combinable and animal feed crops	X	√	√
Grass and forage:			
Grazed	X	[1] X (Deep injected or ploughed down only)	[1] √
Harvested	X	[1] √ (No grazing in season of application)	[1] √

[1] 3 week no grazing and harvest interval applies
[2] 10 month harvest interval applies
√: All applications must comply with the *Sludge (Use in Agriculture) Regulations* (UK SI, 1989) and *DEFRA Code of Practice for Agricultural Use of Sewage Sludge* (DoE, 1996) (both currently under revision)
X: Applications not allowed (except where stated conditions apply)

Table 8.4 Microbiological standards for treated sludge (ADAS, 2001)

Treatment	Escherichia coli	Salmonella
Conventional	2 log reduction $<10^5 \, g^{-1}$ ds	No requirement
Enhanced	6 log reduction $<1000 \, g^{-1}$ ds	Absent in 2 g ds

In addition, the Water Industry has adopted Hazard Assessment Critical Control Point (HACCP) procedures to ensure that sludge treatment processes are effective and comply with the microbiological quality controls (Water UK, 2004).

Other recent work has examined the decay of enteric bacteria in field soil under temperate UK conditions, following the application of conventionally treated sludge, to increase understanding of the fate of enteric microorganisms in the environment (Lang et al., 2003). This showed that numbers of the indicator bacteria, *E. coli*, rapidly decayed in sandy loam soil amended with dewatered digested biosolids. The added population became indistinguishable from the indigenous community within three months (Figure 8.3), well within the harvesting restriction intervals stipulated by the Safe Sludge Matrix, emphasizing the large margin of safety apparent in the land use controls. This work is ongoing and aims to quantify the significance of the ecological mechanisms responsible for pathogen inactivation in sludge-treated soil (Cass et al., 2005).

Microbiological risk assessment (MRA) techniques have been developed to quantify the risks to health from the residual numbers of pathogens in sewage sludge and agricultural soil that may potentially transfer to the food chain (Gale, 2001; 2003; 2005). Enteric infections from bacteria or viruses are the main concern where water supply, wastewater treatment and sludge disposal are of a high standard. The results from MRA show that the risks to human health from disease transmission through sewage

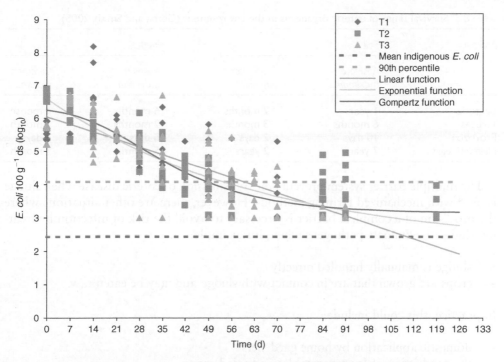

Figure 8.3 Decline in *E. coli* numbers in relation to time following application of dewatered mesophilic, anaerobically digested biosolids to a sandy loam soil at a rate of 10 t DS ha^{-1} and incorporation depth of 10 cm (See also colour plate 6)

Source: Lang et al., 2003

sludge to grazing animals and food crops are very low (Mara and Horan, 2002; Gale, 2005). It has therefore been argued that the strict microbiological standards stipulated in the Safe Sludge Matrix, which are more stringent than those required for foods, are not really justified (Mara and Horan, 2002).

However, the pathogens of concern may vary in different parts of the world. For example, monitoring for the ova of the roundworm *Ascaris lumbricoides* is recommended where infections are prevalent in the population (Feachem et al., 1983). The inactivation of *Ascaris* is an effective indicator of the overall extent of pathogen reduction achieved by a sludge treatment process. The ova of this parasite are among the most resistant to environmental stress compared with other agents responsible for disseminating enteric disease and they remain viable for long periods if released into the environment in sludge (Table 8.5). A word of warning here: Mesophilic anaerobic digestion is favoured by many water utilities for treating sludge and is highly appropriate in many circumstances, not least because it has the advantage of producing biogas that can be used for energy recovery. However, this process is ineffective against the ova of *Ascaris* (Carrington et al., 1991). Therefore, where infections by *Ascaris* and other parasites, that liberate resistant ova or oocysts, are prevalent in the population, alternative treatment processes providing more extreme conditions to eliminate these organisms, or further treatment of the digested sludge (e.g., prepasteurization or composting), should be considered to allow safe reuse in agriculture.

Table 8.5 Survival times of enteric organisms in the environment (Gerba and Smith, 2005)

Pathogen	Soil		Plants	
	Absolute maximum	Common maximum	Absolute maximum	Common maximum
Bacteria	I year	2 months	6 months	I month
Viruses	6 months	3 months	2 months	I month
Protozoa	10 days	2 days	5 days	2 days
Helminth eggs	7 years	2 years	5 months	I month

The multiple barrier system prevents the transmission of enteric disease when sludge is used with mechanized farming methods. However, there are other situations where the provision of a complete barrier is necessary to avoid the risk of infection by treating sludge to eliminate pathogens when, for example:

• sludge is manually handled directly
• crops are grown that are in contact with sludge and may be eaten raw.

In practice, this could include:

• domestic application by home gardeners
• the production of fruit, vegetables and salad crops
• labour-intensive agriculture
• other situations where the adoption of land use restrictions cannot be guaranteed.

These conditions demand a high standard of treatment. Time-temperature exposure is a key principle of sludge treatment processes designed to effectively eliminate pathogens (Figure 8.4). However, sophisticated engineering processes adopted in Europe or the US to achieve this standard of treatment, e.g., thermal drying, may be inappropriate technologies for application in many parts of the world due to their complexity and high cost. Alternative, potentially more practical, approaches under these circumstances include lime addition and composting. Augmentation with lime (calcium hydroxide or oxide) is a simple practical solution and destroys pathogens by heating through exothermic rehydration and by raising the pH of the sludge to >12. However, the process requires the addition of large amounts of lime (up to $550 \, kg \, t^{-1}$ ds), significantly increasing the bulk of material to be managed.

Dewatered sewage sludge requires admixture with a bulking agent to allow aerobic composting action and this can be sourced locally or mature compost may also be used. Invessel and aerated static pile systems are the most effective systems at pathogen inactivation, as all of the composting material is exposed to destructive temperatures (Figure 8.5a), but they have high capital costs. Windrow turning is a lower cost and more practical option and the end-product has good physical characteristics, but the rate of pathogen kill is slower than with the other composting methods. This is because material at the pile edge does not achieve the high temperatures necessary for effective pathogen inactivation and is mixed into the heap when the windrow is turned. Consequently, time-temperature treatment requirements for windrow

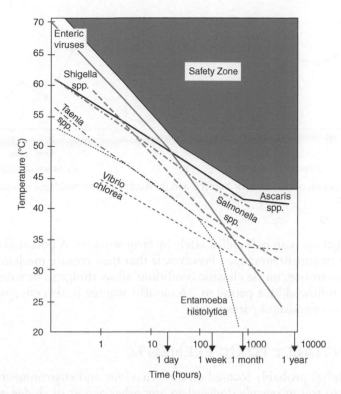

Figure 8.4 **Time-temperature requirements to produce sludge that is virtually pathogen-free**

Source: Strauch, 1991

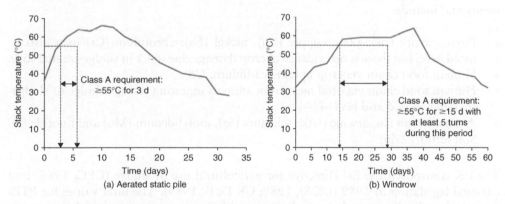

Figure 8.5 **Temperature profiles of composting sewage sludge by (a) aerated static pile and (b) windrow turning, and in relation to US EPA Class A pathogen requirements**

Source: Adapted from Pereira Neto et al., 1987; US EPA, 1993

composting for unrestricted use are more demanding than with invessel systems, but are achievable with careful management and monitoring of the stack temperature as shown in Figure 8.5b. Windrow composting is effective at reducing the numbers of indicator bacteria as well as inactivating resistant *Ascaris* ova (Figure 8.6) producing a

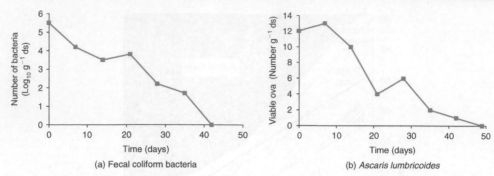

Figure 8.6 Destruction of (a) indicator bacteria and (b) Ascaris ova by windrow composting sewage sludge

Source: Nell et al., 1983

hygienic product that can be handled safely by farm workers. A potential disadvantage to all of these treatment strategies, however, is that they require mechanical dewatering and polyelectrolyte, unless climatic conditions allow sludge to be solar dried. Solar drying sludge followed by a period of >6 months storage is also effective at destroying the viability of resistant parasite ova.

8.4 POTENTIALLY TOXIC ELEMENTS

More research has probably focused on the behaviour and environmental impacts of PTEs applied to soil in sewage sludge than any other aspect of sludge application to agricultural land. The critical environmental end-points for setting soil limits for agricultural recycling sewage sludge vary according to the behaviour of particular elements and include:

- Phytotoxicity (zinc (Zn), copper (Cu), nickel (Ni); chromium (Cr) may also be listed here, but there is no evidence of crop damage due to Cr in sludge-treated soil).
- Human food chain via crop uptake (cadmium, Cd).
- Human food chain via offal meat from animals ingesting sewage sludge or sludge-treated soil (Cd and lead –Pb–).
- Animal health Cu, arsenic (–As), selenium (Se), molybdenum (Mo) and fluoride (F).
- Soil fertility (Zn).

The UK transposed the EU Directive for agricultural use of sludge (CEC, 1986) into national legislation in 1989 (UK SI, 1989; UK DoE, 1989). The limit values for PTEs in agricultural soil were subsequently reviewed following concerns that they may not adequately protect the food chain or soil fertility (MAFF/DoE, 1993a, b). The review concluded that the controls protected human health, but a reduction in the soil limit value for Zn was recommended from $300\,mg\,kg^{-1}$ at pH 6–7 to $200\,mg\,kg^{-1}$ across the pH range 5–7, as a precautionary measure to protect soil fertility. This action was taken based on two field experiments, one in the UK and the other in Germany, which showed that Zn may be potentially toxic to the symbiotic N fixing bacteria, *Rhizobium leguminosarum* bv. *trifolii*. However, there is other evidence suggesting

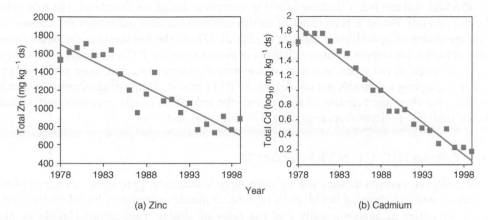

Figure 8.7 Reduction in (a) zinc and (b) cadmium concentrations (untransformed and log10 transformed data, respectively) in sewage sludge from a major sewage treatment works (Nottingham) in the UK during the period 1978–1999

that these apparent effects were linked to artefacts of the trials themselves, such as high soil Cd contents in the case of the UK field experiment at Woburn (Smith, 2000). Long-term research is ongoing at a network of field sites around the British Isles to assess the implications of increasing soil metal concentrations on microbial processes and fertility (Chambers et al., 2004). However, so far, the results concerning potential ecotoxicological effects remain equivocal. Previous work on historically sludge-treated soils showed that the critical microbiological threshold concentration for Zn in soil was above the limit value at 500 mg kg^{-1} (Alloway et al., 1998; Smith, 1994; Smith, 1995; Smith et al., 1999). Furthermore, no evidence of metal toxicity to N fixing *Rhizobium* was found, whereas other soil, environmental and management factors were much more important than the soil metal content in defining the absence of rhizobia from soil (Smith, 1997).

A later review of the scientific evidence relating to the controls on the agricultural use of sewage sludge (Carrington et al., 1998) recommended lowering the soil limit for Pb from 300 to 200 mg kg^{-1} as a precautionary measure to avoid transfers of Pb to offal meat by sheep grazing sludge-amended pasture under exceptional conditions of soil Pb accumulation. The recommendations on Pb and Zn were adopted in the revised *Code of Practice for Agricultural Use of Sewage Sludge* (UK DoE, 1996). More recently, Europe has introduced food quality limits for Cd in grain (ComEC, 2001), with potential implications for the accumulation of this element in soil, and which is also an issue for countries exporting grain to the European market. However, the concentrations of this element in sludge, as with other metals, have dropped dramatically (see Figure 8.7). Consequently, in practice, its concentration in soil is very unlikely to reach a critical value impacting grain quality in relation to European soil limit values (CEC, 1986). Nevertheless, it is emphasized that sludge contamination with all PTEs should be minimized as far as practical by effective discharge controls, to ensure the long-term sustainability of recycling to agricultural land.

In overall assessment, therefore, the potential impacts and consequences of PTEs in sludge and their effects on soil, the environment and human health have been extensively researched and are not a limiting factor to recycling sludge on farmland. For example, WHO recently issued a revised set of health-related chemical guidelines for the use of sewage sludge in agriculture (Chang et al., 2002). Within the European context, the main uncertainty with respect to future changes in limit values for PTEs in sludge-treated soil is likely to arise, not from technical or scientific developments, but from policy enactments regarding acceptable accumulations of PTEs in soil. Supporting ongoing research in this area therefore remains vital to reduce the risk of an overly precautionary stance being taken by policymakers regarding PTE limits for soil.

8.5 ORGANIC CONTAMINANTS

The risk assessments completed to date have concluded that most sludge-applied organics will not increase health risks to either animals or humans, based on their relative toxicities at agronomically realistic rates of sludge application (Jacobs et al., 1987; Wild and Jones, 1991; US EPA, 1992b; ArthurAndersen, 2001). The reason that organic contaminants do not transfer to, or have an impact on, the food chain is related to their behaviour in soil and the presence of only very small concentrations of some of the more potentially toxic compounds in sludge. Organic compounds can be grouped into three broad categories according to behaviour:

- Volatile compounds which are quickly lost to the atmosphere from sludge and sludge-treated soil.
- Compounds which are rapidly mineralized by microorganisms and have little or no persistence.
- Persistent compounds which are strongly adsorbed onto the sludge and soil organic matrix.

Some of the compounds retained in the soil may be relatively persistent in the environment, depending on their cyclical complexity and degree of halogenation, but they are accessible to microorganisms and are slowly destroyed by natural degradation processes in the soil. In an overall assessment, there is a consensus that the risk of human exposure is minimal as there is negligible plant uptake of organic contaminants from sludge-treated soil.

   In spite of the scientific evidence to the contrary, some European countries (e.g., Germany, Denmark and France) decided to regulate the maximum content of certain organic contaminants in sludge. However, because of the absence of direct evidence to support this, compared to controlling heavy metals, for instance, there is no consistency in approach in the regulations introduced for organics in sludge. There has been considerable debate on this issue, but the benefit of the current European Directive on the agricultural use of sludge (CEC, 1986) is that it provides flexibility and permits countries to adopt a more stringent regime of standards, if that is deemed to be necessary under the local, national circumstances, provided the minimum requirements of the Directive are met. In some cases, limits on organic contaminants were therefore introduced, not because this was required technically, but to reduce negative perceptions and increase confidence in the recycling route. The danger, however, is that, should the current Directive be revised, the national limits could become transferred to the new regulations, which would be an inappropriate step given the weight of scientific evidence that limits on organic contaminants are not justified from a technical perspective.

Source and emission controls on persistent organic contaminants were introduced in the 1980s and 1990s in Europe to curb the extent of environmental releases, as concern increased about their widespread occurrence and potential toxicity. This legislation has been effective in reducing environmental emissions and there are several examples illustrating significant reductions in the primary sources of polycyclic aromatic hydrocarbons (PAHs), polychlorinated biphenyls (PCBs), and polychlorinated dibenzodioxins and polychlorinated dibenzofurans (PCDDFs) and this has also reduced inputs to sewage sludge (Figure 8.8). Currently, inputs of these contaminants to sludge are primarily from atmospheric deposition onto paved surfaces and runoff. For example, PCB concentrations in UK sludges have dropped markedly in response to the ban on the production and use of these substances by industry (McIntyre and Lester, 1984; Alcock and Jones, 1993). Controls on combustion and waste incineration have also reduced PCDD/F emissions, and analysis of archived samples of sludge from a major London wastewater treatment works (WWTW) showed that concentrations had declined by >97% in the past 40 years from 166 ng kg^{-1} toxicity equivalent quotient (TEQ) in 1960 to 4.2 ng kg^{-1} TEQ in 1998 (Figure 8.8). Furthermore, the cost of PCDD/F analysis is high and routine, statistically valid monitoring of PCDDFs in sludge is simply impractical.

Historical pollution levels are particularly important in understanding the transfer and cycling of persistent organic pollutants in the environment. Soil is an effective scavenger and sorptive medium for these organics and acts as a long-term and major repository, although biodegradation also takes place. Contemporary remobilization by volatilization from soil and redeposition onto surfaces and consequential transfer to wastewater collection systems represents an important diffuse input to sewage sludge that is difficult to control or eliminate. This process is particularly important for PCBs, which are no longer used, but enter sewage sludge primarily by remobilization

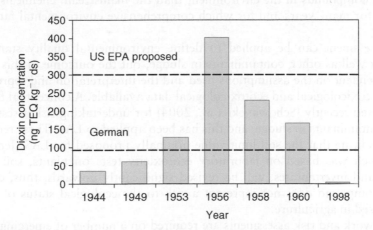

Figure 8.8 Dioxin concentration in a sewage sludge from a major sewage treatment works in West London

Note: both EU and US EPA proposed limits have been retracted

Source: Evans, 1999

from soil and therefore occur in sludge at relatively consistent concentrations irrespective of the mix of industrial or domestic discharges to the wastewater collection system.

Understanding of the behaviour of the major groups of persistent organic pollutants in the environment and the principal emission sources suggests relatively little more can be done to practically reduce their concentrations further in sewage sludge (IC Consultants Ltd, 2001). The increasing amount of scientific investigation also shows there are no significant environmental consequences from PAHs, PCBs or PCDDFs when sludge is used on farmland as a fertilizer. In the light of such developments, and the known physico-chemical behaviour of these compounds in soil, it may be argued that their importance as major pollutants in sludge has significantly diminished and that standards for these substances in sewage sludge are no longer necessary for agricultural application. This view was recently supported by the European Commission (EC) Joint Research Centre (Erhardt and Prüeß, 2001), which recommended that 'organic contaminants in sludge are not expected to pose major health problems to the human population when sludge is reused for agricultural purposes' and that 'it does not make much sense to include PCDD/F, PCBs and PAHs in routine monitoring programmes'.

Emissions of other organic contaminants and their entry into the wastewater system are associated with direct or indirect discharges resulting from their use in commercial and domestic activities. For example, the principal emissions of di(2-ethylhexyl) phthalate (DEHP) occur from the use of finished plastic products, and major domestic inputs to urban wastewater are floor and wall coverings and textiles with polyvinyl chloride (PVC) prints (IC Consultants Ltd, 2001). Numerically, detergent surfactants and residues (linear alkylbenzene sulphonate (LAS) and nonylphenol ethoxylates (NPE)) are the most significant contaminants in sewage sludge and, in a number of European countries, substitution of the most widely used compounds, e.g., LAS, has taken place. However, far less is known about the behaviour and toxicity of the alternative compounds in the environment than the mainstream chemicals that have been used for many years and for which comprehensive environmental fate data are available.

Risk assessment can be applied to define environmental quality standards for organic, as well as other, contaminants in sludge, but the outcome of this process is highly dependent on the assumptions used and the interpretation and representativeness of the toxicological and ecotoxicological data available. A conceptual framework was proposed recently (Schowanek et al., 2004) for undertaking risk assessments for organic contaminants in sludge, and this has been applied to LAS. These recent developments indicate that the soil limit value originally proposed for LAS (Jensen et al., 2001), which was based on laboratory ecotoxicity tests on plants, soil microbial processes and invertebrates, will be revised significantly upwards, thus, confirming that this compound does not represent a risk to the ecological status of soil when sludge is used in agriculture.

Further work and risk assessments are required on a number of emerging groups of organic compounds including: pharmaceuticals and antibiotics, and bodycare products. This does not provide grounds for curtailing land spreading, however, since these enter soil or result in human exposure by other much more important routes than sewage sludge, which should be understood in context, and also investigated. In Europe, action has been taken banning the use of certain persistent and bioaccumulative compounds, such as brominated flame retardants (EP/CEC, 2003), that may represent a threat to human health and the environment and were also detected in sludge. Such source control measures are important and have a positive influence on the quality of sludge for agricultural reuse. However, the implications of this action should also be considered in terms of the properties and fate of alternative compounds that may be used to substitute for the banned substances.

8.6 CONCLUSIONS

Agricultural recycling has many operational, agronomic and economic, as well as environmental benefits that are realized through careful and responsible management. The microbiological quality of sludge is a primary concern and heavy metals are also important, although they represent a longer-term environmental issue with regard to the accumulation of PTEs in soil. International experience of pathogen inactivation and PTE behaviour demonstrates that a pragmatic, technical basis to quality standards to protect the environment and human health can facilitate the safe and beneficial reuse of sludge. The importance of persistent organic pollutants in sludge declines where source controls on production, use and emissions are effective. Bulk industrial and domestic chemicals generally rapidly degrade during biological wastewater and sludge treatment processes and also in soil, and do not present a risk to soil, health or the environment. Emerging contaminants, including pharmaceuticals and bodycare products, in sludge represent a relatively small source of these compounds compared to other environmental exposure routes, but require further evaluation. Effective industrial effluent control on contaminant discharges to the sewer is a major priority to underpin the long-term sustainability of agricultural reuse programmes. Some critical focal points to ensure the secure and safe reuse of sewage sludge in agriculture are listed in Table 8.6.

Table 8.6 Focal points for agricultural use of sewage sludge

Industrial effluent control and screening
Proven treatment processes for stabilization, odour minimization and pathogen destruction
Appropriate use on land (rates of application, nutrient additions, crop types, waiting periods, and sowing and harvesting constraints)
Mechanized agriculture – treatment to significantly reduce pathogens and land use restrictions
Labour intensive agriculture – treatment to eliminate pathogens for unrestricted use
Monitoring protocols and quality assurance systems (HACCP) for sludge treatments and field application operations
Standard analytical methods for PTEs, organic contaminants and microorganisms
Sludge quality monitoring programme for PTEs and microorganisms
Technically based standards for PTEs in sludge-treated soil
Research, extension advice and demonstrations for farmers on use and agronomic value
Management systems
Communication (user, regulator, public, landowners and retailers)
Training and staff competence
Record keeping and reporting

REFERENCES

ADAS, Agricultural Development and Advisory Service. 2001. *The Safe Sludge Matrix. Guidelines for the Application of Sewage Sludge to Arable and Horticultural Crops.* 3rd. edn. (www.adas.co.uk).

Alcock, R.E. and Jones, K.C. 1993. Polychlorinated Biphenyls in Digested UK Sewage Sludges, *Chemosphere*, Vol. 26, pp. 2199–2207.

Alloway, B.J., Hird, A.B., Zhang, P., Chambers, B., Nicholson, F., Smith, S. and Carlton-Smith, C.H. 1998. *The Vulnerability of Soils to Pollution by Heavy Metals – Contract Ref No. CSA 2736.* Final Report to the Ministry of Agriculture, Fisheries and Food.

ArthurAndersen. 2001. *Disposal and Recycling Routes for Sewage Sludge.* Final Report to DG Environment, European Commission. Available at: http://europa.eu.int/comm/environment/waste/sludge/sludge_disposal.htm.

Carrington, E.G., Davis, R.D., Hall, J.E., Pike, E.B., Smith, S.R. and Unwin, R.J. 1998. *Review of the Scientific Evidence Relating to the Controls on the Agricultural Use of Sewage Sludge.* WRc Report No. DETR4415/3 and 4454/4. WRc Medmenham, Marlow.

Carrington, E.G., Pike, E.B., Auty, D. and Morris, R. 1991. Destruction of Faecal Bacteria, Enteroviruses and ova of Parasites in Wastewater Sludge by Aerobic Thermophilic and Anaerobic Mesophilic Digestion, *Water Science and Technology*, Vol. 24, No. 2, pp. 377–380.

Cass, J., Rogers, M. and Smith, S.R. 2005. Enteric Pathogen Decay Kinetics in Biosolids Amended Agricultural Soils, *Proceedings of the 10th European Biosolids and Biowastes Conference and Workshop.* 13–16 November, Wakefield.

CEC, Council of the European Communities. 1986. Council Directive of 12 June 1986 on the Protection of the Environment, and in Particular of the Soil, when Sewage is Used in Agriculture (86/278/EEC), *Official Journal of the European Communities* No L181/6-12.

Chambers, B., Gibbs, P., Mcgrath, S.P., Chaudri, A., Carlton-Smith, C., Godley, A., Becon, J., Campbell, C. and Aitken, M. 2004. Long-term Sludge Experiments: Soil Metal Availability to Crops and Microbes, *Sustainable Land Application Conference*, 4–8 January, Orlando, Florida.

Chang, A.C., Pan, G., Page, A.L. and Asano, T. 2002. *Developing Human Health-related Guidelines for Reclaimed Water and Sewage Sludge Applications in Agriculture.* World Health Organization, Geneva.

ComEC, Commission of the European Communities. 2001. Commission Regulation EC No 466/2001 of 8 March 2001 setting maximum levels for certain contaminants in foodstuffs, *Official Journal of the European Communities* L 77/1-21.

EP/CEC, European Parliament and Council of the European Union. 2003. Directive 2003/11/EC of the European Parliament and of the Council of 6 February 2003 amending for the 24th time Council Directive 76/769/EEC relating to restrictions on the marketing and use of certain dangerous substances and preparations (pentabromodiphenyl ether, octabromodiphenyl ether), *Official Journal of the European Union* L 42/45-46.

Erhardt, W. and Prüeß, A. 2001. *Organic Contaminants in Sewage Sludge for Agricultural Use.* Final Report to European Commission Joint Research Centre. Available at: http://europa.eu.int/comm/ environment/waste/sludge/organics_in_sludge.pdf.

Evans, T. 1999. Biosolids and residuals in Europe. *Water Environment Federation, 72nd Annual Exhibition and Technical Conference on Water Quality and Wastewater Treatment*, 9–13 October.

Feachem, R.G., Bradley, D.J., Garelick, H. and Mara, D.D. 1983. *Sanitation and Disease: Health Aspects of Excreta and Wastewater Management.* John Wiley & Sons, Chichester.

Gale, P. 2001. Microbiological risk assessment. Pollard, S. and Guy, J. (eds) *Risk Assessment for Environmental Professionals.* The Chartered Institution of Water and Environmental Management, London, pp. 81–90.

Gale, P. 2003. Using Event Trees to Quantify Pathogen Levels on Root Crops from Land Application of Treated Sewage Sludge, *Journal of Applied Microbiology*, Vol. 94, pp. 35–47.

Gale, P. 2005. Land Application of Treated Sewage Sludge: Quantifying Pathogen Risks from Consumption of Crops, *Journal of Applied Microbiology*, Vol. 98, pp. 380–396.

Gerba, C.P. and Smith, J.E. 2005. Sources of Pathogenic Microorganisms and their Fate During Land Application of Wastes, *Journal of Environmental Quality*, Vol. 34, pp. 42–48.

Horan, N.J., Fletcher, L., Betmal, S.M., Wilks, S.A. and Keevil, C.W. 2004. Die-off of Enteric Bacterial Pathogens During Mesophilic Anaerobic Digestion, *Water Research*, Vol. 38, pp. 1113–1120.

IC Consultants Ltd. 2001. *Pollutants in Wastewater and Sewage Sludge*. Final Report. DG Environment, European Commission. Available at: http://europa.eu.int/comm/environment/waste/sludge/sludge_pollutants_xsum.pdf.

Jacobs, L.W., O' Connor, G.A., Overcash, M.A., Zabik, M.J. And-Rygiewicz, P., Machno, P., Munger, S. and Elseavi, A.A. 1987. Effects of Trace Organics in Sewage Sludges on Soil-plant Systems and Assessing their Risk to Humans. Page, A.L., Logan, T.J. and Ryan, J.A. (eds). *Land Application of Sludge: Food Chain Implications*. Michigan, Lewis Publishers Inc, Chelsea, pp. 101–143.

Jensen, J., Løkke, H., Holmstrup, M., Krogh, P.H., and Elsgaard, L. 2001. Effects and Risk assessment of linear alkylbenzene sulfonates in agricultural soil. 5. Probabilistic Risk Assessment of Linear Alkylbenzene Sulfonates in Sludge-amended Soils, Environmental Toxicology and Chemistry, Vol. 20, pp. 1690–1697.

Lang, N.L., Smith, S.R., Bellett-Travers, D.M. and Pike, E.B. 2003. Effects of Biosolids Application, Environment and Time on *Escherichia Coli* Population Dynamics in Agricultural Soil. Workshop focused on research at Imperial College London on Recycling – Nutrient Management Issues. *Proceeding of the Joint CIWEM Aqua Enviro Technology Transfer 8th European Biosolids and Organic Residuals Conference*, 24–26 November, Wakefield.

McIntyre, A.E. and Lester, J.N. 1984. Occurrence and Distribution of Persistent Organochlorine Compounds in U.K. Sewage Sludges, *Water, Air, and Soil Pollution*, Vol. 23, pp. 397–415.

MAFF/DoE, Ministry of Agriculture, Fisheries and Food. 1993a. *Review of the Rules for Sewage Sludge Application to Agricultural Land: Food Safety and Relevant Animal health Aspects of Potentially Toxic Elements*. Report of the Steering Group on Chemical Aspects of Food Surveillance. PB 1562. MAFF Publications, London.

MAFF/DoE, Ministry of Agriculture, Fisheries and Food. 1993b. *Review of the Rules for Sewage Sludge Application to Agricultural Land: Soil Fertility Aspects of Potentially Toxic Elements*. Report of the Independent Scientific Committee. PB 1561. MAFF Publications, London.

MAFF, Ministry of Agriculture, Fisheries and Food. 2000. *Fertiliser Recommendations for Agricultural and Horticultural Crops (RB209)*. 7th edn. The Stationery Office, London.

Mara, D.D. and Horan, N.J. 2002 Sludge to land: Microbiological double standards, *Journal of the Chartered Institution of Water and Environmental Management*, Vol. 16, No. 4, pp. 249–252.

Mininni, G., Zuccarello, R.B.D., Lotito, V., Spinosa, L. and Di Pinto, A.C. 1997. A Design Model of Sewage Sludge Incineration Plants with Energy Recovery, *Water Science and Technology*, Vol. 36, pp. 211–218.

Morris, R., Smith, S.R., Bellett-Travers, D.M. and Bell, J.N.B. 2003. Reproducibility of the Nitrogen Response and Residual Fertiliser Value of Conventional and Enhanced-treated Biosolids. Workshop focused on research at Imperial College London on Recycling – Nutrient Management Issues. *Proceeding of the Joint CIWEM Aqua Enviro Technology Transfer 8th European Biosolids and Organic Residuals Conference*, 24–26 November, Wakefield.

Nell, J.H., Steer, A.G. and van Rensburg, P.A.J. 1983. Hygienic Quality of Sewage Sludge Compost, *Water Science and Technology*, Vol. 15, pp. 181–194.

O' Connor, G.A., Sarkar, D., Brinton, S.R., Elliott, H.A. and Martin, F.G. 2004. Phytoavailability of Biosolids Phosphorus, *Journal of Environmental Quality*, Vol. 33, pp. 703–712.

Pereira Neto, J.T., Stentiford, E.I. and Mara, D.D. 1987. Comparative Survival of Pathogenic Indicators in Windrow and Static Pile. de Bertoldi, M., Ferranti, M.P., L' Hermite, P. and Zucconi, F. (eds) *Compost: Production Quality and Use*. Elsevier Applied Science Publishers Ltd, Barking, pp. 276–296.

RCEP, Royal Commission on Environmental Pollution.1996. *Nineteenth Report: Sustainable Use of Soil*. Cm 3165. HMSO, London.

Schowanek, D., Carr, R., David, H., Douben, P., Hall, J., Kirchmann, H., Patria, L., Sequi, P., Smith S. and Webb, S. 2004. A Risk-based Methodology for Deriving Quality Standards for Organic Contaminants in Sewage Sludge for Use in Agriculture – Conceptual Framework, *Regulatory Toxicology and Pharmacology*, Vol. 40, pp. 227–251.

Smith, S.R. 1994. *Effects of Heavy Metals on the Size and Activity of the Soil Microbial Biomass after Long-term Treatment with Sewage Sludge*. Report No. FR 0469. Foundation for Water Research, Marlow.

Smith, S.R. 1995. Soil Microbial Biomass Content of Sewage Sludge-treated Agricultural Soil, *Third International Conference on the Biogeochemistry of Trace Elements*, 15–19 May, Paris.

Smith, S.R. 1996. *Agricultural Recycling of Sewage Sludge*. CAB International, Wallingford.

Smith, S.R. 1997. *Rhizobium* in Soils Contaminated with Copper and Zinc Following the Long-term Application of Sewage Sludge and Other Organic Wastes, *Soil Biology and Biochemistry*, Vol. 29, pp. 1475–1489.

Smith, S.R. 2000. *Rhizobium* in Long-term Metal Contaminated Soil, *Soil Biology and Biochemistry*, Vol. 32, pp. 729–731.

Smith, S.R., Alloway, B.J., and Nicholson, F.A. 1999. Effect of Zn on the Microbial Biomass Content of Sewage Sludge-treated Soil, *Fifth International Conference on the Biogeochemistry of Trace Elements*, 11–15 July, Vienna.

Smith, S.R., Morris, R., Bellett-Travers, D.M., Ferrie, M., Rowlands, C.L. and Bell, N. 2002a. Implications of the Nitrates Directive and the Provision of Fertiliser Advice for the Efficient Agricultural use of Conventional and Enhanced-treated Biosolids Products, *Proceedings of the Joint CIWEM and Aqua Enviro Technology Transfer 7th European Biosolids and Organic Residuals Conference*, 18–20 November, Wakefield.

Smith, S.R., Triner N.G., and Knight, J.J. 2002b. Phosphorus release and Fertiliser Value of Thermally Dried and Nutrient Removal Biosolids, *Water and Environmental Management Journal*, Vol. 16, pp. 127–134.

Strauch, D. 1991. Microbiological Treatment of Municipal Sewage Waste and Refuse as a Means of Disinfection Prior to Recycling in Agriculture, *Studies in Environmental Science*, Vol. 42, pp. 121–136.

Sweet, N., Mcdonnell, E., Cochrane, J. and Prosser, P. 2001. The New Sludge (Use in Agriculture) Regulations, *Proceedings of the Joint CIWEM Aqua Enviro Consultancy Services 6th European Biosolids and Organic Residuals Conference*, 12–14 November, Wakefield, UK.

Thierbach, R.D. and Hanssen, H. 2002. Utilisation of Energy from Digester Gas and Sludge Incineration at Hamburg's Köhlbrandhöft WWTP, *Water Science and Technology*, Vol. 46, pp. 397–403.

UK DoE, UK Department of the Environment. 1989. *Code of Practice for Agricultural Use of Sewage Sludge*. HMSO, London.

UK DoE, UK Department of the Environment. 1996. *Code of Practice for Agricultural Use of Sewage Sludge*. HMSO, London.

UK SI, UK Statutory Instrument. 1989. *The Sludge (Use in Agriculture) Regulations 1989*. No. 1263. HMSO, London.

UKWIR, UK Water Industry Research. 2002. *Pathogens in Biosolids – The Fate of Pathogens in Sewage Treatment*. Report Ref. No. 02/SL/06/6. UKWIR, London.

US EPA, US Environmental Protection Agency. 1992a.*Technical Support Document for Land Application of Sewage Sludge, Volume I.* Eastern Research Group, Lexington.

US EPA, US Environmental Protection Agency. 1992b. *Technical Support Document for Land Application of Sewage Sludge, Volume II.* Eastern Research Group, Lexington.

US EPA, US Environmental Protection Agency. 1993. Part 503-Standards for Use or Disposal of Sewage Sludge, *Federal Register*, Vol. 58, pp. 9387–9404.

US EPA, US Environmental Protection Agency. 1999. *Control of Pathogens and Vector Attraction in Sewage Sludge.* EPA/625/R-92/013. Office of Research and Development, US EPA, Washington DC.

Water UK. 2004 *The Application of HACCP Procedures in the Water Industry: Biosolids Treatment and Use on Agricultural Land.* London, Water UK. Available at: www.water.org.uk/static/files_archive/0WUK_Haccp_guide_FINAL_19_Mar_04.pdf.

WHO, World Health Organization. 1981. *The Risk to Health of Microbes in Sewage Sludge Applied to Land.* EURO Reports and Studies 54. WHO, Copenhagen.

Wild, S.R. and Jones, K.C. 1991. Organic Contaminants in Wastewaters and Sewage Sludges: Transfer to the Environment Following Disposal. Jones, K.C. (ed.) *Organic Contaminants in the Environment.* Elsevier Science Publishers Ltd, Barking, pp. 133–158.

WRC, Water Research Centre. 1999. *Scientific Justification for Agricultural Use of Sewage Sludge in Egypt.* Phase 2 Final Report of the Cairo Sludge Disposal Study. WRc Report No. CSDR014/1. WRc Medmenham, Marlow.

Chapter 9

'Closing the urban water cycle' integrated approach towards water reuse in Windhoek, Namibia

Ben van der Merwe

Environmental Engineering Services, P.O. Box 6373, Ausspannplatz, Windhoek, Namibia

ABSTRACT: Windhoek, the capital of Namibia, is situated in the central highlands of Namibia, approximately 1,600 metres above sea level. The nearest perennial river is approximately 700 km away. The average annual rainfall is approximately 360 mm, whilst the evaporation is approximately 10 times higher. The integrated approach towards water reuse in the city entails water reclamation for potable use, provision of irrigation water for landscaping through a dual pipe system and artificial recharge of the Windhoek Aquifer (water banking) with high-quality treated water sourced from surface sources blended with reclaimed water. In 1968, a direct reclamation system from treated domestic sewage to augment the potable water supply to the city was pioneered. The process was improved over time and the Goreangab Water Reclamation Plant has consistently supplied water of acceptable quality over the years. Due to water scarcity in the region and with few alternative supply sources, the construction of a new water reclamation plant became necessary. This plant incorporated a modified process train, based on the multiple barrier concept to provide high-quality water in accordance with agreed standards. The new plant, with a capacity of 21 000 m³/day, was completed in August 2002, and is capable of providing approximately one-third of the city's water demand until 2012. All the design parameters of the old plant were re-assessed and evaluated in the design of the new plant. The reclamation of water over a period of 37 years has demonstrated that, with proper care and diligence, water of acceptable quality can consistently be produced from mainly domestic effluent. The cost of reclaimed water versus other supply options to Windhoek is very competitive. Although perceived to be very controversial in the rest of the world, the consumers in Windhoek have accepted that reclaimed water forms part of the potable water supply.

9.1 INTRODUCTION

Namibia is located in the south-western part of Africa and is the most arid country south of the Sahara Desert. Windhoek, the capital of Namibia, is situated in the Central Highlands of Namibia, approximately 1,600 m above mean sea level. The average annual rainfall is 360 mm, while the average evaporation is 3,400 mm/a. Namibia's water resources are unevenly distributed throughout the country and there are no perennial rivers within the country's borders. The nearest perennial river, the Okavango, is 700 km north of the city on the north-eastern border.

The estimated population of Windhoek in mid-2005 was 273,000. According to the Namibian Economic Policy Research Unit (1996), Windhoek has 51% of manufacturing, 96% of utilities, 56% of construction and trade, 94% of transport and communications,

82% of financial and business services, and 68% of community and social services in the country. The city produces 47% of value added, and private consumption expenditure in Windhoek comprises 35% of the national total. In contrast, only 8% of production comes from the north where 60% of the population resides and where 33% of private consumption expenditure takes place.

9.2   WATER SOURCES IN WINDHOEK

Windhoek is underlain by an aquifer with an estimated safe yield of $1.73\,\mathrm{M\,m^3}$ per annum. There are 50 production boreholes with a maximum emergency production rate of $5.5\,\mathrm{M\,m^3}$ per annum, which can be maintained over a maximum period of two years. The Avis Dam ($2.4\,\mathrm{M\,m^3}$) and the Goreangab Dam ($3.6\,\mathrm{M\,m^3}$), built on ephemeral rivers near the city, were completed in 1933 and 1959 respectively. The 95% assured safe yield of the two sources is approximately $1.1\,\mathrm{M\,m^3}$ per annum.

In 1969, the government embarked on a scheme of dams on ephemeral rivers further from Windhoek to provide water to the city. The Von Bach Dam ($48.6\,\mathrm{M\,m^3}$) was added in 1970, the Swakoppoort Dam ($63.5\,\mathrm{M\,m^3}$) was completed in 1977, while the Omatako Dam ($43.5\,\mathrm{M\,m^3}$) was completed in 1982. The distances to the various dams are 70, 100 and 200 km respectively. The combined safe yield of the three dams, based on 95% assurance supply, is approximately $20\,\mathrm{M\,m^3}$ per annum. Of this volume, approximately $15\,\mathrm{M\,m^3}$ is available for supply to Windhoek.

9.3   REUSE OPTIONS IMPLEMENTED IN WINDHOEK

In a paper presented to the Institute of Water Pollution Control, dated March 1970, it was indicated that: 'As early as 1954, it was realized that the future economic progress of Windhoek depended on the utilization of reclaimed water' (Stander et al., 1970). During the early 1960s, the Windhoek City Council resolved to run a pilot plant for the reclamation of treated sewage effluent and a pilot-scale plant was erected at the Gammams Plant (Clayton, 2005). During the mid-1960s, following favourable results from the Gammams pilot plant, the Windhoek City Council decided to implement potable reclamation. Continued pressure on water supplies led to a decision to extend the Goreangab Treatment Plant that treated water from the Goreangab Dam. Work commenced in 1967 and was completed in 1968. The conventional Goreangab water treatment plant was extended by the addition of a second treatment train to treat not only the surface water from the Goreangab Dam, but also the final effluent from the Gammams Plant. Gammams had been constructed to treat predominantly Windhoek's domestic effluents, while industrial effluents were diverted for separate treatment at a different plant, situated to the north of the city. Thus, the Goreangab Water Reclamation Plant (GWRP) was born. It had an initial capacity of $3,287\,\mathrm{m^3}$ per day ($1.2\,\mathrm{M\,m^3}$ p/a). The reclamation of potable water from sewage at the GWRP has been ongoing for the past 37 years and is well documented and recognized as the only plant of its kind in the world. The process was improved and the plant extended on various occasions over a period of 30 years to a capacity of $2.9\,\mathrm{M\,m^3/a}$. During the drought of 1996/1997, the old GWRP was extended to increase the capacity to $7,500\,\mathrm{m^3/day}$.

During discussion of the Central Area Water Master Plan of 1992 (a plan to augment water supplies to the Central Areas of Namibia, including Windhoek), it was concluded that implementation of Water Demand Management and the extension of the Goreangab Water Reclamation Plant to provide approximately 35% of the total potable supply to Windhoek were the most feasible interim measures for implementation. Technically, the extension of water reclamation was seen as a readily available source of water for the City of Windhoek. The New Goreangab Water Reclamation Plant (NGWRP) was completed August 2002 with a capacity of 7.6 M m^3/annum.

During eight years between 1968 and 2000, when conventional sources were unable to meet normal demand for the city, the reclamation plant produced from 12% to 18% of the total potable water supply to Windhoek, with highest production in 1997 of 2.8 M m^3 or 18% of the total demand of 15.5 M m^3. It is contemplated that the new plant will be operated continuously to provide potable water to the city, as well as water for the planned artificial recharge project for the Windhoek Aquifer. During 2003, the plant produced 5.23 M m^3 or 26% of the supply of 20.2 M m^3 to Windhoek. The production rate was limited by the restricted availability of raw water for reclamation, technical plant problems, the raw water supply infrastructure and the blending ratio restriction of 35% reclaimed water in the final blended water. In 1992, a dual pipe system was installed to distribute irrigation water for restricted irrigation of sports fields, parks and cemeteries in the city. On completion of the NWGRP the old GWRP was integrated to supply higher quality water for unrestricted irrigation through the dual pipe system. The treatment steps include dissolved air flotation, sand filtration and chlorination with a contact period of 60 minutes for the treatment of irrigation water. The irrigation water supplied for sports fields, parks and cemeteries in the city contributes 6% to 7% of the annual water supply to the city.

A recent study (Central Area Joint Venture Consultants, 2004) concluded that artificial recharge of the Windhoek Aquifer (water banking) is the 'best next' supply option to the Central Area of Windhoek. It contemplates storing excess treated water from surface dams, which would be available during periods of high runoff, as well as reclaimed water, in the Windhoek Aquifer for abstraction during periods of drought. In a study of the required quality of the injection water, it was determined that a blend of 3 to 1 of surface water with reclaimed water would be suitable for injection. Phase 1 of the project was completed in 2004 and a further phase, including the drilling of deeper large-diameter boreholes to increase the size of the water-bank, is expected to be completed by the end of 2008. If completed in full (4 phases), the estimated storage will represent approximately 3 years of supply to the city.

When implemented, this project will be an efficiency improvement of existing sources, since it will improve the use of existing surface sources through the reduction of evaporation losses and allow increased production from the water reclamation plant. One of the biggest benefits is that it will downsize the required capacity of future supply augmentation options. Moderate savings through water demand management initiatives (reduction of leakages on government properties through improved maintenance) could provide approximately 50% of the average volume required for recharge during the next 18 years. This is the only option for indirect reuse water in the Windhoek environment.

Figure 9.1 New Goreangab Water Reclamation Plant under Construction (2000) with the old
Reclamation Plant and Goreangab Dam on the Far Right

Other options implemented to augment supplies include:

• Conjunctive use of water, based on the premise that surface water is used during
  periods of ample supply while groundwater is used as a backup system during
  drought years when surface supplies are limited.
• Transfer of water from surface dams with high evaporation characteristics to
  dams with better basin characteristics.
• Implementation of Water Demand Management, which lowered the unrestricted
  water demand by approximately 30% and reduced the annual increase in consumption to less than 3%.

9.4 FUTURE WATER SUPPLY AUGMENTATION TO WINDHOEK

The long-term solution to secure the water supplies for central Namibia is to link the
Eastern National Water Carrier (the infrastructure system that delivers water from
remote sources to Windhoek and the Central Area) to the Okavango River on the
north-east border, a distance of 700 km. The Okavango Delta in Botswana is an important international wetland that is very sensitive from an ecological perspective and
abstraction from the Okavango River would require negotiation with neighbouring
Botswana. It is essential to develop all local sources to their maximum potential and
to consider alternative water supply sources which could be developed to replace or
defer, for as long as possible, the extension of the Eastern National Water Carrier to
the Okavango River.

Table 9.1 Process configurations and modifications since commissioning

Configuration One (1969)	Configuration Two (1977)	Configuration Three (1980)	Configuration Four (1986)
GAMMAMS WASTEWATER TREATMENT PLANT			
primary settling	primary settling	primary settling	primary settling
biological filters	biological filters	activated sludge	activated sludge
secondary settling	secondary settling	secondary settling	secondary settling
maturation ponds	maturation ponds	maturation ponds	maturation ponds
GOREANGAB WATER RECLAMATION PLANT			
carbon dioxide	lime	chlorine	alum
alum	settling	alum and lime	
autoflotation	ammonia stripping	settling	dissolved air
foam fractionation	primary carbon dioxide		flotation
breakpoint chlorination	chlorine	breakpoint chlorination	chlorine* lime
settling	settling	settling	settling
sand filtration	secondary carbon dioxide	sand filtration	sand filtration
	sand filtration		
carbon filtration	breakpoint chlorination	chlorine	breakpoint chlorination
	chlorine contact	chlorine contact	chlorine contact
	carbon filtration	carbon filtration	carbon filtration
chlorine	chlorine	chlorine	chlorine
blending	blending	blending	blending

Notes: Process modifications in bold blocks.
Chemical addition in shaded blocks.
Chlorine marked * only for intermittent shock dosing.

9.5 VARIOUS PROCESS MODIFICATIONS FROM 1968 TO 1995

The reclamation plant has undergone a process of evolution and improvement since 1968. The different configurations are summarized in Table 9.1 and have been described in greater detail in earlier publications (Haarhoff and Van der Merwe, 1996).

Modifications to configuration four (adapted in 1995) entail the use of ferric chloride in the place of alum, carbon filtration moved to immediately follow sand filtration and breakpoint chlorination after carbon filtration. In 1996, a filter-to-waste facility was added to the sand filters to eliminate possible breakthrough of *Giardia* and *Cryptosporidium* after filter backwash.

9.6 PROCESS DESIGN FOR THE NEW GOREANGAB WATER RECLAMATION PLANT[1]

9.6.1 Summary

The sensitive nature of wastewater reclamation demands unusual attention to the process selection criteria for such a plant. The recent upgrading of the GWRP

[1] Summarised from Haarhoff et al., 1998.

necessitated a complete re-analysis of the process needs and the treatment objectives in the light of new technologies and ever-tighter water quality guidelines. The process selection methodology is summarized below:

• Establishing a raw water quality profile in probabilistic terms.
• Defining final water quality objectives, based on own experience and recent international developments.
• Developing the vague 'multiple barrier' concept into a quantitative tool for comparison of process options.
• Critical analysis of the process train used successfully in the past.
• Evaluation of numerous experimental and pilot trials performed during the period from 1997 until 2000.
• The finalization of the process train for the new reclamation plant.

9.6.2 Raw water quality profile

The raw water data records over the past three years were probabilistically analyzed. Sixty different parameters were measured, of which forty were also duplicated by external laboratories as part of the ongoing quality control programme. The two raw water sources are chemically remarkably similar, both in terms of their mean values as well as the 95th percentile values. Physically they are slightly more dissimilar, with the natural surface water from Goreangab Dam more turbid and algae-infested than the maturation pond effluent from the Gammams Water Care Works. This dissimilarity, however, is small enough to allow essentially the same process design for both sources. The high levels of iron, manganese and ammonia necessitate an early oxidation step in the process train. The occasional high level of nitrate in the water from Gammams is a problem that cannot be treated by conventional physical treatment processes, but requires ion exchange or low cut-off membrane treatment. The occurrence of nitrate is a special problem and should be dealt with at the Gammams Water Care Works.

Due to the high organic loading of the raw water, an organic removal process should form part of the treatment process, as has been the case since the inception of the GWRP. The organic material from Gammams is the refractory material left after biological treatment and is not very amenable to precipitation during normal coagulation. This is supported by the specific UV absorbance (SUVA) values of 2.94 for Goreangab raw water and 1.87 for Gammams raw water, suggesting that only about 20% of the organics would be precipitable.

9.6.3 Determination of treatment objectives

Table 9.2 summarizes the finally adopted water quality of treatment objectives, as determined from an analysis of various publications and the experience gained on the old GWRP. The guidelines for the planning, design and implementation of a water reclamation scheme that were used included the Namibian Guidelines (1991), Rand Water (1994), World Health Organization (1993), National Drinking Water Standards and Health Advisories issued by the United States Environmental Protection Agency (1996) and the Water Research Commission Guide (RSA, 1982).

Table 9.2 reflects the current US EPA 'treatment technique' philosophy that treatment plant operation should be directed towards a log-reduction approach when routine measurements of contaminants cannot be quickly or easily measured. By

Table 9.2 Final water quality criteria and operational results

Parameter	Units	Final Water Specification	Actual Operational Results	
			50% tile	*95% tile*
Physical and organic				
Chemical oxygen demand	mg/L	10–15	6.6	11
Colour	mg/L Pt	8–10	0.5	0.5
Dissolved organic carbon	mg/L	3*	1.7	2.8
Total dissolved solids	mg/L	1000 max or 200 above incoming	838	938
Turbidity	NTU	0.1–0.2	0.05	0.10
UV$_{254}$	abs/cm	0.00–0.06	0.015	0.027
Inorganic	**Units**			
Aluminium	Al mg/L	0.15	0.005	0.05
Ammonia	N mg/L	0.1	0.05	0.18
Iron	Fe mg/L	0.05–0.10	0.01	0.03
Manganese	Mn mg/L	0.01–0.025	0.005	0.015
Microbiological	**Units**			
Heterotrophic plate count	per 1 mL	80–100	0	4
Total coliforms	per 100 mL	0	0	0
Fecal coliforms	per 100 mL	0	0	0
Chlorophyll a	µg/L	1	0.27	2.58
Giardia	per 100 L	0 count/100 l or 5 log removal	0	0
Cryptosporidium	per 100 L	0 count/100 l or 5 log removal	0	0
Disinfection by products	**Units**			
Trihalomethanes	µg/L	20–40	35	57

Note: *A target was set for 3 mg/L dissolved organic carbon (DOC) with a maximum of 5, based on the premise that in the final blended water not more than 1 mg/L DOC originating from sewage water should be present. A maximum blending ratio of 35% to 65% water from other sources was specified.

maintaining a residual concentration C after time t, the link between the Ct-product and the log-reduction of certain microbiological contaminants will ensure compliance with the required log-reduction. The 5-log requirement for the protozoan oocysts is based on the restraint of no more than 1 infection per 10,000 people per year.

9.6.4 The multiple-barrier concept

The term 'multiple barriers' is frequently encountered in relation to wastewater reuse, but is often loosely used as an approximate equivalent of 'safety factor' without clear meaning. It is necessary to differentiate between non-treatment barriers, treatment barriers and operational barriers. Three significant non-treatment barriers have been present in the past, and will continue to be present in the future:

- The diversion of industrial effluents to different municipal drainage areas, and the continuous policing of discharges into the sewerage system draining to the Gammams Wastewater Treatment Plant.

- Rigorous, continuous quality monitoring of raw and treated water, which will detect problems as they arise and allow corrective action before the public can be exposed to any prolonged risk.
- Blending of reclaimed water with water from conventional sources to limit reclaimed water to a maximum of 35% of the blended water supplied to the public.

Operational barriers are treatment processes which are not normally used, but which provide backup or standby for other essential treatment processes. In the case of the NGWRP, for example, a powdered activated carbon facility was planned to provide additional adsorption capacity if the water quality deteriorates during periods of drought to such an extent that the final water quality criteria could not be met.

Treatment barriers are barriers always present in the treatment plant to guard against specific contaminants, such as the barrier of chlorination against bacteria. It should be noted that 'barriers' in this sense generally do not imply absolute dead-stop barriers. The use of settling and filtration, for example, is regarded as a barrier against turbidity, although it does not completely remove all turbidity. The precipitation of DOC during coagulation, as another example, is only partial and therefore far from a complete 'barrier'. Different contaminants respond differently to different treatment methods. Likewise, the consequences of failure are different for each contaminant. By considering all these factors, the following criteria for treatment barriers were eventually adopted:

- For aesthetic parameters, such as turbidity or colour, two complete barriers are required.
- For microbiological parameters, three complete barriers are required.
- As an additional precaution against protozoan oocysts, filters with slow-start and filter-to-waste facilities are required to ensure the required removal as a minimum. (The combination of Dissolved Air Flotation (DAF) with dual-media filtration will give 5 log removal if properly operated).
- For organic parameters, two partial barriers are required.
- For the parameters that determine stability (with respect to corrosion), one barrier is required.
- The minimization of disinfection by-products will be ensured by adherence to practical guidelines, such as using a chemical dosing strategy, which maximizes turbidity as well as organic removal ('enhanced coagulation' in current US vernacular), moving chlorine dosing as far downstream as possible, and using alternative oxidants as far as possible.

The performance data for each individual process on the old GWRP was scrutinized over a three-year period. The deficiencies in the old Goreangab process train could theoretically be rectified by any number of additional processes. Before the final selection was made, numerous experiments were performed at full scale, pilot scale and at bench scale. The most important experimental results are summarized below.

9.6.5 Experiments and pilot studies to determine process design criteria

Two full-scale experiments were performed with activated carbon on the GWRP to determine the efficiency of DOC removal through combined granular activated carbon (GAC) / sand filters and powder activated carbon (PAC). The performance of the GAC filters was excellent, leading to DOC removals down to levels of below 3 mg/L. The breakthrough time, however, was about three months after which the carbon had to be replaced. This rendered it impractical for continuous use and frequent regeneration. The second experiment was performed to assess the efficiency of powdered activated carbon (PAC) adsorption added to the raw water, and to determine whether the deliberate lowering of the pH with acid improved the rate of adsorption. The lowering of the pH had a positive effect on the organic removal efficiency of both the PAC and the DAF process, and, in general, better results were obtained with 20 mg/L PAC with acid than with 60 mg/L PAC without acid. The DOC levels were reduced to a low value of less than 4 mg/L after sand filtration.

An ozone / GAC pilot plant was operated from November 1995, and consisted of an ozonated train and a non-ozonated train. Two different carbons were tested and the ozone dosage was regularly adjusted with varying organic content of the incoming water.

Four different ultra-filtration membranes were tested for varying periods on a pilot scale. These provided useful results on allowable flux rates and backwash frequency. Membrane filtration did not lower the organic content of the water.

9.7 SELECTION OF FINAL PROCESS TRAIN

The resulting treatment train, after consideration of all the foregoing, is shown in Figure 9.2.

The proposed process train provides for the following:

- Two complete barriers against turbidity (DAF/filtration and membrane filtration) as well as the secondary filtration effect of GAC.
- Three complete barriers against microbiological contaminants (ozone, membrane filtration, chlorination).

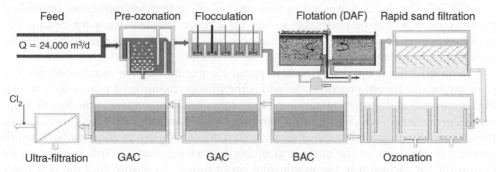

(DAF: dissolved air flotation, BAC: biological activated carbon-filter, GAC: granular activated carbon-filter)

Figure 9.2 Final treatment train for the new Goreangab water reclamation plant

- Four complete barriers against *Giardia* (DAF/filtration, ozonation, membrane filtration and sufficient chlorination).
- Two complete barriers and two partial barriers against *Cryptosporidium* (DAF/filtration and membrane filtration as complete barriers, ozonation and chlorination as partial barriers).
- Four partial barriers against organic material (enhanced coagulation, ozonation, GAC adsorption).
- Pre-ozonation is used on a permanent basis for the beneficial effect on coagulation/flocculation.
- The use of ozone with GAC on a pilot scale has reduced the trihalomethane (THM) formation potential to less than 20 µg/L compared to the less than 70 µg/L THMs actually measured during earlier months, showing the beneficial effects of using ozone in lieu of later chlorination with respect to disinfection by-products (DBPs).
- The PAC facility will be used only when the ozone/GAC facility is not operating at full capacity or during quality deterioration during periods of drought.
- The membrane choice between micro- and ultra-filtration was made only during tender adjudication stage.

9.8 OPERATIONAL EXPERIENCE TO DATE

Enhanced coagulation is practised (considered as best practice) where pre-ozonation is a critical component for floc development and DOC removal, typically 40% at the DAF process.

With effective iron and manganese oxidation across the sand filter media no significant DOC removal is experienced. The main ozonation process ensures the fractionization of organic substances. Biological GAC filters as well as the GAC filters account for 20% DOC removal respectively, which highlights the necessity of the Biological Activated Carbon (BAC) filters as an effective process that extends GAC saturation and reduces operational cost.

9.9 WATER QUALITY AND MONITORING

From the experience gained over the past 15 years, since 1990, with the operation of the old reclamation plant and during the research for the design and operation of the NewGWRP, the decision was taken that instead of testing for numerous individual parameters, it would be preferable to test on a more frequent basis for certain key parameters at process points that represent barriers for those particular parameters or would contribute towards their removal.

The monitoring is divided into three parts. First is the on-line monitoring for key parameters at different processes. Second is the process monitoring part, which is vital for the control and proper operation of a process unit. Third, there is the health monitoring programme, which is aimed at parameters that are still in a state of research or that are mostly measured in concentrations that do not constitute an immediate health concern in the light of the current available knowledge but are still deemed to be important for inclusion in the monitoring programme. To ensure that the treatment unit is operated according to specification, heavy penalties would be levied if certain parameter values are not adhered to at critical control points.

On-line monitoring of *Giardia/Cryptosporidium* and virus: on-line sampling devices have been developed in-house to sample one litre every 15 minutes over a period of seven days for *Giardia* and *Cryptosporidium*. These have proved to be very effective and give a clear indication of plant failure or breakdown of a barrier. This device will now also be used for the concentration of virus onto a cartridge.

On-line monitoring for chemical parameters: pH, turbidity, conductivity, DOC, UV254 absorption, dissolved oxygen (DO), particle counters, ozone concentration and free chlorine concentration is implemented at various points in the process.

Automatic composite samplers are placed at critical control points in the process to collect the desired volumes of samples over a 24-hour period for chemical analysis.

The operational monitoring programme includes:

- Physical and organoleptic: turbidity, colour, pH, conductivity, total dissolved solids (TDS), hardness, corrosiveness.
- Macro elements: potassium, sodium, chloride, sulphate, fluoride, iron, manganese, silica.
- Organics and nutrients: DOC, chemical oxygen demand (COD), UV254 absorption, nitrate, nitrite, ammonia, total Kjeldahl nitrogen (TKN), ortho-phosphate and trihalomethanes (THM).
- Microbiology, biology and parasites: heterotrophic plate count (HPC), total coliform (TC), fecal coliform (FC), *E coli* (Ec), fecal streptococci, clostridium perfringens (viable and spores), coliphage (somatic), *Giardia* and *Cryptosporidium*, chlorophyll a. Begin a new porphyria Monitoring (Research) for health aspects includes:
  - Micro-elements: silver, argon, arsenic, gold, boron, barium, cadmium, copper, chromium, cobalt, mercury, nickel, nead, selenium, tin, zinc.
  - Toxicity and mutagenicity: waterflea lethality (24 + 48 hours), urease enzyme assay, bacterial growth test and ames salmonella mutagenicity.
  - Virus: enteric virus by cytopathogenic effect and polymerase chain reaction (PCR), f-RNA phages.
- Other tests that are or have been part of a research programme or that are conducted at irregular intervals: bromate, oestrogens, oestrogenic activity, inflammatory activity, medical substances, organic pollution index (OPI).

9.10 QUALITY CONCERNS WITH THE PRESENT PROCESS CONFIGURATION

With the present operating procedures at the NGWRP, the following concerns need to be addressed:

- The build-up of TDS in the distribution system. If not addressed, urgently, the water quality of the NGWRP will soon exceed the stipulated TDS value of 1,000 mg/L. This is aggravated during periods of drought with greater production from boreholes with higher TDS values.
- The high concentrations of nitrates emanating from the wastewater treatment plant which cannot deal effectively with the high nitrate loads currently experienced. This will have to be addressed at the wastewater treatment plant by creating longer retention times in the anoxic basins.

Table 9.3 Cost comparison of existing and potential future water sources

Supply source	N$/m³	U$/m³
Conventional water supply sources		
Groundwater (borehole supplies)	1.75	0.25
Surface sources (Namwater supply)	5.75	0.82
*Estimated Tsumeb Karst aquifer supply (unit reference value)	22.50	3.21
*Estimated Okavango River supply (unit reference value)	188.00	26.86
Unconventional water supply sources		
New Goreangab water reclamation plant supply	7.35	1.05
*Estimated artificial recharge supply (under construction)	10.60	1.51
Irrigation water production costs (landscaping)	3.65	0.52

Notes: * Refers to new augmentation schemes. The exchange rate was accepted as N$7.00 = U$ 1.00 for 2007

- During droughts, DOC concentrations are higher due to the deterioration of the water quality from surface sources. An investigation is underway to address this at the Von Bach Water Treatment Plant that supplies water with relatively high DOC values from surface sources.

9.11 COST CONSIDERATIONS

The cost of water produced by the NewGWRP is compared to other supply sources in Table 9.3. The costs for the new supply augmentation options were calculated according to the Unit Reference Value, based on the estimated capital investment, as well as the estimated costs, based on 2007 cost estimates with an 8% return on investment. All the alternative supply augmentation options are based on an accepted level of security of supply. This design goal implies a 1 in 100 year risk that shortfall magnitudes will not exceed 13% during the 18 year planning period for the option.

9.12 PUBLIC ACCEPTANCE OF DIRECT POTABLE REUSE

On 24 November 1968, the *Sunday Tribune* (Johannesburg, South Africa) carried a headline: '**WINDHOEK DRINKS SEWAGE WATER, a purified world first**'. The City Engineer of Johannesburg made the following statement in 1974: 'It had been predicted that by the year 1985 we would all be drinking sewage and by the turn of the century there would not be enough to go around'. In practice, this prediction has not been realized with respect to direct water reuse, or was it realized, in certain places without the consent of the consumers through indirect reuse of water and even with conventional water treatment processes in place?

In the *American Waterworks Association Journal* of January 1968, Dwight F. Metzler and Heinz B. Russelmann, reported on how the City of Chanute, Kansas, survived the severe drought of 1956, on water from the virtually dry Neosho River. They reported:

> The 4,000-sq mi drainage area yielded only the wastes of upstream communities. Much of the river flow past Chanute during the drought was made up of sewage plant effluent, and the city had been using this water for several months without known ill effects.

Windhoek is still the only example of direct potable reuse. This begs the question, why would anyone want to start drinking 'sewage'. It can be questioned why the people of Windhoek have come to accept and are indeed proud of still being the only place in the world where direct potable reuse is practised on a commercial scale. In the case of Namibia, and Windhoek in particular, severe water shortages occur almost every five years. The main reason for instituting water reclamation in 1968 was the fact that there was no alternative supply available.

Perhaps the most important cornerstone of potable reclamation is public acceptance and trust in the quality of this reclaimed water. To maintain public confidence, water quality at the Goreangab Treatment Plant is monitored on an ongoing basis through on-line instrumentation as well as through collection of composite water samples after every process step. In the event of any quality problems (specified target values that are not met), the plant goes into recycle mode and water is not delivered. The final product water is also continuously sampled by way of online instrumentation and composite sampling and is analyzed for the full range of currently recognized contamination parameters, including pathogens, such as viruses, *Giardia* and *Cryptosporidium*.

Contrary to practices with indirect reuse, the history of the feed water to Goreangab is specifically recognized as treated sewage and the treatment process is designed to cope with just that. In that sense, the very aspect of this water that initially made people reluctant to use it, is in fact the aspect that leads to its being adequately treated to ensure its quality.

Over the years, the Goreangab Water Reclamation Plant (GWRP) has been widely publicized locally and internationally. Reclaimed water has become a fact of life in Windhoek, and the average citizen drinks water from the tap on a daily basis. Very few if any water quality complaints have ever been linked to poor quality from the GWRP.

Since 1995, the City of Windhoek has embarked on an active campaign to inform the public about water-saving measures. This has included inviting schools and various public target groups including members of the diplomatic corps to visit the wastewater treatment plant, the laboratory facilities and the reclamation plant. Water pollution, water saving and environmental protection issues have also been introduced into the school curricula at primary and high school levels.

In 1992, a technical committee of experts in the different disciplines of water and wastewater treatment and health aspects was appointed to provide advice during the planning and design stages of the reclamation plant.

In its own news publication, the City of Windhoek makes sure that these facts are well publicized and that the citizens are informed. In eight years, no single consumer has complained about the fact that reclaimed water was supplied. Enquiries received centred more around the possibility of contracting HIV Aids through the medium of reclaimed water. With the aid of science and the available knowledge of the virus, it has always been possible to deal satisfactorily with such enquiries. During 1996, a local newspaper enquired about the possibility of medicinal substances being unintentionally distributed to consumers, probably because at the time this subject was receiving attention in Europe. The City of Windhoek responded by having a European laboratory analyze samples from the treatment train for certain groups of medical substances. The results confirmed that through the maintaining of sound process protocols, the reclamation plant successfully removed all the substances in these groups.

The citizens of Windhoek have, over time, become used to the idea that potable reuse is included in the water provision process. In fact, there is a degree of pride that their city, in many respects, leads the world in direct water reclamation.

9.13 NEW RESEARCH AND DEVELOPMENT OPTIONS

9.13.1 Process-related refinements

From a process optimization point of view, various projects have been identified and are currently in progress to enhance process efficiency and to reduce operational cost. The increased sensitivity of the ultra-filtration membranes to DOC and the fouling phenomena that have been experienced have resulted in the continuous evaluation of the existing chemical cleaning regime. This includes using various blends of caustic soda and sodium hypochlorite or the respective singular use as dictated by the quality of the feed water. Effective wastewater treatment is important to provide a good quality effluent for reclamation, and continued upgrading of the process is required. In future, the use of membrane bio-reactors will be tested to modify the current wastewater treatment plant to improve the effluent water quality, especially through the biological reduction of organic carbons.

9.13.2 Quality control

The microbiology and virus-monitoring programme has been changed to include critical control points according to the Hazard Analysis and Critical Control Points (HACCP) principle, where certain relevant parameters are tested on a more frequent basis to establish direct relationships between different microbiological indicators and virus detected by means of the Polymerase Chain Reaction method. This will determine the risk and removal efficiency. Further developments are the continuous sampling for the concentration of virus on a suitable medium, and the continuous sampling and concentration of parasites on a filter. This has also proved to be a good visual indicator of certain treatment plant mishaps.

The HACCP has also been introduced in the monitoring of the distribution system. This has helped considerably to improve the water quality in the distribution system.

9.13.3 Health

Important issues related to health, especially with the use of ozone, are bromate and THM formation. This research is nearing completion and early indications are that the multiple barrier approach reduces the risk of THM in the final water to the extent that European Water Quality Standards are consistently met. The determination of bromates is part of a current study. Research continues on the prevalence and occurrence in the process water of endocrine disruptive substances as identified and prioritized by the Global Water Research Coalition. The fact that Windhoek does not have a significant industrial sector and that industrial wastewater is separated from domestic wastewater and does not reach the GWRP, means that the components identified for observation are minimized.

Between 2001 and 2003, oestrogenic screening tests and inflammatory activity tests, developed during a research project at a South African University, were conducted on

Windhoek's supply system. Findings proved that oestrogenic activity was removed during reclamation. Inflammatory screening tests suggested a seasonal pattern, which is higher during the rainy season, but the NGWRP is effective in reducing this hazard. Indications are that if very high inflammatory activity is present in the raw water, breakthrough could occur in the final water. This work will be followed by direct determination of oestrogenic substances. Further research on additional screening tests is needed, as this can be used as a cheaper indicator test than testing directly for hormones.

Research on the quantitative characterization of organic substances in the raw water sources and process water, the removal by each process and the use of the information for process optimization was initiated recently.

9.14 CONCLUSION

The 'urban water cycle' was indeed closed in Windhoek through:

- direct water reclamation for potable reuse in 1968
- installation of a dual pipe system to provide irrigation water for landscaping from purified effluent
- and, more recently, the artificial recharge of the Windhoek Aquifer with blended water from surface sources and reclaimed water.

The experience gained in Windhoek using all options to reuse water is important in view of water shortages elsewhere in the world. The reuse of water can increase the total quantity of available water substantially, at an affordable price without resorting to additional, more costly natural water sources.

Water reuse is by far the cheapest option to provide additional supplies. In the case of Windhoek, reclamation for potable reuse is approximately 30% of the cost of additional augmentation schemes from remote conventional supplies.

Potable reuse (reclamation) should not be treated as a technical process in isolation, but should be part of a holistic approach to include the water supply strategy, collection of wastewater, prevention of pollution of wastewater, optimal treatment of wastewater through biological processes, public participation and water demand management. With the development of new technology, as well as the refinement of test methods for water quality, the process is dynamic and should be scrutinized and upgraded to ensure the provision of safe potable water on an ongoing basis.

The NGWRP by far exceeds the performance of the previous reclamation plant and after operating for one-and-a-half years it is good to know that the targets that were set for each treatment unit can be reached consistently. The sage words of Dr. Lucas van Vuuren, one of the pioneers of the Windhoek reclamation system, have successfully withstood the test of the past 40 years:

'Water should be judged by its quality; not its history'

REFERENCES

Central Area Joint Venture Consultants. 2004. *Feasibility Study on Water Augmentation to the Central Area of Namibia*. Report No. NWPC-IP-Tsumeb97-1, Volume 6, Namibia Water Corporation Ltd.

Clayton, A.J. 2005. Personal Communication. July.

Die SuidWester. 1968. Windhoek Daily Newspaper, 22 November.

Haarhoff, J. and Van der Merwe, B.F. 1996. Twenty-five Years of Wastewater Reclamation in Windhoek, Namibia, Water Science and Technology, Vol. 33, No. 10–11.

Haarhoff, J., Van der Walt, C.J. and Van der Merwe, B.F. 1998. Process Design Considerations for Windhoek Water Reclamation Plant. Paper presented at the biannual Conference of the Water Institute of Southern Africa, Cape Town May.

Metzler, F.D. and Russelmann, B.H. 1968. Wastewater Reclamation as a Water Resource, American Waterworks Association Journal, January.

Namibian Economic Research Unit (NEPRU). 1996. Projections and Opinions on Economic and Business Prospects in Windhoek.

Namibian Guidelines. 1991. Guidelines for the Evaluation of Drinking Water for Human Consumption with Regard to Chemical, Physical and Bacteriological Quality. Department of Water Affairs, Republic of Namibia.

Rand Water. 1994. Potable Water Quality Guidelines. Johannesburg, Republic of South Africa, Scientific Services, Rand Water.

Stander, G.J. and Clayton, A.J. 1970. Planning and Construction of Waste Water Reclamation Schemes as an Integral Part of Water Supply. Paper presented to the 1970 Conference of the Institute of Water Pollution Control, Cape Town March 1970.

US EPA. 1996. National Drinking Water Standards and Health Advisories. United States Environmental Protection Agency, Washington DC.

Water Research Commission, Republic of South Africa. 1982. Guide for the Planning, Design and Implementation of a Water Reclamation Scheme.

World Health Organization. 1993. Guidelines for Drinking Water Quality. World Health Organisation, Geneva, Switzerland.

Chapter 10

Reducing risk from wastewater use in urban farming – A case study of Accra, Ghana

Pay Drechsel[1], Liqa Raschid-Sally[1] and Robert Abaidoo[2]

[1]International Water Management Institute (IWMI), Africa Office, PMB CT 112, Accra, Ghana
[2]Kwame Nkrumah University of Science and Technology (KNUST), University Mail Box, Kumasi, Ghana

ABSTRACT: In West Africa, urbanization has outpaced the provision of sanitation infrastructure and services. Only a fraction of cities has sewer systems and most domestic wastewater ends in drains and streams. Vegetable farmers in and around the cities have problems finding (any) unpolluted surface water sources for irrigation. Consequently, most vegetables found on urban markets are highly contaminated. In the case of Accra, every day, about 200,000 urban dwellers consume street food containing a portion of raw vegetables from these sources. As wastewater treatment facilities are unreliable, non-treatment options for health risk reduction at other entry points, namely farms, markets and kitchens, are currently being explored in line with WHO's multiple barrier approach. This research addresses the constraints farmers face in adopting common recommendations, such as using different irrigation methods or changing types of crop; and examines a variety of other options for on-farm risk mitigation. The latter have a high potential to reduce helminth egg counts, while improved washing practices prior to vegetable preparation can reduce the bacterial contamination considerably. Initiatives to safeguard public health have to consider the right balance between the provision of incentives (market channels for safer produce, good price, rewards and recognition) and applying pressure or sanctions to comply with realistic hygiene standards especially in the post-harvest sector.

10.1 INTRODUCTION

Africa's rate of urbanization is estimated to be about 3.5% per annum, which is one of the highest in the world (UN Population Division, 2004). The projection is that by 2030 there will be 41 countries in Sub-Saharan Africa (SSA) with higher urban than rural populations (UN-Habitat, 2001). Urbanization is particularly visible in West Africa, where rural–urban migration is catalyzed by the emergence of a market economy and a relative concentration of investments in urban trading centres (Club du Sahel, 2000). The dynamic expansion of the sprawling conurbation of Lagos in Nigeria, from a population of only 75,000 in 1940 to a megacity with a population of more than 9 million presently, is the most noticeable example. This development poses a major challenge to emerging and expanding cities, not only in view of the provision of urban services, such as shelter, water, energy and sanitation, but also of urban food supply and food security.

As part of urban food supply, the sourcing of perishable crops, such as leafy vegetables, is the most difficult, given the lack of adequate storage and transport facilities. Irrigated urban farming focusing on these crops is the response in many African cities;

and inside major urban centres of West Africa, there may be up to 650 ha under irrigation and often several thousand if you consider the peri-urban fringe (Drechsel et al., 2006). Preferred locations for irrigated farming are open spaces (typically these are unused government land, clan or 'stool' lands or undeveloped private plots) selected because of their proximity to urban water bodies, such as streams, which provide access to a low or no-cost source of water year round and especially in the dry season when it is very profitable to produce these crops (Gaye and Niang, 2002; Faruqui et al., 2004; Obuobie et al., 2006). However, the inadequate sanitation and waste disposal practices common to most West African cities, where less than 10% of the urban households are connected to a piped sewerage and (usually dysfunctional) treatment system (Keraita et al., 2003b; Tandia, 2002) have resulted in pollution of these water sources. Thus, in many instances, in the West African urban context, the type of water used for agriculture is some form of diluted, partially treated or untreated urban wastewater, which includes black water, grey water and stormwater (Drechsel et al., 2006). The use of safe pipe-borne water for irrigation is rare, due to price and/or common supply shortages (Tandia, 2002; Moustier and Fall, 2004). Thus, in the sub-region, in cities such as Dakar, Accra, Tamale, Kano and parts of Ouagadougou, the supply of vegetables largely depends on the availability of once-used water. As the dominant method of watering the crops is overhead (spray) irrigation with watering cans, health risks to farmers and consumers are high (Amoah et al., 2005; Faruqui et al., 2004; Endamana et al., 2003; Hussain et al., 2002). However, irrigated urban agriculture provides livelihoods and has a particular function for urban food supply (Drechsel et al., 2006). Thus, a balanced approach is needed that accommodates the two dimensions and is locally sustainable. The important role that non-treatment options play, to safeguard public health, whilst being locally sustainable, has been recognized by the new WHO (2006) guidelines. Some of these options are currently being tested in Ghana and are discussed in this chapter.

10.2 THE CASE OF ACCRA

Ghana's capital city Accra is located by the Atlantic Ocean, covers an area of about 240 km² and had, in 2000, an estimated population of 1.66 million, with a population growth rate of about 3.4% annually. Beyond the outdated city boundary, the adjoining districts of Ga and Tema are growing at a much faster rate of 6.4 and 9.2%, respectively, leading to urban sprawl and uncontrolled physical expansion beyond the municipal boundary of the Accra Metropolitan Area (AMA). These two districts, together with AMA, form the functional boundaries of the city (known as Mega Accra) covering an area of 1,260 km² and with a population of 2.7 million. With 'Greater Accra' we finally refer to one of Ghana's ten administrative regions, which includes more land than Mega Accra and represents 2.9 million inhabitants or 15.8% of Ghana's population on only 1.36% of its land area (Twum-Baah, 2002).

In AMA, except for the few high and medium class residential areas, the bulk of the population (about 60%) lives in informal settlements or slums in the centre of the city. Large parts of the middle and upper classes have moved to the periphery, occupying the many residential satellite districts that have sprung up. In addition to the physical expansion mentioned above, there has been increased crowding in existing inner-city residential areas. This has resulted in higher occupancy rates in local housing units and infilling of vacant plots, both with significant implications for the provision of sanitation services.

10.2.1 Urban water use and wastewater management

Water use: The total water supplied to Accra is 11.5 million cubic metres per month (4.4 m³/s) – inclusive of 30% leakage losses – by the Ghana Water Company Ltd (GWCL). Water supply coverage to the city is said to be 80% but this does not imply a house connection nor does it reflect the irregularity of actual water flow within the distribution system. In reality, only 45% of the population has a household or at best a yard connection and this category includes the urban rich. The majority who live in the low-income settlements depends on water vendors for their daily needs. Water supply to the city is intermittent and the estimated consumption pattern of an average 55 L per capita and day is influenced by this factor (London Economics, 1999; GWCL, 2006).

Disposal methods for different types of wastewaters: Sanitation service provision in most West African countries is biased towards their larger cities. In Ghana, most of the functional sewer networks, and more than half of all wastewater treatment plants are found in and around Accra collecting and treating wastewater from only about 5–7% of Accra's population. On-site systems (septic tanks) are dominant.

The national census of 2000 (GSS, 2002) showed that in Greater Accra, about 22% of all households have their own or a shared flush toilet (emptying into septic tanks and sewers in the covered areas), 30% have their own or shared (ventilated) pit or pan latrine, 27% have to rely on public toilets, 9% on friends in other houses and 12% on open defecation (bush, drains and beach).

Septic tanks require frequent emptying either due to poor design or the lack of a drainage field, due to the congested conditions in many parts of Accra. Therefore, except for the flush sewered systems, all other forms of sanitation, including many of the septic tanks, have to be emptied and the fecal sludge disposed of. In Accra, there are two facilities designated for this purpose with fecal sludge treatment ponds, but neither is functioning properly, thus, nearly all sludge is currently dumped into the ocean through a 'beach outfall'.

More than 90% of all grey water in Accra is disposed of via stormwater drains, which are discharged into the natural streams draining the city. The natural drainage system here includes streams, ponds and lagoons (e.g., Songo, Korle and Kpeshie). Accra's major wetland, the Korle Lagoon, receives 'fresh' water through the Odaw stream, which is the main urban stormwater drain with a catchment area covering more than 60% of the city. Due to Accra's limited sanitation and solid waste disposal infrastructure, the Odaw and the lagoon receive a vast amount of wastewater as well as solid waste (Biney, 1998; Boadi and Kuitunen, 2002). Since 2007, the Odaw inflow into the lagoon has been diverted, after de-siltation, directly to the sea through a 1-km-long ocean outfall. The lagoon is now only subject to the tidal system without sweet water inflow as long as the diversion is maintained operational.

Most industrial activities in Ghana take place along the coast where wastewater is disposed of directly into the ocean, thus, urban wastewater used in farming mostly derives from domestic sources.

10.2.2 Irrigated urban vegetable farming

In Accra, irrigated urban vegetable production takes place on over seven large sites and in many other small scattered locations. IWMI (unpublished) estimated that, on average, there are about 100 ha under vegetable irrigation in the dry season. Some of the sites have been in use for more than 50 years (Anyane, 1963). There are about 800–1,000 vegetable farmers of whom 60% produce 'exotic' vegetables and 40% indigenous local or traditional ones. Typical exotic crops cultivated are lettuce, cabbage, spring onions, sweet peppers, cauliflower, radish and spinach, while the more traditional crops are tomatoes, okra, garden eggs (eggplant) and hot pepper. In addition, a number of indigenous leafy vegetables, such as ayoyo and alefu, and exotic herbs, such as parsley, coriander, fennel and basil, are also grown (Gbireh, 1999; Obuobie et al., 2006). Plot sizes under cultivation range between 0.02 and 0.05 ha per farmer within the city, and 2.0 ha (maximum) in peri-urban areas. Clean piped water is used only on one site in Accra, by some of the farmers. On all other sites, the main sources of irrigation water are highly polluted streams, that have been often transformed into drains due to wastewater inflows. The pollution levels in these drains vary by location and by season, ranging from raw wastewater to polluted stream water.

The most common method of urban irrigation in the sub-region is manual water fetching and application using a watering can (Keraita et al., 2003a). Perception studies have shown that farmers are well aware of different water qualities they use, but rarely complain, partly because their livelihoods depend on it (Obuobie et al., 2006).

10.2.3 Irrigation water quality

Chemical and bacteriological water quality of irrigation water sources can vary widely, both spatially and temporally, in and around the cities. Water quality can show diurnal, daily and seasonal variations, with clear upstream–downstream gradients around urban centres (Keraita et al., 2003b; Obuobie et al., 2006) and activities. Faruqui et al. (2004) mentioned as an example, the effects of 'laundry days' and 'Friday prayer' on stream water quality. Studies from Ghana, Burkina Faso, Senegal and Cameroon confirm predominantly bacteriological contamination ranging in most cases between 10^3 and 10^6 MPN fecal coliforms per 100 mL (Cissé, 1993; Niang et al., 2002; Armar-Klemesu et al., 1998; Cornish and Lawrence, 2001; Endamana et al., 2003; Faruqui et al., 2004; Sonou, 2001; Mensah et al., 2001; Keraita and Drechsel, 2004), which exceeds common irrigation standards. There are only a few cases which show evidence of water pollution from heavy metals of industrial origin (Cornish et al., 1999; Faruqui et al., 2004; McGregor et al., 2002) but common irrigation standards at agricultural sites are seldom exceeded. Most industrialization within cities is from light to medium industries, which are less polluting, but there are exceptions, such as the examples of Kano (Nigeria) or Ouagadougou (Burkina Faso), which are known for their tanneries and chromium contamination (Binns et al., 2003; Bosshart, 1997), or Kumasi (Ghana) with its Suame suburb that specializes in car repairs, resulting in heavy oil and grease contamination. So in many African cities where wastewater is used, the health risk from pathogen contamination is more immediate.

In addition to pathogens, urban wastewater also contains nutrients that influence water quality. Like the pathogen load, the fertilizer value of wastewater varies with the degree of dilution. It is sometimes debated whether farmers use 'wastewater' more for its water value or for its nutrients. Our surveys show that even in the humid part of West Africa, farmers are looking foremost for 'water' (to irrigate their leafy vegetables every day all year round) and only in selected cases are they consciously considering its (additional) nutrient value.

Amoah et al. (2005, 2006) sampled irrigation water (specifically for coliform levels and helminth eggs) in the main urban farming sites in Accra and Kumasi. In Accra, fecal coliform levels ranged from 10^1 to 10^7 per 100 ml (Table 10.1), the lower values being recorded in Dzorwulu where farmers use clean piped water stored in shallow wells. Farming sites in Korle-bu, La and Marine Drive, where farmers use water from urban drains for irrigation, recorded the highest values. Previous studies carried out in Accra (Armar-Klemesu et al., 1998; Sonou, 2001; Zakariah et al., 1998) also showed that there are hardly any unpolluted water sources available for irrigation. The worst case is the Odaw river/Korle lagoon, which has a highly populated drainage basin where the water is also used on smaller irrigation sites. Its BOD load has been estimated at 132,000 kg/day (Biney, 1998).

10.2.4 Quality of vegetables in urban markets in Accra

The quality of vegetables in urban markets and their marketing channels were investigated for Accra, using lettuce as an example of a leafy vegetable that is grown using wastewater and is consumed raw by urban households. This is the type of vegetable most susceptible to contamination particularly when spray irrigation is practised, given the extensive leafy surface areas exposed.

Figure 10.1 shows the sources of lettuce, and channels of marketing, from source to consumer in Accra. As can be seen, 98% of lettuce arriving in the city goes to places

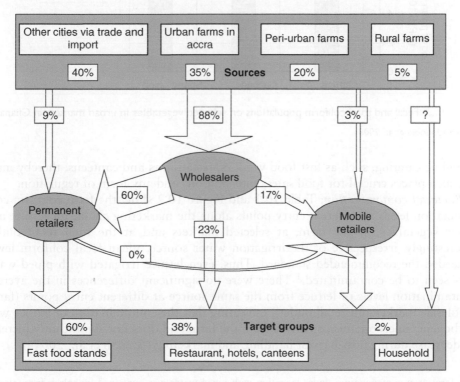

Figure 10.1 **Flow chart of lettuce distribution in Accra**

Source: Henseler et al., 2005

Table 10.1 Mean fecal coliform contamination levels of lettuce at different entry points along the production – consumption pathway of lettuce (Amoah *et al.*, 2007a, modified)

City	Irrigation water source	Statistics	Log fecal coliform levels (MPN* 100 g⁻¹ wet weight)		
			Farm	Wholesale market	Retail
Accra	Drain	Range	3.40–6.00	3.00–6.80	3.00–6.50
	(n = 216)	Geometric mean	4.25	4.24	4.48
	Stream	Range	3.20–5.70	3.10–5.90	3.20–5.50
	(n = 216)	Geometric mean	4.22	4.29	4.37
	Piped water	Range	2.90–4.70	2.90–4.80	2.80–4.50
	(n = 216)	Geometric mean	3.44	3.46	3.32

*MPN, Most Probable Number

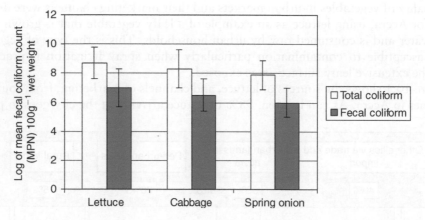

Figure 10.2 Fecal and total coliform populations on selected vegetables in urban markets in Ghana
Source: Obuobie et al., 2006

of public catering, such as fast food vendors, restaurants and canteens, thereby making such places crucial for food safety interventions and any form of regulation.

Bacterial contamination: Table 10.1 and Figure 10.2 show the fecal coliform contamination levels at different entry points along the marketing chain, for lettuce and other vegetables, at the farm, at selected markets and at the final retail outlet. Interestingly, irrespective of the irrigation water source, mean fecal coliform levels exceeded the recommended standard. Thus, even lettuce irrigated with piped water was seen to be contaminated.[1] There were no significant differences in the average contamination levels of lettuce from the same source at different entry points (farm, wholesale market and retail outlet), indicating that the contamination of lettuce with pathogenic microorganisms derives from the farm and does not significantly increase or decrease during post-harvest handling and marketing (Amoah et al., 2007a).

[1]On-farm crop contamination under irrigation with piped water was explained by splash from already contaminated soil (1×10^4 fecal coliform levels per $10\,g^{-1}$ soil in the upper 5 cm) and the 'broadcast' application of inadequately composted poultry manure (Amoah et al., 2005; Drechsel et al., 2000).

Table 10.2 Mean helminth egg populations on selected vegetables in
Ghanaian markets (Obuobie et al., 2006)

Vegetable	No. of samples	Average number of eggs/100 g
Lettuce	27	1.14 a
Cabbage	32	0.42 a
Spring onion	26	2.74 b

The mean difference is significant between *a* and *b* at the 0.5% level.

The same was observed for pipe-water irrigated lettuce, implying that post-harvest contamination is not simply hidden (by high initial contamination levels) but indeed is not significant in our case. The situation may be different as observed for example in Pakistan (Ensink et al., 2007).

A larger impact of post-harvest contamination was expected as environmental sanitation and hygiene conditions, such as availability of sanitation infrastructure at the marketplace, washing practices, and clean display and handling of food, appear to be deficient in Ghana. Only 31% of the markets in Accra are connected to a drainage system, 26% have toilet facilities and 34% are connected to pipe-borne water, as shown by a survey some years ago (Nyanteng, 1998).

Compared to cabbage and spring onions, lettuce showed the highest level of contamination both for fecal and total coliforms (Figure 10.2) with contamination levels ranging between 10^6 and 10^{11} for total coliforms per 100 g lettuce wet weight and between 10^3 and 10^9 for fecal coliforms. These contamination levels are similar to those found in earlier studies on food contamination conducted in Accra (Akpedonu, 1997; Abdul-Raouf et al., 1993). The larger contamination surface offered by lettuce leaves might explain the differences.

Helminth eggs: Table 10.2 shows helminth populations from the vegetables analyzed. The eggs identified include *Ascaris lumbricoides*, hookworm, *Trichostrongylus*, *Schistosoma heamatobium*, *Trichuris trichiura*, *Strongyloides stercoralis* and *Nuaplius larvae*. *S. stercoralis* had a high occurrence and was observed in all samples. *Ascaris lumbricoides* was the most predominant among all the other organisms and was observed in 85% of the contaminated vegetables. This could be attributed to the high level of persistence of *A. lumbricoides*, hence high survival time. The presence of helminths, particularly *A. lumbricoides*, on the vegetables could pose a serious problem because of their low infective dose and low host immunity. The potential threat, however, would depend on egg viability, which, though not yet systematically studied, appeared in preliminary sampling to be low. Cabbage had the lowest counts of helminth eggs and spring onions the highest (p < 0.05). The high helminth egg count in spring onion is surprising considering the low surface area of the leaves compared with lettuce and cabbage.

10.2.5 Numbers of consumers at risk

If we take lettuce as a high-risk vegetable for the reasons explained previously, the largest group of buyers of lettuce are fast food sellers (60% of the total lettuce sold in Accra) and restaurants/hotels/canteens (38% of the lettuce) as seen from Figure 10.1.

The lettuce is served as an ingredient in the salad accompanying common 'urban' fast food meals. First estimations of the total number of urban dwellers eating fast food with lettuce daily, in Accra's streets were 150,000 to 200,000 people (Obuobie et al., 2006). If we include the people eating in canteens and restaurants, we can add another 80,000 consumers. All these people benefit from consuming vegetables from urban agriculture, but at the same time this large group represents the part of Accra's population at risk from food contamination (Amoah et al., 2007a).

This type of fast food is sold throughout the city and is not restricted to a particular type or income group of consumer. Customers are mostly male (70%) and eat this type of food 3–4 times a week. The typical consumer is therefore an urban dweller, living in low (50%) and middle income (38%) neighbourhoods. They are mostly workers in the small-scale private sector, but businessmen and students are also customers. Often, fast food with a lettuce supplement is the cheapest type of food available in town, thus, explaining its popularity, with the poor and their dependants, as stated by Essamuah and Tonah (2004).

In Ghana, food caterers commonly wash their vegetables with tap-water, salt or vinegar, but their washing methods proved to be inadequate for sanitizing the lettuce, as reconfirmed in our surveys and laboratory tests (Amoah et al., 2007b); resulting in contaminated salad sold in Accra streets (Mensah et al., 2002). A perception study conducted in Kumasi (Olsen, 2006) shows however, that street food vendors are convinced that they have eliminated risks for consumers by their practices of washing lettuce. For health campaigns to be effective, greater awareness on invisible risk factors, such as pathogens, is needed.

10.2.6 Risk assessment to farmers and consumers

Epidemiological surveys for health risk assessment for farmers are ongoing in the city of Tamale, Ghana, where farmers use wastewater for vegetable cultivation and fecal sludge for cereals. Stool samples are being tested for helminth ova and fecal coliforms. The large number of confounding factors makes comprehensive assessment of the risks for consumers very difficult. However, QMRA estimates are available for farmers and consumers, based on the pathogen density in wastewater and irrigated vegetables. The annual risks of infections associated with diluted wastewater reuse, for farmers and consumers, were 10^{-3} to 10^{-4} for Ascaris, and 10^{-2} for Rotavirus, compared to the WHO tolerable health risk of 10^{-4} per person per year for the two pathogens. For fecal sludge reuse, farmers' risk of Ascaris infection ranged from 10^{-2} to 10^{-4} while it was negligible for rotavirus infections due to fecal sludge treatment (Seidu et al., 2008a).

Amoah et al. (2007a) and Seidu et al. (2008b) identified that the farm where irrigation takes place is the major source of pathogens, but based on the occurrence of extreme contamination values on lettuce samples, deduced that the highest risk of infections as calculated by QMRA is in wholesale or retail markets. The authors, however, stressed that the estimated risks of infection need to be validated against follow-up data obtained from epidemiological investigations coupled with studies on the effect of different health risk barriers, such as hand washing. Indeed, empirical results from health perception studies among wastewater and pipe water irrigating farmers in Accra and Kumasi revealed no noteworthy health burden (Obuobie et al., 2006; Gbewonyo, 2007), which does not correspond to the findings of the QMRA risk assessment studies on farmers.

10.3 RISK REDUCTION MEASURES

In a country like Ghana, where wastewater treatment is inadequate, complementary non-treatment approaches as recommended by WHO (2006) are required. Two projects of the Consultative Group on International Agricultural Research (CGIAR) Challenge Programme on Water and Food (CPWF) were started in 2005 to study corresponding interventions at the farm, market and kitchen level. In late 2006, WHO, FAO and IDRC recognized this work and funded two follow-up projects testing the implementation potential of the new WHO guidelines in Kumasi and Tamale, both in Ghana.

The studies in Ghana focus on the following non-treatment options as described in the new WHO guidelines and in Drechsel et al. (2002):

10.3.1 Explore alternative farmland, tenure security and safer water sources

Authorities could reduce farmers' and consumers' health risks easily if they have the opportunity to provide the concerned farmers with safer water sources in existing or alternative locations. In Ghana, the Ministry of Food and Agriculture started a related programme of borehole drilling on urban farming sites. Groundwater in Accra is, unfortunately, too saline on many sites for this option to succeed, but it may be possible in other cities. An associated issue is the provision of tenure security which is usually absent in urban farming on open spaces: where farmers feel secure about land they are more likely to invest labour and capital in irrigation infrastructure to reduce pathogen contamination. Examples are investments in drip or furrow irrigation (which reduces crop–water contact), or in 'stream-barriers' or on-farm storage reservoirs which could provide basic treatment by way of sedimentation of worm eggs and pathogen die-off, while at the same time balancing irrigation water supply with demand.

BOX 10.1 The example of Cotonou

Moving farmers to alternative farming sites with safer water sources is a possibility, provided farmers are willing to move. An interesting case where this was initiated has been reported from Benin. Following a multi-stakeholder process facilitated by the Institut Africain de Gestion Urbaine (IAGU), the cities of Cotonou and Seme-Kpodji, supported by the Ministries of Agriculture, Interior, Finances, and the State Ministers Council, agreed to allocate about 400 ha of farmland for use by urban and peri-urban farmers. The site is located about 20 km from Cotonou towards Porto-Novo and has shallow non-saline groundwater, which can easily be lifted by treadle pump for all-season irrigation. About 1,000 farmers declared their interest to move to this site and more than 100 have done so, to date. However, there are constraints, such as the required subsidies for building houses and other necessary infrastructure. This initiative had a triple objective of addressing tenure insecurity, access to safe water and supporting the need for providing perishable vegetables to cities. However, it also relocates farming out of the cities, which might have been a fourth objective. The success of this initiative will depend, among others, on the impact of increased transport costs on farmers' profits.

Source: Drechsel et al., 2006

10.3.2 Promote safer irrigation methods

In many parts of the world, and Sub-Saharan Africa in particular, most urban farmers use watering cans, which serve the dual purposes of water carrying and application, and require only a small financial outlay. However, this method increases crop contamination, especially of leafy vegetables through spraying of droplets on the leaves. Irrigation techniques which apply water at a lower height or to the root zone (such as a wastewater adapted drip irrigation kit) are much safer and as an additional benefit, might use less water and labour. Reducing crop contamination by ceasing irrigation a few days before harvest is another common recommendation, which, however, does not work for every crop and under every climate. Box 10.2 shows some examples of on-farm tested interventions in Ghana.

BOX 10.2 Alternative techniques to reduce crop contamination

Ongoing research by the University of Kumasi (KNUST) and IWMI shows that simple changes in pond construction, water collection and application techniques; can significantly reduce the number of helminth eggs. Avoiding the use of fresh poultry manure and allowing it to mature and stabilize over a dormant period, is a further means of effectively reducing crop contamination along this contamination pathway. Participatory on-farm research targeting farmers' perception of such innovations and related changes in labour allocation is critical for understanding which 'best practices' with their corresponding risk reduction rates have the highest adoption potential in a given context (Keraita et al., 2007c, d). For the institutionalization of safer practices, the extension services have to be fully involved in any project and simultaneously trained.

Cessation of irrigation before harvesting, as part of a multiple-barrier approach, is a very simple and effective measure. In Kumasi, Ghana, on average, 0.65 log units for indicator fecal coliforms and 0.4 for helminth eggs were removed on $100\,g^{-1}$ of fresh weight lettuce, for each non-irrigated day in the dry season. However, due to the hot and humid weather, a corresponding daily loss of 1.4 tons/ha of fresh weight of lettuce was recorded. Four to five days of cessation before harvesting might be the best option, but resulted in a yield loss of about 25% which is significant for the farmers. A compromise with higher adoption potential would be a cessation of two days which corresponds to a yield reduction of 10%. This, however, still requires the consent of farmers, and traders, who have to agree on the days for collecting the vegetables at the farm-gate, and adhere to these (Keraita et al., 2007a). This shows that under the given climate, cessation has to be complemented by other methods for health risk reduction for a more significant cumulative impact. However, in cooler climates, such as Addis Ababa, where lettuce is watered only thrice a week and not twice a day, as in Ghana, cessation will have a higher potential for risk reduction with less yield loss.

Field trials were also conducted to compare simple and low-cost irrigation methods, such as watering cans, low-cost drip kits and furrow irrigation, to evaluate their effectiveness in terms of both crop productivity and contamination. The drip kits recorded the lowest levels of contamination, with an average fecal coliform reduction of 4 log units per $100\,g$ of lettuce wet weight compared to watering cans. However, the disadvantages were clogging of pipes, lower cropping densities and interference with other routine activities, such as weeding. Slightly changing the preferred practice of using watering cans, by capping the spout and irrigating from less than 0.5 m high, showed an average reduction of 2.5 log units fecal coliforms and 2.3 helminth eggs per $100\,g^{-1}$ lettuce compared to the usual practice of uncapped spouts lifted high to access all parts of the beds (a low height reduces splash and lettuce contamination via soil particles).

Source: Keraita et al., 2007b

10.3.3 Influence the choice of crops grown

When irrigation projects are centrally managed or where there is an appropriate market demand for alternative crops, it is possible to introduce and enforce crop restrictions to ensure that wastewater is not used to irrigate high-risk crops, such as leafy vegetables that are eaten raw. Research in Mexico, Chile and Peru has shown that this is only successful when the crops allowed under the restrictions are of similar profitability for the farmer (Danso and Drechsel, 2003; Drechsel et al., 2002; Cornish and Lawrence, 2001; Gockowski et al., 2003).

It is doubtful that such an approach would be successful in the context of Ghana, where exotic leafy vegetables are highly profitable, and where in addition, restrictions would be difficult to enforce in the informal irrigation sector. A certification programme for 'safer crops' with related marketing channels and awards for innovative farmers are examples of possible incentives to effect this change. These efforts have to be strongly supported by awareness campaigns and the media, to increase market demand for safer crops. Although it is unlikely that all farmers will change their practices, some will, which offers the consumer alternatives. However, poor urban dwellers might not have the choice.

10.3.4 Avoid post-harvest contamination

Health risk reduction measures should not only focus on improving irrigation water quality and reducing crop contamination at the farm level. Transport, markets and restaurants are potential points of additional contamination. Even in a case such as Ghana, where post-harvest contamination appeared to be low (see above), markets and especially kitchens are critical places for vegetable decontamination and risk reduction as described below. Another example is the common practice in West Africa where market women wash the crops they have just harvested themselves, with the local (contaminated) irrigation water, to get rid of soil and dirt. This has to be avoided by creating awareness of health risks and providing alternative water sources for washing purposes.

10.3.5 Assist post-harvest decontamination

An important option for complementary risk reduction is vegetable washing and disinfection at household level and at food outlets. This is, in fact, a common practice in the West African sub-region, but investigations showed that it is rarely done correctly. It does not require large financial outlays and has a high potential for large-scale risk reduction where pathogen contamination is likely, be it from wastewater irrigation or post-harvest handling. In West Africa, various washing methods and disinfectants are used, but the techniques often proved to be very ineffective (Amoah et al., 2007b). Ongoing research in Ghana tried to improve existing practices. Current recommendations target chlorine tablets, potassium permanganate and *Eau de Javel*[2] (Table 10.3). Concentrated vinegar solutions (one part vinegar in five parts water) proved to be

[2]This term is commonly used in French speaking countries, to indicate, liquid bleach used for household washing purposes including toilets.

Table 10.3 Impact of common and improved washing practices on fecal coliform levels (Amoah et al., 2007b; modified)

Method	Log reductions	Comments
Water in a bowl	1–1.4	• Increased contact time improves the efficacy of cold water considerably • Not very efficient compared to washing with other sanitizers • Not very effective for helminth eggs if washing has to be done in a bowl of potable water (because of accumulation and re-suspension) • Increasing the temperature does not significantly increase its efficacy
Running tap water	1–2.2	• Comparatively more effective than washing in a bowl, especially for helminth egg removal • Increased impact with increased contact time • Limited application potential due to absence of running water (taps) in poor households
Salt (NaCl) solution	1.4–2.1	• Salt solution is a better sanitizer compared to potable water at an appropriate concentration and 2 min contact time • Efficacy improves with increasing temperature and increasing concentration, however, high concentration have a deteriorating effect on the appearance of some crops, such as lettuce
Vinegar	1–>4.0	• Very effective at higher concentration but this may result in vegetables smelling of vinegar • To achieve a high efficacy at lower vinegar concentration, the contact time should be increased • Efficacy is improved even at low concentration if carried out at a higher temperature
Potassium permanganate	1.2–2.5	• More effective at higher concentrations (200 ppm) and also with increasing temperature (3 log units) and contact time • Higher concentration colours washed vegetables purple which requires more water for rinsing or may raise questions about negative health impacts
Washing detergent (OMO)™	1.6–2.3	• Significant reductions could be achieved. As OMO™ contains surfactants which could affect health, thorough rinsing is required • Perfumes used might affect consumer's acceptance • People's perception that soap is not to be consumed could affect its use
'Eau de Javel' (chlorine bleach)	2.1–3.1	• Effective but content and concentrations vary without proper labeling. Potential health risk if overdosed; but widely used in most Francophone West African countries • Effect of higher dosages on efficacy not tested in this study
Chlorine tablets	2.3–2.7	• Effective but not commonly used in most West African countries • Effect of higher concentrations on efficacy not tested in this study

effective but more costly, and can be adopted only by mid-class restaurants. Evaluating not only the efficiency of these methods under local conditions, but also their cost, and the perception of food caterers is crucial to understanding possible drivers of adoption and the upscaling potential.

10.3.6 Improve institutional coordination to develop integrated policies

In the context described above, it is clear that effective risk reduction requires collaboration between different ministries and institutions dealing with the concerned farmers, traders and food caterers. Case studies from around the world, including Ghana, show that sanitation, agricultural, environmental and health legislations and interventions are usually vested in various agencies. There is often an overlap of roles and responsibilities as well, leading to confusion. Furthermore urban and peri-urban agriculture, or informal irrigation in general, was, until recently, not officially recognized in Ghana's agriculture sector, thus, posing difficulty for effecting changes. A response to this is dialogue and action through multi-stakeholder platforms. These are vital to find mutually satisfactory solutions with a high potential for institutionalization, e.g., of the new WHO guidelines. Such a process is currently supported in Accra by the network of Resource Centers on Urban Agriculture and Food Security (RUAF) through IWMI, its focal point for Anglophone West Africa (Obuobie et al., 2006).

10.4 CONCLUSIONS

In the recent past many countries have started recognizing the value of irrigated urban and peri-urban agriculture. Ghana has also joined this trend, by integrating informal irrigation and irrigated urban and peri-urban agriculture into its designated[3] national irrigation policy, a process strongly supported by IWMI and FAO (Obuobie et al., 2006). The leading argument in support of this recognition is the vast extent of the informal irrigation sector in Ghana, compared to the formal one, its impact on livelihoods, and the cities' dependency on urban production of perishable vegetables. The policy, however, also addresses the concerns of the authorities about the use of polluted water for irrigation, and recommends applied research in risk reducing practices. This development reflects a subregional shift towards more progressive and accommodative approaches to address the common reality of water pollution and (peri)urban irrigation with polluted water than applied in the past (Drechsel et al., 2006).

Changes in irrigation methods, timing and types of crops are commonly suggested methods, but may not always be possible for the farmers (Drechsel et al., 2002). However, the trials reported here show that especially helminth counts can be reduced significantly with simple on-farm measures. This can be further improved by good washing practices prior to vegetable preparation (at home or in restaurants and food stalls). Typical fecal coliform concentration as analyzed in Ghana, on irrigated lettuce (10^6 fecal coliform counts per 100 g fresh weight) could be reduced with the tested

---

[3]Awaiting cabinet approval (as of February 2008).

interventions at the following levels:

- farm level by 2–4 log10 units
- market level by at least 1 log10 unit
- kitchen level by 2–3 log10 units.

All options tested require both training and behaviour change. To increase the sustainability of any change it is important to analyze, as early as possible, local factors and opportunities that might support or constrain their adoption. A certification programme for 'safer crops' combined with exclusive marketing channels, or awards for innovative farmers might be possible incentives for change. Farmers with safe produce could then be linked directly to large consumer establishments, such as hotels or supermarkets. There can also be designated selling points for 'safe food' in markets for the general public. These efforts have to be supported by awareness campaigns to increase the demand for safer crops.

Other incentives could be institutional support from government agencies, such as providing extension services, training, loans and improved tenure security. Perhaps good media publicity will also encourage farmers to adopt practices for safer vegetable production. Although it is unlikely that all farmers will change their practices, at least some farmers might, thus, reducing the general risk to the public.

The promotion of good washing practices prior to vegetable preparation (at home or in restaurants and food stalls) will remain an additional crucial risk barrier which simultaneously addresses on-farm and post-harvest contamination. Health initiatives and campaigns broadcasted through radio, TV, or newspapers have to consider the right balance between the provision of incentives (rewards, recognition in tourist guides, etc.) and gradual enforcement to comply with hygiene standards.

Although increased awareness and behaviour changes can be effective means for risk reduction in the short term, a combination of non-treatment options and functional wastewater treatment services, should remain the long-term goal.

REFERENCES

Abdul-Raouf, U.M., Beucht, L.R. and Ammar, M.S. 1993. Survival and Growth of E. Coli on Salad Vegetables, *Applied Environmental Microbiology*, Vol. 59, No. 7, pp. 1999–2006.

Akpedonu, P. 1997. *Microbiology of Street Foods from a High Density Community in Accra*. Noguchi Memorial Institute for Medical Research, University of Ghana (mimeo).

Amoah, P., Drechsel, P. and Abaidoo, R.C. 2005. Irrigated Urban Vegetable Production in Ghana: Sources of Pathogen Contamination and Health Risk Elimination, *Irrigation and Drainage*, Vol. 54, pp. 49–61 (special issue).

Amoah, P., Drechsel, P., Abaidoo, R.C. and Ntow, W.J. 2006. Pesticide and Pathogen Contamination of Vegetables in Ghana's Urban Markets, *Arch. Environ. Contam. Toxicol.*, Vol. 50, No. 1, pp. 1–6.

Amoah, P., Drechsel, P., Abaidoo, R.C. and Henseler, M. 2007a. Irrigated Urban Vegetable Production in Ghana: Microbiological Contamination in Farms and Markets and Associated Consumer Risk Groups, *Journal of Water and Health*, Vol. 5, No. 3, pp. 455–466.

Amoah, P., Drechsel, P., Abaidoo, R. and Klutse, A. 2007b. Effectiveness of Common and Improved Sanitary Washing Methods in West Africa for the Reduction of Coli Bacteria and Helminth Eggs on Vegetables, *Tropical Medicine & International Health*, 12 suppl. 2, pp. 40–50.

Anyane, S.L. 1963. Vegetable Gardening in Accra, *The Ghana Farmer*, Vol. 1, No. 6, pp. 228–230.

Armar-Klemesu, M, Akpedonu, P., Egbi, G. and Maxwell, D. 1998. Food Contamination in Urban Agriculture: Vegetable Production Using Wastewater. Armar-Klemesu M, Maxwell D (eds.) *Urban agriculture in the Greater Accra metropolitan area*. Final report to IDRC (Project No. 003149), University of Ghana, Noguchi Memorial Institute, Accra, Ghana.

Biney, C.A. 1998. The Threat of Pollution to the Coastal Zone of the Greater Accra Metropolitan Area, Ghana, *Ghana Journal of Science*, Vol. 31–36, pp. 47–54.

Binns, J.A., Maconachie, R.A. and Tanko, A.I. 2003. Water, Land and Health in Urban and Peri-urban Food Production: The Case of Kano, Nigeria, *Land Degradation and Development*, Vol. 14, pp. 431–444.

Boadi, K.O. and Kuitunen, M. 2002. Urban Waste Pollution in the Korle Lagoon, Accra, Ghana, *The Environmentalist*, Vol. 22, pp. 301–309.

Bosshart, S. 1997. *Analyse de L'état Environnemental du Maraîchage à Ouagadougou*, Rapport définitive du stage professionnel. ETHZ, ITS, EIER: Ouagadougou (mimeo).

Cissé, G. 1993. *Impacts Sanitaire de la Reutilization des Eaux Usées en Agriculture Dans le Contexte Sahelien*. MSc Thesis; École Polytechnique Fédérale de Lausanne, EIER, Ouagadougou (mimeo).

Club du Sahel. 2000. *Urbanization, Rural-urban Linkages and Policy Implications for Rural and Agricultural Development: Case Study from West Africa*, SAH/DLR (2000)1, Paris.

Cornish, G. A. and Lawrence, P. 2001. *Informal Irrigation in Peri-urban Areas: A Summary of Findings and Recommendations*, DFID's Water KAR Project R7132, Report OD 144, HR Wallingford, Wallingford, UK, p. 54.

Cornish, G.A., Mensah, E. and Ghesquire, P. 1999. *Water Quality and Peri-urban Irrigation: An Assessment of Surface Water Quality for Rrigation and its Implication for Human Health in the Peri-urban Zone of Kumasi, Ghana*, Report OD/TN 95 September 1999. DFID's Water KAR (Knowledge and Research) Project R7132, HR Wallingford, UK, p. 44.

Danso, G. and Drechsel, P. 2003. The Marketing Manager in Ghana, *Urban Agricultural Magazine*, Vol. 9, p. 7.

Drechsel, P., Abaidoo, R.C., Amoah, P. and Cofie, O.O. 2000. Increasing Use of Poultry Manure in and Around Kumasi, Ghana: Is Farmers' Race Consumers' Fate? *Urban Agricultural Magazine*, Vol. 2, pp. 25–27.

Drechsel, P., Blumenthal, U.J. and Keraita, B. 2002. Balancing Health and Livelihoods: Adjusting Wastewater Irrigation Guidelines for Resource-poor Countries, *Urban Agriculture Magazine*, Vol. 8, pp. 7–9.

Drechsel, P., Graefe, S., Sonou, M. and Cofie, O.O. 2006. *Informal Irrigation in Urban West Africa: An Overview*. IWMI Research Report 102. IWMI, Colombo. Available at: www.iwmi.cgiar.org/pubs/pub102/RR102.pdf.

Endamana, D., Kengne, I.M., Gockowski, J., Nya, J., Wandji, D., Nyemeck, J., Soua, N.N. and Bakwowi, J.N. 2003. *Wastewater Reuse for Urban Agriculture in Yaoundé, Cameroon: Opportunities and constraints*.

Ensink, J., Tariq, M. and Dalsgaard, A. 2007. Wastewater-irrigated Vegetables – Market Handling Versus Irrigation Water Quality, *Tropical Medicine and International Health*, Vol. 12 suppl. 2, pp. 2–7.

Essamuah, E. and Tonah, S. 2004. Coping with Urban Poverty in Ghana: An Analysis of Household and Individual Livelihood Strategies in Nima/Accra, *Legon Journal of Sociology*, Vol. 1, No. 2, pp. 79–96.

Faruqui, N.I., Niang, S. and Redwood, M. 2004. Untreated Wastewater Use in Market Gardens; a Case Study of Dakar, Senegal. Scott, C., Faruqui, N.I. and Raschid, L. (eds) *Wastewater Use in Irrigated Agriculture: Confronting the Livelihood and Environmental Realities*. IWMI-IDRC-CABI, Wallingford, pp. 113–125.

Gaye, M and Niang, S. (eds) 2002. *Epuration des Eaux Usées et L'agriculture Urbaine*. Etudes et Recherches, ENDA-TM, Dakar, Sénégal.

Gbewonyo, K. 2007. *Wastewater Irrigation and the Farmer: Investigating the Relation Between Irrigation Water Source, Farming Practices, and Farmer Health in Accra, Ghana*. Harvard College, Cambridge, Massachusetts (unpublished thesis).

Gbireh, Z.A. 1999. *Development of Urban and Peri-urban Irrigation for Food Security Within and Around Accra Metropolitan Area*. FAO Technical Report. (unpublished).

Gockowski, J., Mbazo'o, J., Mbah, G. and Moulende, T.F., 2003. African Traditional Leafy Vegetables and the Urban and Peri-Urban Poor, *Food Policy*, Vol. 28, No. 3, pp. 221–235. Available at: www.iwmi.cgiar.org/waterpolicybriefing/files/wpb17.pdf.

GSS. 2002. Ghana Statistical Service 2002 census data.

GWCL. 2006. Ghana Water Company Limited. Internal Data.

Henseler, M., Danso, G. and Annang, L. 2005. *Lettuce Survey*. Project Report. Lettuce Survey Component of CP51, CGIAR CPWF Project 51. Unpublished report, IWMI, Ghana.

Hussain I., Raschid, L., Hanjra, M.A., Marikar, F., van der Hoek, W. 2002. *Wastewater Use in Agriculture: Review of Impacts and Methodological Issues in Valuing Impacts*. (With an Extended List of Bibliographical References). Working Paper 37. International Water Management Institute, Colombo, Sri Lanka.

Keraita, B. and Drechsel, P. 2004. Agricultural Use of Untreated Urban Wastewater in Ghana, Scott, C., Faruqui, N.I. and Raschid, L. (eds) *Wastewater Use in Irrigated Agriculture: Confronting the Livelihood and Environmental Realities*. IWMI-IDRC-CABI, Wallingford, pp. 101–112.

Keraita, B. and Drechsel, P. 2007. Safer Options for Wastewater Irrigated Urban Vegetable Farming in Ghana, *LEISA* Vol. 23, No. 3, pp. 26–28.

Keraita, B., Danso, G. and Drechsel, P. 2003a. Urban Irrigation Methods and Practices in Ghana and Togo, *Urban Agriculture Magazine*, Vol. 10, pp. 6–7.

Keraita, B., Drechsel, P. and Amoah, P. 2003b. Influence of Urban Wastewater on Stream Water Quality and Agriculture in and Around Kumasi, *Ghana, Environment & Urbanization*, Vol. 15, No. 2, pp. 171–178.

Keraita, B., Konradsen, F., Drechsel, P. and Abaidoo, R.C. 2007a. Reducing Microbial Contamination on Lettuce by Cessation of Irrigation Before Harvesting, *Tropical Medicine & International Health*, Vol. 12, suppl. 2, pp. 8–14.

Keraita, B., Konradsen, F., Drechsel, P. and Abaidoo, R.C. 2007b. Effect of Low-cost Irrigation Methods on Microbial Contamination of Lettuce, *Tropical Medicine & International Health*, Vol. 12, suppl. 2, pp. 15–22.

Keraita, B., Drechsel, P., Agyekum, W. and Hope, L. 2007c. In Search of Safer Irrigation Water for Urban Vegetable Farming in Ghana, *Urban Agriculture Magazine*, Vol. 19, pp. 17–19.

London Economics. 1999. *Ghana Urban Water Sector, Willingness and Ability to Pay, Demand Assessment and Tariff Structure Study*. In association with John Young & Associates.

McGregor, D., Simon, D. and Thompson, D. 2002. *Peri-urban Natural Resources Management at the Watershed Level: Kumasi, Ghana*. Final Technical Report for DFID Project R7330. CEDAR-IRNR, Royal Holloway University of London, London.

Mensah, E., Amoah, P., Abaidoo, R.C. and Drechsel, P. 2001. Environmental Concerns of (Peri-) Urban Vegetable Production – Case Studies from Kumasi and Accra. Drechsel, P. and D. Kunze (eds) *Waste Composting for Urban and Peri-urban Agriculture – Closing the rural-urban nutrient cycle in Sub-Saharan Africa*. IWMI/FAO/CABI, Wallingford, pp. 55–68.

Mensah, P., Yeboah-Manu, D., Owusu-Darko, K. and Ablordey, A. 2002. Street Foods in Accra, Ghana: How Safe are They? *Bulletin of WHO*, Vol. 80, pp. 546–554.

Moustier, P. and Fall, S.A. 2004. Les Dynamiques de L'agriculture Urbaine: Caractérisation et Évaluation. Smith, O.B., Moustier, P., Mougeot, L.J.A. and Fall, S.A. (eds). *Développement durable de l'agriculture urbaine en Afrique francophone. Enjeux, concepts et méthodes*. Centre de coopération internationale en recherche agronomique pour le développement (CIRAD), Montpellier, and International Development Research Centre (IDRC), Ottawa, pp. 23–43.

Niang, S., Diop, A., Faruqui, N., Redwood, M. and Gaye, M. 2002. Reuse of Untreated Wastewater in Market Gardens in Dakar, Senegal, *Urban Agriculture Magazine*, Vol. 8, pp. 35–36.

Nyanteng, V.K 1998. Draft Summary Report on Food Markets and Marketing in the Accra Metropolis, *Food Supply and Distribution to Accra and its Metropolis. Workshop – Proceedings*, Accra, Ghana, 13–16 April 1998.

Obuobie, E., Keraita, B., Danso, G., Amoah, P., Cofie, O.O., Raschid-Sally, L. and Drechsel, P. 2006. *Irrigated Urban Vegetable Production in Ghana: Characteristics, Benefits and Risks*. IWMI-RUAF-IDRC-CPWF, IWMI, Accra, Ghana, p. 150. Available at: www.cityfarmer.org/GhanaIrrigateVegis.html.

Olsen, M. 2006. *Risk Awareness of Street Food Handlers and Consumers*. Lettuce survey. Project Report. Lettuce Survey Component of CP51, CGIAR CPWF Project 51. Unpublished report, University of Copenhagen, Denmark.

Seidu, R., Amoah, P., Heistad, A., Strenstrom, T.-A. and Drechsel, P. 2008b. A Quantitative Microbial Risk Assessment of Reclaimed Water Irrigation in Accra: Exploring the Effects of Water Quality and Marketing Points on Health Risks, *Journal of Water and Health* (submitted).

Seidu, A.R., Drechsel, P., Amoah, P., Löfman, O., Heistad, A., Fodge, M., Jenssen, P. and Stenström, T.-A. 2008a. *Quantitative Microbial Risk Assessment of Wastewater and Faecal Sludge Reuse in Ghana*. Paper presented at the 33rd WEDC International Conference, Accra, Ghana, 2008.

Sonou, M. 2001. Peri Urban Irrigated Agriculture and Health Risks in Ghana, *Urban Agriculture Magazine*, Vol. 3, pp. 33–34.

Tandia, M. 2002. Urban Farming in a Context of Water Scarcity. The Case of the Tel Zaatar Market Gardening Site in Nouakchott (Mauritania), *African Cities* (Special Issue), December 2002: 11–12.

Twum-Baah, K.A. 2002. Population Growth of Mega-Accra: Emerging Issues. Mills-Tettey, R. and Adi-Dako, K. (eds) *Visions of the City. Accra in the 21st Century*. Woeli Publishing Services, Accra, pp. 31–38.

UN Population Division. 2004. *World Urbanization Prospects: The 2003 revision*. United Nations. New York, US, Department of Economics and Social Affairs. Available at: www.un.org/esa/population/publications/wup2003/WUP2003Report.pdf.

UN-Habitat. 2001. Cities in a globalizing world. *Global Report on Human Settlements 2001*. UNCHS. Earthscan Publications, London.

WHO. 2006. *Guidelines for the safe use of wastewater, excreta and grey water: Wastewater use in agriculture (Volume 2)*. WHO, Geneva, p. 219.

Zakariah, S., Lamptey, M.G. and Maxwell, D. 1998. Urban agriculture in Accra: A descriptive analysis. Armar-Klemesu, M. and Maxwell, D. (eds) *Urban Agriculture in Greater Accra Metropolitan Area*. Final Report to IDRC. Centre file: 003149. Noguchi Memorial Institute for Medical Research, University of Ghana.

Chapter 11

Drinking water – potential health effects caused by infiltration of pollutants from solid waste landfills

Tuula Tuhkanen

Tampere University of Technology, Institute of Environmental Engineering and Biotechnology, Finland

ABSTRACT: Waste disposal by landfill have caused health and environmental concern. People who live near landfill sites may be exposed to chemicals released into the air, water and soil. Local surface water and groundwater can become contaminated and these may endanger the potable water supply or water for recreational use. Thus, the landfill can cause a health risk for local residents and particularly for children. Concentrations of toxic substances in groundwater near the landfills often exceed drinking-water norms, but not as consistently as with surface water. The health risk depends on the quality and quantity of the waste material in the landfill, the engineering structure of protective barriers and the hydrogeological conditions. Risk is high in the case of landfills situated near groundwater reservoirs used as potable water sources.

11.1 INTRODUCTION

Landfilling is one of the main methods of municipal and industrial solid waste disposal. Nowadays, landfill management requires effective and rigorous control regarding quality standards for wastewater discharged into the water bodies and surrounding environment. For example, the Council of Finland enacted legislation (VnP 861/97) stating that all the leachates and runoff waters must be controlled and treated by all who own and maintain the landfill. This has led to centralized and controlled landfill management. The infiltration of pollutants into the soil and groundwater can be controlled by the proper placement of landfill and engineering solutions. Isolation layers can prevent the infiltration of intestinal water and runoff as well as other types of leakages into the surrounding environment (La Grega et al., 1994).

Legislation and technical guidelines control the management of modern landfills. The present European Union policy on hazard classification and landfill disposal of waste material requires tight control over the release of contaminants into the environment. Hazardous Waste Directive 1991689/EC defines a set of 14 properties to be used in hazard waste classification (Official Journal of the European Communities, 2000). According the EU Landfill Directive 1999/31/EC, the waste acceptance criteria impose requirements for waste classification and quality monitoring (Official Journal of the European Communities, 2003). The basic requirements for waste classification are source and origin of the waste, information about the waste-producing process, information on its composition and on its leaching behaviour and appearance.

Hazardous waste cannot be disposed of in municipal landfills, but municipal landfills can contain hazardous household products, such as paints, garden pesticides,

Figure 11.1 A site for snow disposal in Finland at the end of the summer in the middle of the 1980s

pharmaceuticals, photographic chemicals, detergents, personal care products, fluorescent tubes, waste oil, heavy metal contained in batteries, wood treated by dangerous substances, waste electronic and electrical equipment, and discarded CFC-containing equipments (Slack et al., 2005). The water balance of the entire waste bank is controlled, to reduce the risk of emissions from the hazardous compounds getting into the environment. The situation in developed countries has previously been unsatisfactory, as simply getting rid of waste was the standard method used in the past and unfortunately still is in some developing countries. In developing countries, waste disposal management is not a top priority, even today. Meanwhile, old landfills in developed countries can contain hazardous industrial waste, as the inspection of the material dumped into the waste sites has often been lax. For example, the dumping of hazardous waste, such as waste oils, in waste heaps, or their infiltration into the soil, was standard practice until the 1980s in Finland. According a Finnish survey, 72 landfills have been located in areas of significance to the country's drinking-water supply (Loikkanen, 1984), most of them small ones that have not been used for years.

Due to the uncontrolled disposal of municipal and industrial waste into the environment, a large number of landfills contain hazardous material from previous disposals. Leachates from these landfills contain high and variable concentrations of organic and inorganic compounds. There is also the possibility of pathogens developing if waste of human origin has been dumped.

Because of the improper isolation of old landfills, the waste material from the leachates can penetrate the underlying soil and groundwater.

11.2 POLLUTANTS IN LANDFILL LEACHATES

Leachate can be defined as a liquid that has percolated through solid waste and extracted dissolved material. In most landfills, leachate comprises the liquid that has entered the landfill from external sources, such as drainage, rainfall, groundwater, water from underground springs and the liquid produced during composting of waste. The quality and quantity of contaminants in leachate are influenced by numerous factors, including refuse age, moisture loading, type of composition, as well as the landfilling technique, the presence of sludges and ashes, and seasonal weather conditions (Schultz et al., 2002; Haapea et al., 2002). The leachate contamination of groundwater is less likely from modern landfills because of engineered barriers and leachate collection and treatment.

Water infiltrating into waste will first be absorbed by solid waste material. When the waste becomes saturated the water moves through the waste by gravity and becomes contaminated with waste constituents through a partitioning process. The concentration of contaminants depends on the contaminate concentration of the leachable material in the waste, the ease of mass transfer, surface area, contact time, pH, column depth and the infiltration rate (Vaajasaari, 2005).

The quality of leachate is affected by the composition of the waste matter. Organic and inorganic contaminants of landfill leachates are released from waste due to successive biological chemical and physical processes. The decomposition of leachates can be divided into two phases. The first one is aerobic decomposition, the duration of which is rather short. The second is the anaerobic phase, during which methane gas and water are generated. The water formed during the anaerobic stage has a high content of molecular organic compounds (humic-like substances) and organic acids. There is a difference between wastewaters from fresh and matured landfills. The organic matter content of a new landfill is usually higher and the water is more biodegradable. In matured landfills, the BOD/COD ratio is low since the biodegradable fraction of the organic matter has been decomposed.

Besides organic matter, which can be measured as BOD, COD or TOC some inorganic compounds in landfill leachates can also be considered macropollutants (Christensen et al., 2001). They occur in relatively high concentrations, such as iron, manganese, calcium, sodium, potassium, ammonia/ammonium and nitrate.

The environmental impact and risks caused by hazardous waste disposal in Finnish landfills were assessed on the basis of a five-year study of 43 landfills. The main interest was in the chemical quality and toxicity of surface water runoffs from municipal waste landfills. The effect of the infiltration of the leachate plume in the quality of groundwater was also monitored. The average concentration of organic and inorganic compounds in municipal and industrial landfill leachates are listed in Table 11.1. Also, the microbiological quality is assessed by the standard tests.

11.3 THE EXPOSURE PATHWAYS AND MECHANISMS

Following the disposal of waste into the environment, exposure to hazardous substances can occur outside, in the vicinity of the landfill. Contaminants, which have

Table 11.1 The chemical and microbiological quality of landfill leachate of municipal and industrial landfills in Finland (Assmuth et al., 1990)

Parameter	Municipal landfill			Industrial landfill		
	Min	Mean	Max	Min	Mean	Max
pH	2.8	7.0	8.6	5.0	7.0	8.7
Conductivity, mS/m	4.6	200	820	5.6	69	580
Suspended solids, mg/L	11	39	130	4	56	550
Cl⁻, mg/L	5	220	1,800	4	47	270
Total hardness, mmol/L	0.2	4.3	17.3	0.5	2.1	9.9
Iron, mg/L	0.3	26	150	0.3	8.9	86
NH_4-N, mg/L	0	46	340	0	1.7	9.8
COD, mg/L	40	400	2,200	42	360	1,300
TOC, mg/L	0	180	590	27	62	140
BOD_7, mg/L	1	29	97	8.7	170	490
Fecal streptococci, pfu/100 mL	0	135	65,800	10	32	93
Thermotolerant coliforms,	0	230	720,000	2	34	150
Coliforms, cfu/100 mL	5	4,600	17,200	–	–	–
As, µg/L	<0.1	5.6	23	<1	35	760
Cd, µg/L	<0.01	3.6	70	<0.1	3.4	10
Co, µg/L	<6	36	100	<6	65	260
Cr, µg/L	0.5	16	80	<1	150	7,000
Ni, µg/L	3.3	19	59	1.8	640	3,200
Pb, µg/L	<1	4.2	40	<0.5	2.7	10
Zn, µg/L	10	1,400	110,000	380	3,600	
Fe, µg/L	250	26,000	150,000	12,000	86,000	
Mn, µg/L	50	1600	8,800	40	1700	81,000
K, µg/L	1,400	74,000	310,000	40,000	650,000	

been released from the waste material in the landfill, move and respond to a number of natural and man-made factors. The release of contaminants takes place in one of the three environmental states: liquid, solid or gas. This presentation will concentrate on the transportation of chemicals in the water phase, surface water and particularly in groundwater. The volatilization of compounds into the atmosphere or the transportation of hazardous matter by soil particles can be a secondary pollution pathway for drinking-water sources.

Water is an excellent solvent for several organic and inorganic compounds in the environment. Water solubility is one of the most important factors controlling the fate of compounds in groundwater – the more hydrophilic the compound is, the further it can be transferred. The contaminants move in the subsurface in response to advection, hydrodynamic dispersion and molecular diffusion. Some compounds are degraded biologically or abiotically or can be adsorbed into the soil matrix. If the attenuation capacity of the soil is exceeded, the aquifer can become contaminated. The concentration of hazardous substances in groundwater near (about 200 m from) landfills is presented in Table 11.3. The data is based on the study by Assmuth et al. (1990).

After hazard identification, the second stage of a risk assessment consists of estimating the exposure to the chemicals by the population at risk. It is important to know how the contaminant can migrate (for example, via groundwater) to the potential receptor (here the user of the groundwater). In the exposure assessment, both short-term and

Table 11.2 The average and maximum values of organic compounds in Finnish landfill leachates (Assmuth et al., 1990)

Compound	Mean	Max
Halogenated aliphatic compounds, µg/L		
Dichloromethane	520	5,700
1,2-Dichloroethane	55	680
Trans-1,2,-Dichloroethene	30	850
Trichloroethane	5.6	88
Tetrachloroethane	3.3	110
Aromatic hydrocarbons, µg/L		
m- and p -Xylenes	90	2,000
Toluene	53	1,500
Ethyl benzene	48	980
O-Xylene	18	430
Naphthalene	2.3	9
Benzene	0.58	1.2
Phenolic compounds, µg/L		
p-Cresol	60	810
m-Cresol	13	94
o-Cresol	5.2	67
2,4,6-Trichlorophenol	0.82	6
2,4,5-Trichlorophenol	0.59	8.6
2,3,4,-Tetracholrophenol	0.25	5.6
Pentachlorophenol	0.15	3
Pesticides, µg/L		
Dichloprop	120	230
MCPA	13	25
Mecoprop	11	23
Lindane	0.43	15
Dieldrin	0.058	1.1
Eldrin	0.028	0.54
DDD	0.17	2.2
DDE	0.077	3.4
DDT	0.023	0.23
PCDD/Fs, pg/L		
OCDF	6,800	13,000
1234678-HPCDD	250	460
HxCDD	78	88
2378-TeCDD	6.7	13
Other organic compounds		
PCBs, µg/L	0.49	3.8
Mineral oils, µg/L	81	220
CCl_4 – extractables, µg/L	890	3,600
TOC, mg/L	150	590

OCDF: Octa-chloro-dibenzofureane
1234678-HPCDD: 1,2,3,4,6,7,8 –hepta-chloro-dibenzodioxine
HxCDD: Hexachlorodibenzodioxine
2378-TeCDD: 2,3,7,8- tetra-chloro-dibenzodioxine
TOC: Total Organic Carbon

Table 11.3 Quality of groundwater in the vicinity (<200 m) from the landfill. Average and maximum values compared to drinking-water standards

Parameter	Mean	Max	WHO	EU
pH	6.5	7.9		
Conductivity, mS/m	100	960		
Suspended solids, mg/L	120	210		
Cl⁻, mg/L	73	770		250
Total hardness; mmol/L	2.3	17		
NH_4^-N, mg/L	3.1	58		
COD, mg/L	390	1,400		
BOD_7, mg/L	16	42		
Fecal streptococci, cfu/100 mL	2	10	0	0
Thermotolerant coliforms, cfu/100 mL	22	310		
Coliforms, cfu/100 mL	270	960		
Elements, µg/L				
As, µg/L	14	170	10	10
Cd, µg/L	1.8	3.0	3	5
Cr, µg/L	7.1	45	50	50
Cu, µg/L	35	180	2,000	200
Co, µg/L	24	80		
Ni, µg/L	16	48	20	20
Pb, µg/L	18	240	10	10
Zn, µg/L	130	290	5,000	
Fe, mg/L	14	120	0.3	0.2
Mn, mg/L	1.2	7.7	0.4	0.05
Na, mg/L	60	430		200
K, mg/L	20	120		
Halogenated aliphatic hydrocarbons, µg/L				
Dichloromethane	140	1,600	20	
1,2-Dichnoroethane	23	180	30	3
Trans,1,2,-Dichloroethene	84	440	50	
Trichloroethylene	5.7	12	70	10
Tetrachloroethylene	2.1	5.3	40	
Aromatic hydrocarbons, µg/L				
Toluene	17	140	700	
m,p-xylenes	0.16	0.38	500	
o-xylene	0.21	0.5		
Naphthalene	12	68		
Ethyl benzene	0.44	2	300	
Phenolic compounds, µg/L				
Cresols	85	1150		
3,4,5-TCP, trichlorophenol	0.47	1.8		
PCP, pentachlorophenol	0.31	1.1	9	
2,4,6-TCP, trichlorophenol	0.15	0.32		
Other, µg/L				
PCBs, Polychlorinated biphenyles	9.6	9.9		
AOCl, Absorbable organic chlorine	69	410		
Mineral oils	99	370		

long-term exposure should be considered. Also, the sensitive sub-groups, such as children and pregnant women, are of special concern. Tree exposure routes occur when people are in contact with water; the main route is ingestion of drinking water, but inhalation and dermal contact are also possible exposure pathways. Some compounds in the water are so volatile that the exposure can take place through inhalation in the shower, for example. Also, dermal contact is possible in showers, baths and swimming pools.

The average ingested dose of drinking water is 2 litres per day per person but this can greatly vary from individual to individual, based on diet. The worst-case scenario via drinking water is based on the consumption of 2 litres of drinking water per day during a lifetime. The drinking-water guidelines are based on the assumption that there will be no significant adverse health effect among drinking-water consumers. The judgement of safety – or what is an acceptable level of risk in particular circumstances – is a matter in which society as a whole has a role to play (WHO, 2003). In developed countries, the acceptable risk is very low, about one in a million during a lifetime. However, in developing countries, the quantity of water is sometimes more important than its quality. Any improvement in access to safe water should be considered as the target in less-developed conditions.

11.4 CASES

There have been several cases involving a possible correlation between exposure via contaminated drinking water and health effects. The most infamous case is perhaps Love Canal in the 1970s in the US in which a wide EC BIOMED programme, EURO-HAZON, found the risk of congenital anomalies to be greater among babies whose mothers live within 3 kilometres of a landfill site.

There should be further investigation as to whether the risk association is a causal one or not. There could be a socio-economic factor behind the finding. Families with lower socio-economic levels are still living in communities near to landfill sites (Dolk et al., 1998).

11.5 CONCLUSIONS

In accordance with the EU landfill Directive, Finnish legislation on landfills and waste taxes have raised the cost of waste disposal, to encourage environmentally favourable forms of waste disposal and recovery. The number of landfills is decreasing and more attention is being paid to the environmental effects of the old and abandoned landfills, which no longer meet stricter requirements concerning the sealing of the landfill bases. Nevertheless, old hazardous material which is interspersed with the municipal waste represents a constant risk to the environment, and more especially to the quality of groundwater.

REFERENCES

Assmuth, T., Poutnane, H., Strandberg, T., Melanen, M., Penttilä, S. and Kalevi, K. 1990. *Environmental Impacts of Hazardous Waste Landfills*. National Board of Waters and Environment – Report A, 64. p. 214.

Christensen, T.H., Kjeldsen, P., Bjerg, P.L., Jensen, D.L., Christensen, J.B., Baun, A., Albrechtsen, H.-J. and Heron, G. 2001. Biogeochemistry of Landfill Leachate Plumes, *Applied Geochemistry*, Vol. 16, No. 7, pp. 659–718.

Dolk, M., Vrijheid, M., Amstrong, B., Abramsky, L., Bianchi, F., Garne, E., Nelen, V., Robert, E., Scolt, I.E.S., Stone, D. and Tenconi, R. 1998. Risk of Congenital Anomalies Near Hazardous Waste Landfill Sites in Europe: The EUROHAZON Study, *The Lancet*, Vol. 352, No. 8, pp. 423–427.

Haapea, P., Korhonen, S. and Tuhkanen, T. 2002. Treatment of Industrial Landfill Leachates by Chemical and Biological Methods: Ozonation, Ozone + Hydrogen Peroxide and Biological Post-Treatment for Ozonated Water, *Ozone Science and Engineering*, Vol. 24, pp. 369–378.

La Grega, M.D., Buckingham, P.L. and Evans, J.C. 1994. *Hazardous Waste Management and Remediation*. McGraw Hill, p. 1157

Loikkanen, S. 1984. *The effect of Human Activities into the Quality of Groundwater*. (In Finnish). University of Kuopio, Report. 22 s.

Official Journal of European Communities. 2000. 2000/532/EC. Commission Decision of 3 May 2000 Replacing Decision 94/3/EC Establishing a List of Waste Pursuit to Article 1(a) of Council Directive 75/442/EEC on Waste and Council Decision 94/904/EC Establishing a List of Hazardous Waste Pursuit to Article 1(4) of Council Directive 91/689/EEC on Hazardous Waste, Official Journal L 226, 06/09/2000 (As Amended by Commission Decisions 2001/118/EC (OJ L 47, 1-31), 2001/119/EC (OJ 47, 32) and 2001/573/EC (OJ L 203, 18-19).

Official Journal of European Communities. 2003. 2003/33/EC. Commission Decision of 19 December 2002 establishing criteria and procedure for acceptance of waste at landfills pursuant to Article 16 of and an Annex II to Directive 1999/31/EC. Official Journal L11. 16/01/2003.

Schultz, E., Vaajasaari, K., Joutti, A. and Ahtiainen, A. 2002. Toxicity of Industrial Wastes and Waste Leaching Test Eluates Containing Organic Compounds, *Ecotoxicology and Environmental Safety*, Vol. 52, pp. 238–255.

Slack, R.J., Gronow, J.R. and Voulvoulis, N. 2005. Household Hazardous Waster in Municipal Landfills: Contaminants in Leachates, *Science of the Total Environment*, Vol. 337, pp. 199–137.

Vaajasaari, K. 2005. *Leaching and Ecotoxicity Test as Methods for Classification and Assessment of Environmental Hazards of Solid Wastes*. Tampere University of Technology, Tampere, Publication 540.

VnP 861/97, *Legislation of Soil Protection; Landfills*. Available at: www.finlex.fi/fi/laki/alkup/1997/19970861

WHO. 2003. *Guidelines for Drinking Water Quality*. 3rd, edn. Available at: www.who.int/water_sanitation_health/dwq/gdwq3/en/index.html (accessed on 16 May 2006).

Chapter 12

Exploding sewers: the industrial use and abuse of municipal sewers, and reducing the risk – the experience of Louisville, Kentucky US

Gordon Garner

Vice President, Water Business Group, CH2MHILL, USA

ABSTRACT: Industrial chemicals, including petrochemicals, pesticides, solvents, pharmaceuticals and other process chemicals, can have catastrophic consequences for public health when discharged into sewer systems. Risks may be reduced via industrial pretreatment requirements and monitoring programmes. Louisville, Kentucky, US experienced a series of serious industrial chemical catastrophes in the late 1970s and early 1980s, which led to the development of programmes now considered 'best practice' models for risk reduction. The catastrophic incidents are reviewed, the priority pollutants are listed, and these risk reduction programmes are described.

12.1 INTRODUCTION

Of all the risks associated with the water cycle, the use, storage and transport of industrial chemicals (including petrochemicals, pesticides, solvents, pharmaceuticals and other process chemicals) present the greatest risk for catastrophic damage to public health and the environment. Louisville, Kentucky, US experienced a series of serious industrial chemical catastrophes in the late 1970s and early 1980s, which led to progressive legislation and comprehensive regulation of how industrial chemicals were used, stored and transported in the city.

Many of the programmes developed from the disasters were used as models. The United States Environmental Protection Agency (US EPA) developed industrial pretreatment regulations and 'superfund' legislation requiring industry to reduce risk and assume liability for the chemicals and accidents that occur.

The Louisville and Jefferson County Metropolitan Sewer District (MSD) is the agency responsible for the wastewater collection and treatment system that was affected by these chemical incidents. The incidents are described and the progressive programmes that MSD developed are explained. These programmes are now considered 'best practice' models and are used by many large cities in the United States to reduce the risk from chemical use, storage and transport.

12.2 THE HEXA-OCTA INCIDENT

On March 17, 1977, employees at the Morris Forman Wastewater Treatment Plant in Louisville, Kentucky noticed a strong chemical odour that made them sick. It was the beginning of an environmental incident that would set legal precedent in the United States. It took more than a week to identify the substance: a mixture of hexa-chloropentadiene and octachlorocyclopentene, quickly abbreviated to 'hexa' and 'octa'. These highly toxic chemicals are used in pesticides. The contaminated waste-water treatment plant was shut down on March 29, twelve days later. With the treatment plant offline, 378,000 m³/d of untreated wastewater was discharged into the Ohio River.

The US Army sent teams wearing protective gear into the sewers to find the source of the chemicals. The Federal Bureau of Investigations (FBI) joined the investigation. On June 7, a federal grand jury charged Donard E. Distler, president of Kentucky Liquid Recycling, and two of his employees with dumping toxic chemicals into the sewers.

The chemicals, as the courts later found, were wastes that had been sent by tanker truck to Distler's company for disposal. Distler's company had dumped the chemicals down a manhole directly into the sewers in western Louisville.

The treatment plant was shut down for nearly three months while the contaminated material was removed – three months during which 378,000 cubic metres of raw sewage were discharged into the river each day. It took another two years to remove the contaminated material from the sewer lines – years during which the raw sewage from these lines was shunted around the plant and into the river.

In September 1979, the month the clean-up ended, Distler was found guilty in court. This was the first time in the United States that an individual was convicted in a trial of federal criminal charges of polluting a waterway. He was sentenced to two years in prison and fined US$ 50,000. After appealing his case, all the way to the US Supreme Court, he was sent to prison in early 1982.

In January 1983, the companies that had originated the waste – Velicol Chemical Corp. of Chicago, Illinois and Chem-Dyne Corp. of Hamilton, Ohio – agreed to pay Metropolitan Sewer District (MSD) US$ 1.9 million for the medical costs of employees who had exposures from the clean-up and the costs of cleaning up the sewers and the treatment plant.

12.3 THE SEWER EXPLOSIONS

Shortly after 5:15 am on Friday, February 13, 1981, two women on the way to work at a hospital drove under the railroad overpass on Hill Street. There was a gigantic blast, and their car was hurled into the air and onto its side.

At the same time, a police helicopter was heading toward the downtown area when the officers saw an unforgettable sight: a series of explosions, 'like a bombing run,' erupting along the streets of the oldest part of Louisville and through the University of Louisville campus.

After the explosions, more than 3 km of Louisville streets were pockmarked with craters, where manholes had been. Several blocks of Hill Street had fallen into the collapsed 3.6-metre diameter sewer line. Miraculously, no one was seriously hurt. Homes and businesses were extensively damaged and some families had to be evacuated. Louisville was in the newspaper headlines and on national broadcast news for several days.

The cause of the explosion was traced to the Ralston-Purina soybean processing plant, located south-east of the University of Louisville campus. Thousands of litres of a highly flammable solvent, hexane, had spilled into the sewer lines. The fumes from the hexane created an explosive mixture, which lay waiting in the larger sewer lines. As the two women drove under the railroad overpass, a spark from their car apparently ignited the gases.

Several blocks of Hill Street soon became an open trench, as crews cleared away the debris and prepared to replace the sewer line. The trench remained open throughout the summer while the work continued. (The stench was so bad that MSD tried using huge blocks of restroom-type deodorant to mask the odour – without success.)

It took 20 months to repair the sewer lines, and another several months to repair the work on the streets.

Ralston-Purina pleaded guilty to four counts of violating federal environmental laws, and paid a fine of US$ 62,500. In February 1984, the company agreed to pay MSD more than US$ 18 million in damages. Many millions more were paid to other government agencies and private individuals who suffered damage eventually totalling over US$ 45 million.

As for the soybean plant, Ralston-Purina used more than US$ 2 million in city industrial bonds to rebuild it in 1983 – and then sold it in 1984.

12.4 INDUSTRIAL WASTE AND HAZARDOUS SPILLS

Industrial waste had been a problem for generations in Louisville, since the days when wastes from meat-packing plants clogged the sewers and acid discharges from distilleries eroded concrete pipes. The hexa-octa dumping in 1977 and the sewer explosion of 1981 were Louisville's worst examples of the damage caused by industrial waste. There were many smaller spills and discharges that damaged the sewers, disturbed the treatment process and polluted the streams.

The problem, in nearly every case, was that industrial spills and even intentional discharges were usually hard to anticipate, hard to detect and hard to trace to their sources.

In 1981, MSD launched a programme to require industries to obtain permits to discharge wastewater into sewers. The permits also limited the amounts of a wide range of substances to be allowed into the wastewater. At the time of this programme's launch, 46 out of the 148 affected companies failed to comply with the initial deadline. In the early 1980s, the Environmental Protection Agency added its muscle to the programme.

Accidental spills were a more difficult matter. Gasoline, acids, solvents, ammonia, ingredients for plastic and a wide assortment of chemicals were being spilled onto the ground, onto the streets, and into the sewers and streams. No single agency was in charge of dealing with spills and no single agency even kept track of them. There was no enforcement authority to deal with the offenders. Risks to public health and safety were acute.

In September 1984, MSD prepared a detailed report on the problem. MSD suggested a complete overhaul of the system for reporting and dealing with spills, and requirements that businesses and industries using hazardous materials prepare detailed plans for spill clean-up and containment.

It was the first example of MSD identifying a public need, proposing a solution and taking responsibility for a major new programme. It was also one of the nation's first comprehensive local government chemical spill prevention programmes. It led to changes in how industry in the US managed and assumed responsibility for the chemicals they used and the accidents that caused damage to the public and the environment.

12.5 ABOUT THE LOUISVILLE AND JEFFERSON COUNTY METROPOLITAN SEWER DISTRICT (MSD)

MSD provides sanitary sewers, stormwater drainage and flood protection services to over 200,000 user accounts. Each year, MSD adds approximately 4,000 users and 110 km of sanitary sewers to a system that is almost 4,800 km long.

MSD serves a population of about 700,000 throughout the 619 km² of Jefferson County. This includes 6 regional wastewater treatment plants, 19 small treatment plants and roughly 300 pumping stations.

The Morris Forman Wastewater Treatment Plant (MFWTP) is the oldest and the largest wastewater treatment plant in the Louisville MSD service area. It has a treatment design capacity of 454,000 m³/d and a wet-weather capacity of over a million cubic metres per day. The collection system serves approximately 500,000 people with over 4,000 km of pipe and numerous major industries with significant process waste discharges. The combined sewer area encompasses 101 km², which is about one-third of the MFWTP service area. It serves approximately 324,000 people with 1,078 km of combined sewer.

Louisville, Kentucky is located on the Ohio River and has been a longtime manufacturing centre for rubber products, automobiles and trucks, paint and pigments, and food processing. It has 121 significant industrial users (SIU) that discharge untreated or pretreated wastewater to the municipal MSD system. An SIU has the potential to damage the wastewater collecting and treatment system and the environment.[1]

The hexa-octa incident and the sewer explosion are not the only serious chemical incidents to happen in Louisville. They are indicative of the risks that cities have from industrial use, storage and transport of chemicals and chemical wastes.

MSD meets or exceeds all requirements of the EPA that are required by the Clean Water Act (CWA). More than 1,500 other US cities must meet these requirements: those with wastewater plants treating more than 18,900 m³/d, or small wastewater plants that have significant individual industrial users. Basic requirements of any pretreatment programme involve the prohibition of certain chemicals or wastes with characteristics that would harm the wastewater collection and treatment system, human health or the environment. These are categorized in Table 12.1, Specific prohibitions of industrial chemicals or waste to wastewater systems, and Table 12.2, Priority pollutants.

---

[1]Significant Industrial User (SIU): any industrial user that discharges an average of 94 m³/d or more of process wastewater to the municipal system (excluding sanitary, non-contact cooling and boiler blowdown wastewater); contributes a process wastestream which makes up 5% or more of the average dry-weather hydraulic or organic capacity of the treatment plant; or is designated as an SIU on the basis that the industrial user has a reasonable potential for adversely affecting operation or for violating any pretreatment standard or requirement.

Table 12.1 Specific prohibitions of industrial chemicals or waste to wastewater systems

The following shall not be introduced into a wastewater collection and treatment system:

- Pollutants which create a fire or explosion hazard, including but not limited to, waste streams with a closed cup flashpoint of less than 60 degrees Celsius.
- Pollutants which will cause corrosive structural damage, but in no case discharges with pH lower than 5.0, unless the works is specifically designed to accommodate such discharges.
- Solid or viscous pollutants in amounts which will cause obstruction to the flow resulting in interference.
- Any pollutant, including oxygen-demanding pollutants, released in a discharge at a flow rate and/or concentration which will cause interference.
- Heat in amounts which will inhibit biological activity in the treatment plant resulting in interference, but in no case heat in such quantities that the temperature of the treatment plant exceeds 40 degrees Celsius, unless alternative temperature limits are approved.
- Petroleum oil, non-biodegradable cutting oil, or products of mineral oil origin in amounts that will cause interference or pass through.
- Pollutants which result in the presence of toxic gases, vapours, or fumes in a quantity that may cause acute worker health and safety problems.
- Any trucked or hauled pollutants, except at designated discharge points.

Source: Louisville and Jefferson County Metropolitan Sewer District Wastewater Discharge Regulations

Table 12.2 US EPA designated priority pollutants

Pollutant	Pollutant
Antimony	Arsenic
Beryllium	Cadmium
Chromium III	Chromium VI
Copper	Lead
Mercury	Nickel
Selenium	Silver
Thallium	Zinc
Cyanide	Asbestos
2,3,7,8-TCDD Dioxin	Acrolein
Acrylonitrile	Benzene
Bromoform	Carbon Tetrachloride
Chlorobenzene	Chlorodibromomethane
Chloroethane	2-Chloroethylvinyl Ether
Chloroform	Dichlorobromomethane
1,1-Dichloroethane	1,2-Dichloroethane
1,1-Dichloroethylene	1,2-Dichloropropane
1,3-Dichloropropene	Ethylbenzene
Methyl Bromide	Methyl Chloride
Methylene Chloride	1,1,2,2-Tetrachloroethane
Tetrachloroethylene	Toluene
1,2-Trans-Dichloroethylene	1,1,1-Trichloroethane
1,1,2-Trichloroethane	Trichloroethylene
Vinyl Chloride	2-Chlorophenol
2,4-Dichlorophenol	2,4-Dimethylphenol
2-Methyl-4,6-Dinitrophenol	2,4-Dinitrophenol
2-Nitrophenol	4-Nitrophenol

(*Continued*)

Table 12.2 Continued

Pollutant	Pollutant
3-Methyl-4-Chlorophenol	Pentachlorophenol
Phenol	2,4,6-Trichlorophenol
Acenaphthene	Acenaphthylene
Anthracene	Benzidine
BenzoaAnthracene	BenzoaPyrene
BenzobFluoranthene	BenzoghiPerylene
BenzokFluoranthene	Bis2-ChloroethoxyMethane
Bis2-ChloroethylEther	Bis2-ChloroisopropylEther
Bis2-EthylhexylPhthalate	4-Bromophenyl Phenyl Ether
Butylbenzyl Phthalate	2-Chloronaphthalene
4-Chlorophenyl Phenyl Ether	Chrysene
Dibenze[a,h]Anthracene	1,2-Dichlorobenzene
1,3-Dichlorobenzene	1,4-Dichlorobenzene
3,3'-Dichlorobenzidine	Diethyl Phthalate
Dimethyl Phthalate	Di-n-Butyl Phthalate
2,4-Dinitrotoluene	2,6-Dinitrotoluene
Di-n-Octyl Phthalate	1,2-Diphenylhydrazine
Fluoranthene	Fluorene
Hexachlorobenzene	Hexachlorobutadiene
Hexachlorocyclopentadiene	Hexachloroethane
Ideno1,2,3-cdPyrene	Isophorone
Naphthalene	Nitrobenzene
N-Nitrosodimethylamine	N-Nitrosodi-n-Propylamine
N-Nitrosodiphenylamine	Phenanthrene
Pyrene	1,2,4-Trichlorobenzene
Aldrin	alpha-BHC
beta-BHC	gamma-BHC (Lindane)
delta-BHC	Chlordane
4,4'-DDT	4,4'-DDE
4,4'-DDD	Dieldrin
alpha-Endosulfan	beta-Endosulfan
Endosulfan Sulfate	Endrin
Endrin Aldehyde	Heptachlor
Heptachlor Epoxide	Polychlorinated Biphenyls PCBs:
Toxaphene	

Source: United States Environmental Protection Agency

The three main programmes that MSD has implemented include:

- A pretreatment programme, Permitting and Pretreatment Compliance.
- Chemical Spill Prevention and Response – Hazardous Materials Incident Response Team.
- Sampling and Monitoring to reduce risk.

These programmes will be described in some detail.

12.6 REASONS FOR DOING PERMITTING AND PRETREATMENT COMPLIANCE PROGRAMMES

MSD has defined its reasons for implementing the aggressive programmes to regulate the use, storage and transport of industrial chemicals and wastes. This is communicated to regulated industries, MSD's users, elected officials and state and federal regulators. These reasons include:

Risk reduction: Monitoring industrial sites, sewers and surface streams helps to protect public health and water quality by finding and backtracking problems to the source. Wastewater treatment plants are biological systems that can be harmed by unpermitted discharges. Risk of accidents from some chemicals is very high.

Customer service: Investigating and mitigating spills and other unpermitted discharges helps protect public health and safety. Permitting and monitoring industrial and commercial discharges allows all users to share equitably in the sewer service and treatment resources that MSD provides. All these activities aid the protection and enhancement of a cleaner water environment in streams, and quality of life in the community.

Regulatory compliance: The Pretreatment Programme is required under State and Federal US regulations to inspect and monitor the 121 Significant Industrial Users (SIUs) in Louisville. MSD goes beyond the minimum regulatory requirements and monitors all industrial and commercial establishments that have in the past, or could in the future, adversely affect service to public health or the environment, or that incur a disproportionate share of the cost of operating MSD's treatment system.

The Permitting and Pretreatment Compliance Programmes issue Industrial User Permits and monitor compliance with those permits and with the wastewater/ stormwater discharge regulations. Industrial and commercial sites are inspected regularly, monitored as needed or required for the purpose of determining compliance with the local regulations, the site permit, or state and federal regulations. They are assisted in returning to compliance if a problem arises. MSD strongly encourages pollution prevention and works with a University of Louisville Pollution Prevention Centre to assist industries in reducing waste and discharges to the sanitary sewers.

There are six minimum requirements for a pretreatment programme.

1. *Legal Authority.* The municipal wastewater treatment system (POTW) must operate pursuant to legal authority enforceable in Federal, State, or local courts.
2. *Procedures.* The POTW must develop and implement procedures to ensure compliance with pretreatment requirements.
3. *Funding.* The POTW must have sufficient resources and qualified personnel to carry out the authorities and procedures specified in its approved pretreatment programme.
4. *Local limits.* The POTW must develop local limits or demonstrate why these limits are not necessary.
5. *Enforcement response plan.* The POTW must develop and implement an ERP that contains detailed procedures indicating how the POTW will investigate and respond to instances of industrial non-compliance.
6. *List of significant industrial users.* The POTW must prepare, update and submit to the Approval Authority a list of all significant industrial users.

12.7 COMPONENTS OF THE PERMITTING AND PRETREATMENT COMPLIANCE PROGRAMME

The basic components are as follows.

- Commercial/Industrial Process Plan Review
- Permits
- Unusual Discharge Requests
- Industrial Inspections
- Sampling and Monitoring
- Compliance and Enforcement.

These components are each described in detail below.

12.7.1 Commercial/industrial process plan review

Any new or upgraded industrial or commercial facility must go through a plan review process. The aim of the review, as it relates to proposed commercial and industrial facilities, is as follows:

- To provide Pretreatment (PT) regulatory guidance and assistance.
- To ensure that adequate pretreatment equipment and facilities are being used by the commercial/industrial community.
- To ensure that applicable required PT equipment/facilities are constructed in the appropriate location to maximize PT capabilities.
- To ensure that all process waste stream connections, floor drains, floor sinks and other discharge points are identified/accounted for and addressed.
- To ensure that existing facilities that modify or revise their facilities are identified and provided with the same regulatory guidance.
- To provide a mechanism for MSD to update its commercial/industrial database and assure industrial users (IU) are identified and addressed.
- To identify IUs (industrial users) and refer them to the Permit and Hazardous Spills Prevention Section, for review.

The programme is focused on identification of targeted industries, as well as field inspection of these facilities during construction. Identification through the review process can prove beneficial to pretreatment and monitoring facilities, since necessary improvements are less costly to include at the initial planning and installation stage than after the facility is built and in operation.

12.7.2 Permits

The process used by MSD to identify and categorize new industrial discharges includes the following:

- inspections
- referrals
- new user data

- regular phone book and newspaper review
- citizen/public referrals
- plan review for new development.

Once new industries are identified, MSD formally reviews the proposed processes to determine regulatory requirements that may need to be established. The assessment process includes a commercial/industrial process plan review.

Permits issued to the industrial dischargers are classified as Significant Industrial User (SIU) Permits, General Permits, Hauled Waste Permits or Unusual Discharge Permits.

A discharger is classified as an SIU if it is an industrial discharger which:

- is subject to US EPA Categorical Pretreatment Standards
- discharges a base amount per day that is viewed as significant
- contributes a process waste stream which makes up 5% or more of the average dry-weather hydraulic or organic capacity of the treatment plant; or
- is any other industrial use that is designated as such by MSD on the basis that the industrial user has a reasonable potential for adversely affecting MSD's treatment works operation or for violating any pretreatment standard or requirement or is a potential threat to the public health and environment.

12.7.3 Unusual discharge requests (UDR)

MSD's pretreatment programme issues permits to dischargers with ongoing (continuous) discharges. Additionally, short term, 'one-time' discharges may be discharged to the sewer system through the UDR programme. This programme is designed to allow MSD to have sufficient control over the type and characteristics of the wastewater being discharged and to ensure that contaminants are not allowed to enter the sewer system which could potentially cause problems at the treatment plant (such as upset conditions, pass through, or violation of water quality limits). Through the UDR programme, wastewater is discharged to an on-site sewer connection. Typical types of UDRs include underground storage tank excavations, bad batches of product, car wash holding tanks and purge water from monitoring wells. Below is a description of the general process that is followed by MSD when a request for a UDR is made.

Initial Request. Companies which have a batch of wastewater that needs to be disposed of, but is not covered by any other MSD discharge permit, may telephone MSD and describe their request.

Processing of Requests. MSD has developed a UDR Review Committee which typically meets on a weekly basis to review the requests for discharge. This committee generates guidance regarding parameters for analysis, discharge constraints and/or approval/denial of UDRs.

Parameters for Analysis. Based on the description of the request to discharge, MSD will determine what parameters need to be analyzed. This decision will typically be made at the weekly committee meeting. However, MSD attempts to facilitate the UDR process to meet user needs.

Some specific information regarding sampling parameters for underground storage tank (UST) excavations – a common request – follows.

As a minimum, MSD requires that wastewater from all USTs must be analyzed for the following parameters:

- total benzene, toluene, ethyl benzene, xylene (BTEX)
- total petroleum hydrocarbon (TPH or Oil and Grease Hydrocarbon)
- petroleum aromatic hydrocarbon (PAH)
- total lead
- closed cup flashpoint.

12.7.4 Industrial inspections

MSD's service area has been divided into four areas. One pretreatment inspector is assigned to each area (or 'beat'). The primary responsibility of the inspector is to routinely inspect the industrial facilities in their 'beat' to ensure that the pretreatment systems are properly operated and maintained.

In addition, each inspector is assigned one of the regional wastewater treatment plants. The inspector is called to investigate whenever an unusual condition is observed by the treatment plant operators.

The inspector is also an available resource to the users. The inspector reviews and enters the self-monitoring data; develops and evaluates loading data; performs slug control inspections; reviews permit applications; develops facility maps; assists with special projects if associated with the industries or the treatment plant; enforces local and Federal regulations; assists legal staff in the enforcement area; identifies pollution prevention targets for industry; conducts quarterly meetings with operations personnel to review performance; conducts sampling; and tracks correspondence and milestones of compliance.

The inspectors receive extensive formal training regarding proper pretreatment inspection protocol.

12.7.5 Sampling and monitoring

The US EPA establishes test procedures for industrial chemicals. It is MSD's responsibility to ensure the use of trained sampling personnel, the use of protocol established by regulation, and appropriate data management and sampling methodology. The industries are sampled for wastewater quality; some are also monitored for flow. Permitted industries monitor and report their results a minimum of twice per year and the US EPA requires MSD to inspect and sample the industry discharge at the plant site, at minimum, once per year. MSD's Pretreatment Programme adheres to these requirements, but uses a risk-based methodology, which considers many variables, to determine the frequency of sampling and inspection. Higher risk industries and those with a poor spill history are inspected more frequently.

Sampling may be done for metals, cyanide, ammonia, organics, conventional pollutants, toxicity, surfactants and pH.

Some industries are continuously monitored for Lower Explosive Limits (LELs). MSD conducts additional Quality Charge tests and compliance sampling events throughout the year at various locations. Conventional pollutants (BOD, TSS and flow) are used for a programme to charge industries fees for high-strength wastes which are discharged into the municipal wastewater system.

12.7.6 Compliance and enforcement

MSD has its legal authority through the Wastewater Discharge Regulations (WDRs). The WDRs define the terms of compliance and resulting enforcement for industries discharging to the collection system. Compliance is determined by comparing the results from a combination of self-monitoring and MSD sampling to the discharge limitations. The primary enforcement issue is when an industry fails to meet its discharge limits. Other issues which require enforcement activity include failure to submit required reports, failure to meet the requirements of a compliance schedule and the release of prohibited discharges. MSD has developed and implements an EPA-approved Enforcement Response Plan (ERP).

When an industry violates the specific discharge limitation for a given parameter, a Notice of Violation (NOV) is issued and the industry is required to resample within 30 days for the violating pollutant. If the results of the additional sampling exceed the discharge limits, the industrial user may meet the criteria for significant non-compliance (SNC). A discharger is considered to be in SNC if it meets one of the following criteria:

- Chronic violations of wastewater discharge limits, defined here as those in which 33% or more of all of the measurements taken during a six-month period exceed (by any magnitude) the daily maximum limit or the average limit for the same pollutant parameter.
- Technical Review Criteria (TRC) – violations, defined here as those in which 33% or more of all of the measurements for each pollutant parameter taken during a six-month period equal or exceed the product of the daily maximum limit or the average limit multiplied by the applicable TRC (TRC = 1.4 for BOD, TSS, fats, oil and grease, and 1.2 for all other pollutants except pH).
- Any other violation of a pretreatment effluent limit (daily maximum or longer-term average) that MSD determines has caused, alone or in combination with other discharges, interference or pass-through (including endangering the health of wastewater plant and collection system personnel or the general public).
- Any discharge of a pollutant that has caused imminent endangerment to human health, welfare or to the environment, or has resulted in the exercise of emergency authority to halt or prevent such a discharge.
- Failure to meet, within 90 days after the schedule date, a compliance schedule milestone contained in wastewater discharge permit or other order issued hereunder for starting construction, completing construction or attaining final compliance.
- Failure to provide, within 30 days after the due date, required reports, such as baseline monitoring reports, 90-day compliance reports, periodic self-monitoring reports and reports on compliance with compliance schedules.
- Failure to accurately report non-compliance.
- Any other violation or group of violations which MSD determines will adversely affect the operation or implementation of the local pretreatment programme.

12.8 CHEMICAL SPILL PREVENTION AND RESPONSE – THE HAZARDOUS MATERIALS INCIDENT RESPONSE TEAM

Because of the significant potential for industrial discharges and spills to affect the public health and safety, MSD created and funded a special team to respond to incidents.

This team responds to all hazardous materials spills in Louisville and implements the Hazardous Materials Spill Control Plan (HMPC) Programme, to ensure compliance with the Hazardous Materials Ordinance (HMO). Industrial and commercial sites that use, store and transport hazardous materials are inspected regularly and monitored, if needed, to determine compliance with the HMO and the approved HMPC.

Hazardous materials may be considered generally as those substances in a quantity and form which, if released, can be harmful to life, property or the environment. The range of materials encompassed by this definition includes explosives, flammable and combustible liquids and solids, poisons, oxidizing or corrosive materials, and compressed gases. Of particular concern, is the potential harm to life posed by hazardous chemicals.

In recent years, hazardous material spills have caused, or threatened to cause, serious problems to the environment of Louisville. New hazardous substances are continuously being developed and each year larger volumes are produced and transported through the community. In response to this, Louisville adopted an ordinance to prevent serious harm to the environment and reduce the likelihood of a problem from a hazardous material spill.

The ordinance requires that businesses which have hazardous materials on-site must submit a plan for each business site as to how they will respond in the event of a spill of that hazardous material. An HMPC must be submitted by any business that manufactures, uses or stores hazardous materials in minimum designated quantities at their business location. Guidance as to what constitutes a hazardous material has been provided by a master list which has been defined by the US EPA (Table 12.2).

For businesses that have certain minimum quantities of hazardous materials on their business site, an HMPC Plan must be submitted to MSD. Businesses that must submit plans are generally defined by a Standard Industrial Classification (SIC) code in the ordinance.

Examples of businesses that are expected to submit plans include gas stations, manufacturing facilities, hospitals and medical laboratories, cleaning establishments, pest exterminators, and state and local government offices that handle hazardous materials.

Businesses that fall into the designated industrial groups which do not have a hazardous material on-site may request an exemption from the submission of an HMPC Plan by completing an exemption form. The ordinance specifically exempts residents who have hazardous materials on-site for their own personal use, consumer product and food stuff manufacturers who are covered under the food and drug act, agriculture operations that are handling chemicals only for purposes of their application on a farm and instances where hazardous materials do not exceed reportable quantities and the administrating agency sees no danger to public health.

Those businesses that do have to submit a HMPC Plan are contacted by letter by MSD concerning the schedule for submission. Businesses that believe they should be exempt from a submission need to submit an exemption form as described in the letter they receive. Nearly 5,000 businesses in the community have received a letter, to date.

The HMPC Plan is submitted on forms that are provided by MSD. Once submission has occurred, a joint review is conducted by MSD and the Fire Department that has jurisdiction in the area of the business. Plans that are considered to be deficient are returned to the business for correction and resubmission.

Once approved, businesses are responsible for implementing their plan including initiation of a training programme for employees within their business. If there are any changes to the plan, businesses must give prompt notice of these changes to the administering agency.

A major emphasis of the ordinance is that spills that occur by businesses be promptly reported to the '911' US emergency telephone number. In the event of a failure to provide such notification, a fine of up to US$ 5,000 may be levied. The ordinance also provides for penalties of up to US$ 1,000, which may be levied in the event of a spill occurring due to negligence or because of previous spill occurrences by the same business. The purpose of these fines is to emphasize again the importance of protecting the environment and surrounding residents from potential harm from hazardous material spills.

An Appeals Board has been created to allow businesses to appeal if they disagree with either the agency interpretation of the adequacy of an HMPC Plan submission, or want to challenge a fine or penalty that has been levied. The Appeals Board is made up of representatives from industry, regulatory agencies and the general public. The Appeals Board hears cases and makes recommendations to MSD for final action.

Hazardous chemical incidents can range in magnitude from very minor spills causing no adverse health affects to major incidents with the potential to affect hundreds of people. The purpose of this ordinance is to emphasize the responsibilities of those businesses which handle hazardous materials in protecting the environment from adverse damage. The ordinance is a continuing effort by local government to improve the environment of Louisville and prevent disasters, such as the hexa-octa incident and the sewer explosion.

12.9 SAMPLING AND MONITORING TO REDUCE RISK – THE COLLECTION SYSTEM MONITORING PROGRAMME

The third component of MSD's successful programme for reducing risk from the use, storage and transport of industrial chemicals is collection systems monitoring and data management. In order to be proactive in its approach to wastewater treatment, MSD implemented a system for sampling and monitoring flows in the wastewater system. From this effort, a new programme entitled the Collection System Monitoring Programme emerged.

Seventeen sites have been selected for flow monitoring, and data collection takes place within four treatment plant areas. These four treatment plants are the only ones within Jefferson County that have permitted industrial dischargers. The system includes:

- permanent flow monitors at the 17 approved sites
- periodic sampling at the 17 approved sites (at least quarterly).

Sampling is tentatively scheduled to be conducted quarterly. The flow and mass loading data is used for many purposes, including system trend analysis and detection of chemicals that indicate possible violations from industry.

This programme is in addition to random manhole monitoring, explosivity detection, and other system sampling and monitoring done by MSD.

12.9.1 Data management and computerization

MSD has a database which provides a central location for housing data collected and tracked under the Pretreatment Programme.

Each permitted IU is assigned a unique ID number within the system. All information pertaining to the IU is linked with this ID number including outfalls. Self-monitoring and MSD compliance data are entered. MSD's discharge limitations are used to track the compliance status of IUs and prompt enforcement actions.

There are other aspects of the Pretreatment Programme that also utilize tracking data, such as, inspection and enforcement actions, permitting, unusual discharge requests, plan review, hazardous materials spill response and control plans, emergency response incidents, and sampling events.

MSD is always searching and developing new and innovative uses for technology to streamline Pretreatment Programme functions and make jobs easier. Currently, MSD has been evaluating ways to increase the use of computers in the field, to make data collection more efficient by allowing staff to perform real-time data entry. Inspectors have portable computers with access to detailed maps, industry HMPC plans and other critical information, to allow a fast and effective response during emergencies. Many potential problems have been prevented by MSD's aggressive monitoring and tracking programmes.

12.10 CONCLUSIONS: NEED FOR STRONG LOCAL PROGRAMMES TO REDUCE RISK

Louisville's experience tracks that of many other cities that have industries that use, store and transport hazardous chemicals and chemical wastes. It is local populations and the local water environment that are most at risk when accidents happen or when an industry intentionally dumps a waste product. It falls upon the local governments and water and wastewater utilities to develop programmes that will protect the public and the environment.

The Louisville programmes are good examples of the kinds of programmes that local governments can implement.

REFERENCES

USA, Kentucky. Louisville and Jefferson County Metropolitan Sewer District. Jan. 2006. *Wastewater/Stormwater Discharge Regulations*. Available at: www.msdlouky.org/insidemsd/pdfs/MSDWDR2005.pdf.

USA, Kentucky. Louisville and Jefferson County Metropolitan Sewer District. Jan. 2004. *History of MSD*. Available at: www.msdlouky.org/aboutmsd/history.htm.

USA United States Environmental Protection Agency. 2007. *National Pollutant Discharge Elimination System – Pretreatment Program Website*. Available at: http://cfpub.epa.gov/npdes/ index.cfm.

Chapter 13

Lessons learned: a response and recovery framework for post-disaster scenarios

Darren Saywell

Development Director, International Water Association, United Kingdom

ABSTRACT: Responses to man-made and natural disasters have frequently been disjointed, inappropriate and inefficient, as a result of a lack of perspective of 'bigger picture' actions and processes. These experiences, in particular those of the 2004 Asian tsunami disaster response, highlight the need for a framework for disaster response, and the lessons learned from them provide the basis for recommended actions in post-disaster scenarios. The objective for the International Water Association (IWA) – the leading professional association dealing with all aspects of water management – in preparing a 'Response and Recovery Framework' was to provide orientation and guidance to its members seeking to participate in disaster response efforts. The framework aims to illustrate the progression of disaster response for the water and sanitation sector, including priorities and actions to be taken during various stages of response. From a meta-analysis of lessons learned, the basic structure of a typical response covers immediate aftermath (0–7 days); short term (next 3 months); and medium to long term (next 3–12 months). This framework provides a broad overview of the priority actions and emphasis during these phases, and is augmented by boxed material and references highlighting real cases.

13.1 INTRODUCTION

13.1.1 Background

The Asian tsunami disaster of 26 December 2004 was unprecedented in its breadth and scale of impact. It left behind a trail of devastation that affected 13 countries on two continents, leaving hundreds of thousands dead and a larger number of survivors without food, shelter, water or sanitation facilities. It also posed, to governments and relief agencies, the task of mounting an equally unprecedented disaster response. The need for improved response methodologies and approaches has been exacerbated by the apparent increase in frequency and magnitude of disaster events, such as the 2006 cyclone that affected approximately 30 million Bangladeshis.

13.1.2 Rationale

The rationale for developing a response and recovery framework for post-disaster scenarios stems from the following considerations:

- The lack of an integrated approach to disaster response. In the wake of the tsunami, the international community showed an overwhelming demonstration of support. However, because a disaster on this scale has rarely been seen, few individuals or organizations have had experience in preparing a comprehensive response.

Consequently, many found themselves at a loss for what types of action to take or when to take them, and this in turn often resulted in inappropriate types of aid, duplication of efforts and sometimes even more harm than help, despite the good intentions behind them.

● The need to synthesize the experiences and lessons learned by other organizations. Many organizations have documented their responses to the Asian tsunami and other disasters, but there has been no systematic meta-synthesis of these experiences.

13.1.3 Objectives

The objective of the Response and Recovery Framework is to assist agencies working in post-disaster scenarios. While the motivation for this document was the Asian Tsunami Disaster, the broader aim is to provide guidance that can be applied more generally to natural disasters as a whole, including earthquakes, hurricanes, cyclones, forest fires and floods. More specifically, this document seeks to:

● Reflect on key lessons from previous disaster responses, particularly those from the Asian tsunami disaster.
● Provide an overview and timeline of priority activities in the aftermath of a disaster.
● Provide a broader understanding of the goals of a disaster response and the context in which it takes place.

Though the intended primary audience of this document is the IWA and its members, it may also serve as a guide for others working in the sector.

13.1.4 Methodology

This framework is based on a synthesis of the lessons learned from the experiences of a range of agencies in post-disaster scenarios. This was accomplished through:

● A review of over 70 press releases, presentations, situation updates and reports from more than 20 major NGOs, UN agencies, and other public and private organizations working in the Asian tsunami disaster, as well as documents on disaster response in general; a selection of these is listed in the bibliography provided.
● A meta-analysis of these responses through a distilling of similarities and parallels between the documented responses as well as common concerns.

The results of this analysis provide the foundation for the actions recommended in the framework.

13.1.5 General principles

The main lessons to be drawn from past experiences and emphasized in the framework are:

● Ensure that relief activities and donations are in response to need. Often, aid is provided regardless of whether or not it is needed, or is inappropriate with regard to timing or cultural setting, leading to wasted time, effort and money. The need to conduct ongoing field level assessments that include local community consultation was unanimously emphasized in documentation and evaluations of past responses.

- Coordinate response plans and activities among all participants in the response. Following the tsunami, an outpouring of support and aid ensued, leading to duplication of efforts and inefficient allocation of resources. To ensure the most effective use of resources, a clear lesson was the need for organizations and governments of affected countries to better coordinate, communicate and work together to implement field level responses.

13.2 RESPONSE AND RECOVERY FRAMEWORK

13.2.1 General guidelines

There are a few principles that should frame the response as a whole, to maximize the impact of relief and rehabilitation efforts. Throughout the response, the IWA and its members should seek:

- Compliance of donations and relief activities with internationally accepted standards such as the 'Sphere Humanitarian Charter and Minimum Standards in Disaster Response' (see Box 13.1).
- Coordination with other organizations and channelling of aid through a coordinating agency (typically the United Nations (UN), see Box 13.2 for listing and brief descriptions of UN agencies and their functions).
- Awareness of local cultural conditions, needs, capabilities and expectations.

BOX 13.1 International standards for emergency relief

The Sphere Humanitarian Charter and Minimum Standards in Disaster Response is a set of universal minimum standards in the core areas of humanitarian assistance that aim to improve the quality of assistance provided to people affected by disasters and enhance accountability on the part of aiding organizations.

- The Sphere Minimum Standards contains standards and guidelines pertaining to water, sanitation, and hygiene promotion that can be found at: www.sphereproject.org/ handbook/hdbkpdf/hdbk_c2.pdf.
- The full Sphere Handbook (2004 Revised Edition) can be downloaded at The Sphere Project's website: www.sphereproject.org/.

Alternatively, it may be ordered through Oxfam Publishing by sending an order form (found at: www.sphereproject.org/handbook/orderform.pdf) to:

Oxfam, c/o BEBC Distribution,
PO Box 1496, Parkstone, Poole, Dorset, BH12 3YD
Tel: +44 (0) 1202 712933 • Fax: +44 (0) 1202 712930
Email: oxfam@bebc.co.uk

The Code of Conduct for The International Red Cross and Red Crescent Movement and NGOs in Disaster Relief is a set of professional standards that is meant to guide the way NGOs work in disaster situations.

- The full code can be found online at: www.ifrc.org/publicat/conduct/.

BOX 13.2 Selected United Nations (UN) agencies and their functions

The agencies of the United Nations play a central role in disaster response, both in terms of providing aid and relief and in logistics and coordination of activity. It is important for those participating in disaster responses to be aware of the functions of the UN agencies, to coordinate and plan their efforts accordingly, and to avoid duplication of efforts. Below are selected UN agencies that were involved in the post-Tsunami response and/or are coordination points during disasters.

● *Office for the Coordination of Humanitarian Affairs (OCHA)* – The mission of the OCHA is to reduce human suffering and material destruction caused by natural disasters and emergencies, by mobilizing and coordinating the collective efforts of the international community, in particular those of the UN system. Its main functions are to advise the Secretary General on emergencies and recommend appropriate actions.[a]

● *Humanitarian Information Centres and Partners (HIC)* – The HICs are managed by OCHA and support the coordination of humanitarian assistance through the provision of information products and services.[b]

● *Food and Agriculture Organization (FAO)* – The FAO's mission is to raise levels of nutrition and standards of living, and to improve the efficiency of production and distribution of food and agricultural products. In relief operations, it focuses on the provision of agricultural inputs, such as seeds and veterinary services. This includes working closely with NGOs active in this field, and, in some countries, with UNICEF.

● *United Nation's Children's Fund (UNICEF)* – To ensure a 'first call for children,' UNICEF mobilizes both political will and material resources to help developing countries. UNICEF's niche in emergencies is its role as an advocate for children. Its general resources budget is over US$ 500 million, 25 million of which is allocated to its Emergency Programme Fund.[1]

● *United Nations Development Programme (UNDP)* – UNDP is the United Nations largest provider of grant funding for development and the main body for coordinating UN development assistance. UNDP provides logistic, communications and other support for the activities of the United Nations Emergency Relief Coordinator and to UN Disaster Management Teams. UNDP's annual budget is US$ one billion. Five per cent is allocated for disaster preparedness. The UNDP Resident Coordinator has the right to allocate up to US$ 50,000 for emergency needs to a country in a disaster situation.

● *United Nations High Commissioner for Refugees (UNHCR)* – UNHCR's main task is to provide protection and assistance to refugees and to seek permanent solutions to the problems of refugees. UNHCR's operations can be classified into two categories:
   a.   Protection of refugees from further persecution and violence, including being forced back into areas from which they have fled, while helping lay foundations for lasting solutions to refugee problems.
   b.   Meeting the physical needs of refugees – UNHCR works to supply refugees with food, water, health care, shelter and sanitation.

● *United Nations Joint Logistics Centre (UNJLC)* – The UNJLC's mission is to complement and coordinate the logistics capabilities of cooperating humanitarian agencies during large-scale emergencies. The UNJLC aims to collectively identify and eliminate logistics bottlenecks of common interest to the humanitarian community to avoid wasteful competition among Agencies. Related to this task, the UNJLC plans, prioritizes and de-conflicts relief movements when available infrastructure capacity is limited. Through this process, the UNJLC advises on the most efficient transport modes and performs movement control functions. The UNJLC also frames logistics-related policy issues affecting humanitarian logistics operations.

(Continued)

BOX 13.2 (*Continued*)

The UNJLC acts as a platform for gathering, collating, analyzing and disseminating information required by Agencies to optimize logistics planning and management.

[a]'Humanitarian Information Centres'. www.humanitarianinfo.org/abouthics.html.
[b]IFRC. 'Improving Coordination'. p.21.

BOX 13.3 Responses in depth, timeline/chronology of response activities

0–7 days			7–60 days	3–12 months
Initial assessment of situation	Appeal for aid & alerting of network	Identification of partners & formation of collaborative agreements	Ongoing collaboration and coordination with other organizations, agencies, and governments. Holding of periodic sector reviews/ evaluations	
	Statement of purpose & focus		Supporting website development	
			Mobilization of resources – Matching and targeting of volunteers and assets to impacted areas	
			Identification of local organizations and agencies that need help	Twinning of communities/utilities via IWA CEO/utility networks
			Assembly, posting, and circulation of technical guidelines	Holding of training/briefing programmes for volunteers and NGOs
				Holding of 'training of trainers' programmes
			Creation of online forum	Development of mechanisms and tools to monitor outcomes of reconstruction
				Creation of water safety and public information campaigns
				Enhancement of disaster preparedness

13.2.2 Immediate aftermath (0–7 days)

The focus of activities in the immediate aftermath of the disaster should be on beginning to formulate a plan of action and identifying and communicating with other actors. While the overall framework has been organized by time period, the activities in each section have been broken down according to the nature of the activity. For a more detailed chronology of activities, refer to Box 13.3.

13.2.2.1 Preliminary actions

- Identification of partners and formation of collaborative arrangements.
- Rapid impact and needs assessments in targeted areas, to identify the extent of damage and need for immediate relief, in particular regarding clean water and waste disposal (see Box 13.4 for more details on needs assessments). The results of these assessments will form the basis for setting priorities and appeals for immediate aid.

BOX 13.4 Needs assessments

Assessment teams should be multidisciplinary, with expertise and experience in the areas of public health, water and sanitation, nutrition and food security, as well as including a logistician and a local official or community member.[2] Whenever possible, joint assessments with other organizations/agencies should be conducted, and have standardized assessment methods and tools.[3]

Timetable for needs assessments:

	Ongoing assessment approach for rapid-onset natural disasters					
Pre-disaster		**Post-disaster**				
	When possible	**First 10 hours**	**12–36 hours (& then, as needed)**	**7–15 days**	**30–60 days**	**3 months**
Assessment type	Forecasting & early warning	Disaster (early) notification	More detailed disaster needs assessment	Ongoing monitoring of situation and needs		
Information needs	**Collect and disseminate early warning information** (especially for flash floods, tsunamis, storms, volcanoes, forest fires, etc.)	**Alert headquarters** – Disaster type, date – number of casualties reported – number of properties damaged and type of damage – Immediate emergency priority needs (e.g., search & rescue, first aid)	**Assessment team** – Disaster magnitude – Geographic area affected – Detailed assessment of needs and resources and other responders – Define intervention for ensuing weeks and months (if necessary)	**Continued monitoring and assessment** – Ongoing situation, response and needs – Changes in status and needs – Need for longer-term assistance and rehabilitation (after 3 months) – Plan of action for assistance to continue past 3 months		

Source: International Federation of Red Cross and Red Crescent Societies, 'Disaster Emergency Needs Assessment', Disaster Preparedness Training Manual. p. 10–11.

As the response progresses, the content of the needs assessments will evolve:

Immediate Term: Within the first few days post-disaster, an assessment is needed to determine the areas affected, the extent of the damage done to water and sanitation sources and systems, and the immediate water supply and waste disposal needs. These could be in the form of rapid assessment sample surveys of infrastructure or on-site visual assessments, conducted by small teams on the ground in country.

[2]Heijnen, Han Antonius. 'Report Panel 2.8 Water, Sanitation, Food Safety and Environmental Health'. WHO Conference on the Health Aspects of the Tsunami Disaster in Asia. Phuket, Thailand. 5 May 2005. www.who.int/hac/events/tsunamiconf/presentations/2_8_water_sanitation_food_safety_environ_health_heijnen_doc.pdf. p.1.

[3]Houghton, Rachel. 'Tsunami Emergency Lessons from Previous Natural Disasters'. ALNAP. www.alnap.org/pubs/pdfs/tsunamibriefing05.pdf.

(Continued)

BOX 13.4 (*Continued*)

Several organizations have compiled guidance and checklists for rapid assessments:

Water, Engineering and Development Centre (WEDC):
http://wedc.lboro.ac.uk/publications/pdfs/es/ES16CD.pdf
http://wedc.lboro.ac.uk/WHO_Technical_Notes_for_Emergencies/13-Emergency_sanitation-planning/13-Emergency_sanitation-planning.htm

World Health Organization (WHO): http://w3.whosea.org/en/Section23/Section1108/info-kit/Rapid_needs_Assessment_guidelines.doc

The Sphere Project: www.sphereproject.org/handbook/hdbkpdf/hdbk_c2.pdf (Appendix 1, pp. 89–92)

13.2.2.2 Public notification

- Statement of short-term purpose and task and activity focus.
- Appeals for aid and alerting of members and others in databases/network.

13.2.3 Short term (next 60 days)

In the following 60 days, emphasis should be placed on the identification of needs and on the smooth provision and support of emergency aid and relief to survivors. This requires not only the provision of need-based material aid to survivors, but also proper training and support for workers and volunteers on the ground. Towards the end of this period, assessments of longer-term needs should be conducted.

13.2.3.1 Initial evaluation of impacted areas

- Ongoing execution of needs assessments to inform planning and resource allocation. These should be more detailed than the initial rapid needs assessments, in order to frame appropriate response plans and programme designs (see Box 13.4).
- Identification and contacting of local partner organizations and relief agencies to which the experience, expertise, and assets of members can be made available.
- Network identification and strengthening in target countries.
  - professional associations
  - utilities
  - university/students.

13.2.3.2 Mobilization of human, capital, and physical resources

- Mobilization of human, physical and financial support for the provision of short-term measures for water supply, wastewater treatment and hygiene education.
- Facilitating of networking between volunteers and the needs of target communities and relief agencies.
- Broadcast of requests for equipment and or supplies and identification of surplus utility resources and targeting to impacted communities.
- Development of/direction to donations fund.

13.2.3.3 Technical support/guidance

- Training of volunteers in necessary, high priority tasks such as water quality testing, chlorination process, etc. (see Box 13.5).

- Creation of an online forum for advice and offers of assistance by members. This will act as a type of technical advisory service and may also include a Q&A Service, fact sheets and links to other practical and relevant online resources.
- Assembly and posting of technical guidelines and best practices documents on-line; circulation and advocacy to governments and relief agencies on the ground.

BOX 13.5 Technical support

It is widely recognised that water supply and sanitation concerns are of the highest priority in post-disaster situations, and therefore relief agencies often rush to the scene with generous aid and volunteers. However, especially in the water and sanitation sector, volunteers and NGOs on the ground often do not have the experience or expertise required for building infrastructure, or in issues of water quality protection and monitoring, hygiene promotion and waste management, or even background knowledge of the basics of water and sanitation issues. It is therefore important that they have access to and are provided with expertise and the necessary technical guidance and support through manuals and guidelines and/or training programmes.

Examples of technical guidelines/manuals:
World Health Organization (WHO):
www.who.int/water_sanitation_health/hygiene/envsan/ technotes/en/index.html
www.who.int/water_sanitation_health/hygiene/emergencies/emergencies2002/en/

Water, Engineering and Development Centre (WEDC):
wedc.lboro.ac.uk/specialist-activities/interests.php?area=1&Emergency-water-supply-and-sanitation

Oxfam:
www.oxfam.org.uk/what_we_do/emergencies/how_we_work/manuals.htm

Pan American Health Organization: www.paho.org/english/DD/PED/te_albe.htm

Examples of training programmes
IRC Briefing Programmes:
www.irc.nl/page/16091
www.irc.nl/page/16266

WEDC:
wedc.lboro.ac.uk/education/in_country.php

For more general information on learning and NGOs:
ALNAP's Review of Humanitarian Action in 2003: Synthesis of Findings of Evaluation Reports from 2003:
www.alnap.org/rha2003

BOND:
www.Bond.org.uk/pubs/ol.htm

Evaluating Humanitarian Action using the OECD-DAC Criteria
March 2006
www.alnap.org/pubs/pdfs/eha_2006.pdf
Two Lessons Learned papers on the South-Asian Earthquake 2005:
'*Learning from Previous Earthquakes*' **and** '*Learning from Previous Recovery Operations*'.
www.alnap.org/lessons_earthquake.htm

13.2.3.4 Supporting actions

- Collaboration and coordination of efforts with other organizations and agencies. A coordinating agency should be identified/established from the outset.
- Sector coordination meetings to share information, plan joint assessments and interventions, discuss emerging problems or gaps in response.
- Coordination meetings with governments, NGOs, UN agencies and sectoral working groups. These types of coordination are enhanced and more effective when there are pre-existing established lines of communications between groups and/or part of a disaster preparedness plan (Box 13.6).
- Supporting website development.
- Data assembly at the national level through member-led efforts that support short-term targeting and longer-term needs profiles for reconstruction.

BOX 13.6 Disaster preparedness

An essential component of long-term capacity building is disaster preparedness. Disaster preparedness is defined as the 'readiness to reduce the impact of disasters, and where possible, predict and even prevent disasters occurring'.[4] Building this capacity allows communities to respond and recover quicker and more effectively to future disasters; the World Bank estimates that for every dollar spent on risk reduction, $7 is saved on relief and repairs.[5] Important components of disaster preparedness include:

- Development of an early warning system. For more information, see *The International Early Warning Programme (IEWP)*: www.unisdr.org/ppew/whats-ew/basics-ew.htm.
- Annual disaster preparedness reviews, including inspection of assets and reconfirmation of availability and functioning.[6]
- Formulation of national and local frameworks of preparedness and response policies and strategies, especially regarding disruptions in water and sanitation services, hazardous waste spills and explosions, and other incidents that threaten to expose the population to environmental health risks.[7]
- Development of a framework of communications between NGOs and government and pre-disaster collaboration and coordination agreements (see example below).

NGO-government collaboration in the Asian tsunami disaster

In Sri Lanka, sector coordination is institutionalized informally through a network of government agencies, NGOS and sector support institutions, led by the National Water Supply and Drainage Board. The group meets quarterly, supports action-research, develops policies and exchanges sector experiences. This entity allowed for a rapid coordination and networking, cooperative action and understanding of needs. A similar emerging effort in India, and the existence of a roster of emergency-mitigation trained engineers under RedR- India allowed rapid deployment of additional expertise to support relief efforts in water supply, sanitation and environmental health. Pre-crisis existence of these mechanisms has certainly enhanced coordination

[4]International Federation of Red Cross and Red Crescent Societies. 'What We Do, Disaster Preparedness'. www.ifrc.org/what/disasters/dp/index.asp.
[5]Houghton, R. *Tsunami Emergency, Lessons from Previous Natural Disasters*, ALNAP, p.9.
[6]Heijnen, Han Antonius.
[7]Ibid.

(Continued)

BOX 13.6 *(Continued)*

and speeded up deployment of national experts. In Indonesia, a similar set-up was established during the Tsunami crisis. These coordination mechanisms have assisted government and NGO development partners who were already working in the country before the emergency to harness and guide the inputs of new humanitarian aid agents. The existence of similar mechanisms in Bangladesh, in both emergency response and the water and sanitation sector, has, over the last few years, allowed for faster and more effective responses to emergencies. A functioning coordination mechanism will also allow capacity building for emergency preparedness, including prior agreement on local standards and approaches for water and sanitation, taking into account national expertise and the SPHERE standards.

Source: Report Panel 2.8 Water, Sanitation, Food Safety and Environmental Health. www.who.int/hac/events/tsunamiconf/presentations/2_8_water_sanitation_food_safety_environ_health_heijnen_doc.pdf. p.3.

13.2.4 Medium term (next 3–12 months)

As involvement moves into the third month and beyond, aid efforts should transition into supporting the rehabilitation and reconstruction of communities' infrastructure, with an associated emphasis on capacity building within the affected communities.

13.2.4.1 Reconstruction assistance

- Twinning of communities/utilities via IWA Chief Executive Officer (CEO)/utility networks.
- Continued recruitment and targeting of volunteers with expertise in water quality testing, assessment and other technical skills to affected communities.
- Holding of briefing programmes/workshops for volunteers and NGOs on water solutions, sanitation systems designs and hygiene promotion strategies.
- Working with local water authorities to create sanitation and waste management programmes.

13.2.4.2 Capacity building – implementing institutions

- Holding of national/regional workshops on development/restoration of a sustainable water and sanitation system with the aim to increase experience exchange among developing countries in similar conditions, to build and strengthen existing capacity and to improve access to appropriate expertise.
- Enhancement of disaster preparedness strategies and plans.

13.2.4.3 Capacity building – local/community level

- Creation of water safety, hygiene promotion, and other health promotion and public information campaigns.
- Development and distribution of guidance/training manuals and materials on critical subject matter.

BOX 13.7 Reviews and evaluations

Reviews and evaluations of response plans and activities are an important component of an effective and efficient disaster response. The purposes of these reviews are manifold:

- To assess the effectiveness and appropriateness of the disaster response.
- To identify weaknesses in the current response and make changes accordingly.
- To enhance learning and accountability of relief agencies.
- To garner lessons learned to improve future responses.

These objectives may be met through the undertaking of the following types of reviews:

- strategic review
- real-time evaluation (RTE)
- after action review.

For more information and guidance on conducting reviews and evaluations, the following organizations have published their experiences and guidelines:

Food and Agriculture Organization of the United Nations (FAO).
www.reliefweb.int/rw/RWB.NSF/db900SID/EVIU-6D4ELC?OpenDocument

ALNAP (Tsunami Evaluation Coalition): www.tsunami-evaluation.org/The+TEC+Thematic+Evaluations/

TEC Synthesis Report: www.tsunami-evaluation.org/NR/rdonlyres/0B60502D-167D-478C-82DF-1961FCB48A8A/0/ExSum.pdf

United Nations Office for the Coordination of Humanitarian Affairs:
ochaonline.un.org/webpage.asp?Page=777

Department for International Development:
www.dfid.gov.uk/aboutdfid/evaluation.asp

- 'Training of trainers' programmes for local engineers and community members on the operation and maintenance of water and sanitation facilities, utilities and systems, water quality and sanitation conditions surveillance, and hygiene promotion.

13.2.4.4 Data collection/evaluations

- Creation of profiles of water-related reconstruction needs in target countries through member-led efforts at the local level.
- Development of mechanisms and tools, as well as standardized indicators, to monitor the outcomes of the reconstruction activities.
- Holding of periodic sector reviews/evaluations; dissemination of good practices and lessons learned as they emerge (Box 13.7).

13.3 CONCLUSION

As illustrated by the aftermath of the Asian tsunami, a disaster response requires extraordinary dedication of resources and commitment on the part of all participants.

Although the IWA and many of its members do not provide ground level assistance in disaster relief and recovery, they can play crucial roles in the response through the activities outlined in the framework, including:

• Matching and providing their expertise and technical assistance to those who need it. This includes the development and dissemination of technical guidelines and manuals and best practices, and the holding of training programmes. The IWA and its members may also play a consultative role in the recovery plans during the reconstruction and rehabilitation stages.
• Provision of water and sanitation supplies and equipment.
• Serving as a conduit between members/volunteers and requested aid.
• Collaborating with other relief agencies and governments to plan joint assessments and interventions.

This framework has aimed to provide a context for these key activities, as well as provide other supporting actions in a step-wise fashion for organizations in the water and sanitation sector during a post-disaster period. It is not intended to be a rigid rule, but rather a general guide, flexible enough to be applied and/or adapted to any disaster. It will be published on the IWA Tsunami Response web page found at: www.iwa-conferences.org/templates/dynamic/Conferences/conferenceB. aspx?ObjectId=224458.

For further help and advice during a disaster response, members may contact the IWA HQ, the local national committee structure, and, where necessary, the specialist. Contact information can be found on the IWA's homepage at: www.iwahq.org.

REFERENCES

Though not a comprehensive list of the materials reviewed or available on disaster response, provided below is a selected list of references and documentation of responses for further referencing.

Calvi-Parisetti, P. 2004. *Workshop of Lessons Learnt on the National and International Response to the Bam Earthquake, Kerman, Islamic Republic of Iran, 14–15 April 2004, Report*. UN Office for the Coordination of Humanitarian Affairs, Geneva, Switzerland. Available at: www.reliefweb.int/library/documents/2004/ocha-irn-15apr.pdf.

Clasen, T. and Smith, L. 2005. *The Drinking Water Response to the Indian Ocean Tsunami Including the Role of Household Water Treatment*, Water, Sanitation and Health Protection of the Human Environment, World Health Organization, Geneva. Available at: www.who. int/household_water/research/DW_response_tsunami.pdf.

Harvey, P. and Reed, B. *Emergency Sanitation – Planning. WHO Technical Note for Emergencies No. 13*. The Water, Engineering, and Development Centre (WEDC). Available at: http://wedc.lboro.ac.uk/WHO_Technical_Notes_for_Emergencies/13-Emergency_sanitation-planning/ 13-Emergency_sanitation-planning.htm.

Harvey, P., Baghri, S. and Reed, B. 2002. Chapter 16: Rapid assessment and priority setting. *Emergency Sanitation, Assessment and Programme Design*. Loughborough University, Water, Engineering & Development Centre. ISBN 184 380005 5. Available at: www.lboro.ac.uk/departments/cv/wedc/publications/es/ES00CD.pdf.

Haws, N. 2005. Water for people and long-term development for the Asian tsunami area. *Water Conditioning & Purification*, 26–27. April 2005. Available at: www.waterforpeople.org/upload/0405wfp.pdf.

Heijnen, H.A. 2005. *Report Panel 2.8 Water, Sanitation, Food Safety and Environmental Health*. WHO Conference on the Health Aspects of the Tsunami Disaster in Asia. Phuket, Thailand. (5 May 2005) Available at: www.who.int/hac/events/tsunamiconf/presentations/2_8_water_sanitation_food_safety_environ_health_heijnen_doc.pdf.

Houghton, R. 2005. *Tsunami Emergency Lessons from Previous Natural Disasters*. ALNAP. Available at: www.alnap.org/pubs/pdfs/tsunamibriefing05.pdf.

International Federation of Red Cross and Red Crescent Societies. Disaster Emergency Needs Assessment. *Disaster Preparedness Training Programme*. Available at: www.ifrc.org/cgi/pdf_dp.pl?disemnas.pdf.

— Improving Coordination. *Disaster Preparedness Training Programme*. Available at: www.ifrc.org/cgi/pdf_dp.pl?impcoor.pdf.

International Water Management Institute. 2005. *IWMI's Tsunami Relief Effort in Sri Lanka*. (22 Jun 2005). Available at: www.iwmi.cgiar.org/tsunami_appeal/20050105.pdf.

IRC International Water and Sanitation Centre. 2005. *Tsunami Briefing Programme*. (7 Feb 2005). Available at: www.irc.nl/page/16091.

— *Tsunami causes devastation: how IRC helps*. (22 February 2005). Available at: www.irc.nl/page/15675.

— *Water supply and emergencies (ch 24)*. (4 January 2005) TP40_Chapter24.pdf.

Joint Assessment Team from the Asian Development Bank (ADB), the United Nations, and the World Bank. 2005. Annex 7 – Water supply, sanitation and solid waste management sectors. *Tsunami Impact and Recovery, Joint Needs Assessment World Bank-Asian Development Bank-UN System*. Available at: www.un.int/maldives/ TsunamiImpactandRecovery.pdf.

Ministry of Foreign Affairs, Danida. 2005. *Tsunami follow-up: Water and Sanitation in Asia, Overview of Current and Planned Rehabilitation Assistance to the Water and Sanitation Sector*. Available at: www.danishwaterforum.dk/newdwf/dwf2/doc/DWF Research/Tsunami-m%C3%B8de/2-HN_Tsunami.ppt.

OXFAM. 2005. *How do we do it?* (20 June 2005). Available at: www.oxfam.org.uk/what_we_do/emergencies/country/asiaquake/howwework.htm.

— *How we work in emergencies: Oxfam's response to the Tsunami Crisis*. (20 June 2005). Available at: www.oxfam.org.uk/what_we_do/emergencies/country/asiaquake/howwework.htm.

— *Learning the Lessons of the Tsunami – One Month On*. (25 January 2005). Available at: www. oxfam.org/eng/pdfs/doc050125_tsunami_externalbulletin.pdf.

— (20 March 2006). *Addressing water and sanitation needs in camps in Tamil Nadu* Available at: www.oxfam.org.uk/what_we_do/emergencies/country/asiaquake/reports/water_tamil_nadu220205.htm.

— *Oxfam's response to short-term needs: providing clean water* (3 February 2006). Available at: www.oxfam.org.uk/coolplanet/teachers/tsunami/shortterm.htm.

Red Cross. 2005. *The Tsunami: looking back at an emergency operation*. (16 December 2005). Available at: www.icrc.org/Web/Eng/siteeng0.nsf/iwpList317/ 6F1CCD5C6C5A711EC12570D9003F5686.

RedR India. 2005. *Rapid Sector Review and Capacity Strengthening (Sanitation, Hygiene, and Water Supply)*. *Tsunami Affected Areas – Tamil Nadu, A Debrief Report of Assessment Mission 14 Feb – 20 Feb 2005*. Available at: www.redr.org/india/TsunamiWatsanReport.pdf.

Sarana, L. 2005. *The First 30 Days: Organizing Rapid Response (Indonesian Red Cross/IRC)*. WHO Conference on the Health Aspects of the Tsunami Disaster in Asia. Phuket, Thailand. (5 May 2005). Available at: www.who.int/hac/events/tsunamiconf/presentations/2_7_first_30_days_sarana_doc.pdf.

Snoad, N. 2005. *Logistics, Information Technology, and Telecommunications in Crisis Management*. WHO Conference on the Health Aspects of the Tsunami Disaster in Asia.

Phuket, Thailand. (5 May 2005). Available at: www.who.int/hac/events/tsunamiconf/presen-tations/ 2_18_logistics_it_telecoms_snoad_ppt.pdf.

The Sphere Project. 2004. *The Sphere Project Humanitarian Charter and Minimum Standards in Disaster Response*. Rev. Ed. Geneva, Switzerland. Available at: www.sphereproject.org/ handbook/hdbkpdf/hdbk_full.pdf.

UNICEF. *Indian Ocean Earthquake and Tsunami UNICEF Response at Six Months Update*. Available at: www.unicef.org/emerg/disasterinasia/files/Tsunamiat6report16june.pdf.

— *Quick tsunami response prevented water-borne epidemics*. Available at: www.unicef. org/emerg/disasterinasia/24615_25680.html.

— *Learning lessons from tsunami response*. Available at: www.unicef.org/emerg/ disasterinasia/24615_25934.html.

— *Water and sanitation in emergencies*. Available at: www.unicef.org/wes/index_emergency.html.

— *India. Tsunami response* (video clip). *Bringing clean water and sanitation to children in Tamil Nadu, India*. Available at: www.unicef.org/5048l_indiawatersanitation.ram.

— *Maldives. Tsunami response* (video clip). *Desalinating seawater, harvesting rainwater and promoting hygiene*. Available at: www.unicef.org/5176l_maldive.ram.

— *Sri Lanka* (video clip). *The provision of safe drinking water to Tsunami-affected families saves lives*. Available at: www.unicef.org/5183l_srwaterenglish.ram.

— *Our response – Water & Environmental Sanitation (Indonesia)*. Available at: www.unicef. org/indonesia/wes_2880.html.

— Joint Meeting of the Executive Boards of UNDP/UNFPA, UNICEF and WFP United Nations. *Transition from relief to development, focusing on natural disasters*. 20–23 Jan. 2006. Available at: www.unicef.org/about/execboard/files/06-Joint_Mtg_item2.pdf.

UN/ISDR. *What's Early Warning, Basics of Early Warning*. Platform for the Promotion of Early Warning. Available at: www.unisdr.org/ppew/whats-ew/basics-ew.htm.

— *Evaluation and strengthening of early warning systems in countries affected by the 26 December 2004 tsunami* (updated March 2006). Available at: www.unisdr.org/ppew/tsunami/ news-events/tsunami-brochure.pdf.

— *Evaluation and strengthening of early warning systems in countries affected by the 26 December 2004 tsunami* (March 2006). Progress Report. Available at: www.unisdr.org/ppew/ tsunami/news-events/TEWS-progress-report.pdf.

— International Strategy for Disaster Reduction (ISDR). 29 March 2006. Available at: www. unisdr.org/unisdr/eng/country-inform/indonesia-keydoc.html.

United Nations Office for the Coordination of Humanitarian Affairs. 2005 *Situation Report No. 22 280105 -South Asia (tsunami)*. (28 January 2005). Available at: http://ochaonline.un.org/ DocView.asp?DocID=2838.

Walden, V. 2005. *Challenges for water, sanitation and hygiene promotion interventions in the immediate aftermath of the tsunami: An Acehnese perspective*. WHO Conference on the Health Aspects of the Tsunami Disaster in Asia. Phuket, Thailand. (5 May 2005). Available at: www.who.int/hac/events/tsunamiconf/presentations/2_8_water_sanitation_food_safety_ environmental_health_walden_doc.pdf.

World Health Organization. 2004. *Rapid Needs Assessments for Water, Sanitation and Hygiene*. (December 2004). Available at: http://w3.whosea.org/en/Section23/Section1108/ info-kit/Rapid_needs_Assessment_guidelines.doc.

— *Three Months after the Indian Ocean Earthquake-Tsunami*. (21 Jun. 2005). Available at: www.who.int/hac/crises/international/asia_tsunami/3months/report/en/index.html.

— *WHO Conference on the Health Aspects of the Tsunami Disaster in Asia*. Phuket, Thailand. (4–6 May 2005). Available at: www.who.int/hac/events/tsunamiconf/en/.

WHO. 2005. *Health Action in Emergencies, Including Response to Earthquakes and Tsunamis of December 2004*. Available at: http://w3.whosea.org/LinkFiles/Health_Ministers_Meeting_ WP3-HMM-health_action_in_emergencies.pdf.

— *Solid waste management in emergencies.* Available at: http://wedc.lboro.ac.uk/WHO_
Technical_Notes_for_Emergencies/7%20-%20Solid%20waste%20management%20in%
20emergencies.pdf.
— *Emergency sanitation – technical options.* Available at: http://wedc.lboro.ac.uk/WHO_Technical_
Notes_for_Emergencies/14%20-%20Emergency%20sanitation%20-%20technical.pdf.
— *Technical Notes for Emergencies.* Available at: http://wedc.lboro.ac.uk/WHO_Technical_
Notes_for_Emergencies/.

- The International Federation of Red Cross and Red Crescent Societies has created
  a website on disaster preparedness that includes publications, training manuals,
  and links to other resources: www.ifrc.org/what/disasters/dp/index.asp.
- The WHO's technical guidelines include a section outlining components of emergency preparedness and response plan for the water and sanitation sector: www
  .who.int/hac/techguidance/training/Planning%20emergency%20preparedness%
  20and%20response.pdf.
- FEMA Recovering from disaster.
  www.fema.gov/areyouready/recovering_from_disaster.shtm.
- DERA
  www.disasters.org/.
- PLAN 12 months after
  www.plan-international.org/pdfs/12Months.pdf.

More detailed descriptions of the responses of other organizations to the Asian
Tsunami Disaster can be found at the following websites:

General documentation:

- World Health Organization: www.who.int/hac/crises/international/asia_tsunami/
  3months/report/en/index.html.
- WHO Conference on the Health Aspects of the Tsunami Disaster in Asia
  www.who.int/hac/events/tsunamiconf/en/.
- The Drinking Water Response to the Indian Ocean Tsunami, Including the Role of
  Household Water Treatment www.who.int/household_water/research/DW_
  response_tsunami.pdf.
- Indonesian Red Cross: www.who.int/hac/events/tsunamiconf/presentations/2_7_
  first_30_days_sarana_doc.pdf.
- Oxfam – Learning the Lessons of the Tsunami – One Month On www.oxfam.
  org/eng/pdfs/doc050125_tsunami_externalbulletin.pdf.
- Tsunami Impact and Recovery, Joint Needs Assessment World Bank-Asian
  Development Bank-UN System, Annex 7 – Water Supply, Sanitation and Solid
  Waste Management Sectors www.un.int/maldives/Wpr;d%20Bank%20Report%
  20-%20A7.pdf.
- Asian World Bank (22 April 2005) Tsunami report http://siteresources.worldbank
  .org/INTTSUNAMI/Resources/tsunamireport-042205.pdf.

Short-term responses:

- International Water Management Institute (IWMI): www.iwmi.cgiar.org/
  tsunami_appeal/20050105.pdf.
- IRC International Water and Sanitation Centre: www.irc.nl/page/15675.

Medium- to long-term responses:

- Water for People: www.waterforpeople.org/upload/0405wfp.pdf.
- UNICEF (this report is organized by country and then further broken down into sector responses): www.unicef.org/emerg/disasterinasia/files/Tsunamiat6report-16june.pdf.
- Oxfam (this site contains a compilation of different reports and articles on OXFAM's response to the tsunami): www.oxfam.org.uk/what_we_do/emergencies/country/asiaquake/archive.htm#.
- How do we do it? www.oxfam.org.uk/what_we_do/emergencies/country/asiaquake/howwework.htm.

Lessons learned from other disasters:

- Workshop on lessons learned on the national and international Response to the Bam earthquake: www.reliefweb.int/library/documents/2004/ocha-irn-15apr.

Chapter 14

Managing urban water risks: managing drought and climate change risks in Australia

J.M. Anderson

Afton Water Solutions P/L, 1 Cumbora Circuit, Berowra, NSW 2081, Australia

ABSTRACT: Climate change due to global warming will reduce river flows across much of Australia and reduce the yield of existing urban and rural water systems. These reductions are additional to yield reductions that have occurred from the allocation of more water for environmental flows under national water reforms, and the severity of the 2001–2007 drought in southern Australia. Australian water authorities are adapting to these changes by implementing extensive water-savings programmes and by developing new water sources including water reuse, stormwater and desalination as well as traditional river and groundwater sources. This chapter describes an Australian case study and discusses how stochastic analysis of system scenarios can help identify drought and climate change risks, and the economic benefits of water conservation and water reuse.

14.1 INTRODUCTION

A safe, reliable water supply is vital to community wellbeing, public health and economic prosperity.

A majority of urban water systems around the world depend on water sources which are fed by rainfall. In some cases rainfall is highly variable, varying not only seasonally but also varying widely from year to year. Most urban water systems include some form of water storage to maintain supply when rain-fed source waters are insufficient to meet the demand for water. Storage may be in the form of tanks and reservoirs, surface water storages or groundwater storages.

The capacity, or 'yield', of an urban water system and storage can be calculated in terms of the annual demand that could be supplied from the water sources through a repeat of the historical pattern of rainfall and flows. It is usual to define system yield in terms of a level of service which includes measures of the reliability of the supply and the security against droughts.

This chapter discusses the management of drought risks, the extent of climate change and other impacts on the yield of urban and rural water systems. Potential adaptation strategies are discussed. These strategies include major commitments to water savings and to water recycling.

The chapter also discusses how stochastic methods of water system analysis are being enhanced to provide better management of supply risks. The enhancements include incorporation of cost functions in the analyses. A case study is presented showing how water conservation and water recycling can reduce the cost of water restrictions and drought contingency works, and compensate for projected climate change impacts.

14.2 MANAGING DROUGHT RISKS

A reliability and risk approach is commonly used to manage drought risks in urban water systems. In major droughts, the practice has been to impose restrictions on discretionary water uses, such as garden watering, to ensure that supply for essential household and commercial needs can continue uninterrupted. In recent years, restrictions have become less effective in reducing demands during droughts because recent increases in the efficiency of household and commercial water use have reduced the volume of discretionary water use.

In some systems, drought contingency supplies may be commissioned to supplement normal sources and reduce the duration and severity of restrictions during droughts. Examples of contingency supply options include groundwater wells, desalination and cartage from more distant sources. Such sources are typically more expensive than normal sources but may be preferable to more severe and prolonged restrictions.

A level of service for drought security can be defined in terms of the reliability (per cent of the time) that the system can supply 100% of water needs, and the frequency, duration and severity of restrictions. Restrictions should not occur too frequently, should not last too long and should not be too severe. Severe restrictions associated with the depletion of storages to very low levels can cause political anxiety in the community over the risks of running out of water.

In Australia, which has highly variable rainfall, system behaviour and yield are analyzed typically over a 100-year period of analysis using historical and synthesized data. Systems are generally sized to achieve 95 to 97% reliability of full supply, with security achieved through a drought management plan that incorporates a pre-planned schedule of restrictions and commissioning of drought contingency works. The drought management plan needs to ensure that essential supplies can be maintained in the event of a drought that is worse than any drought in the period of the historical records. Some utilities deal with this problem by providing additional storage reserves. Other utilities devise a drought management plan such that, regardless of the starting storage level, essential supplies could be maintained through a repeat of the worst drought sequence on record by means of restrictions and drought contingency works. The framework for urban water resource planning in Australia has been documented by Erlanger and Neal, 2005.

Many larger urban water utilities in Australia are supplementing yield analyses using historical rainfall and streamflow data with analyses using stochastic data generation techniques to provide multiple replicate analyses. The stochastic analyses do not give any better estimate of system yield, but provide a better understanding of how to manage the system in extreme drought events. Stochastic analyses are also useful in understanding how a system will respond to rising demands, addition of new works, or changes in available water resources.

14.3 ADAPTING TO CLIMATE CHANGE IMPACTS

14.3.1 Climate change forecasts

Climate change associated with global warming is likely to have a major impact on the reliability and security of urban and rural water systems worldwide. The 4th Assessment

Report of the WMO/UNEP Intergovernmental Panel on Climate Change reports that global climate models indicate that average global temperature is likely to rise by between 1.8°C and 6.4°C by 2100 (IPCC, 2007). Even with good progress on abatement strategies, the temperature rise will most likely be 3.0°C or more over this century. The global climate models suggest that significant shifts in rainfall patterns may occur.

In Australia, global warming is likely to cause significant reductions in rainfall across much of southern and eastern Australia (Hennessey et al., 2004). The modelling indicates that flows in Australian rivers will decline and droughts will become more frequent and more prolonged as a result of global warming. There has been a significant change in rainfall patterns in Australia in the last 20 years. Streamflows in the period 1950 to 1990 were about 25% higher than the long-term average. In many catchments, streamflows since 1990 have been less than half the long-term average, and flows in 2006 were the lowest on record. In parts of Australia the current (2001–2007) drought is considerably more severe than any previous event in the short, 120-year historical record of reliable rainfall measurements. It is not yet clear whether the current drought is within the range of variation to be expected based on historical rainfall records, or is a manifestation of a step-change in climate. Current studies of paleo-climate measures, such as limestone deposits may help clarify this question.

Climate change impacts will pose major challenges in maintaining adequate water supplies for urban and rural water needs in Australia. The Murray–Darling Basin, which extends over 15% of the Australian continent, is Australia's prime food-growing region. Climate modelling indicates that streamflows in the basin are likely to fall by 20% or more. Corresponding rises in river salinity are likely.

14.3.2 Modelling of impacts

To demonstrate the potential impacts of climate change, drought and other factors, the capacity (yield) of an urban water supply system supplied by a coastal river in eastern Australia has been modelled. System behaviour has been modelled using 122 years of streamflow records generated from the correlation of streamflow and rainfall data. System capacity is defined as the average annual demand, D, which can be supplied with 95% reliability (restrictions not applied in more than 5% of months), and able to supply restricted demands through a repeat of any historic drought event starting with the storage at the restriction level (55% of full storage) and not falling below 5% storage. The system model incorporates climate variable demands and three levels of restrictions. System capacity results are presented in the form of non-dimensional graphs in which system capacity D (given as a percentage of average annual streamflow, Qm) is expressed as a function of the storage S (also given as a percentage of average annual streamflow, Qm).

Curve A in Figure 14.1 shows the estimated system capacity prior to the current drought, and before any allowance for climate change and other impacts.

14.3.3 Water reforms and environmental flows

Even prior to the current drought it was recognized that there was a need to undertake a programme of water reforms to correct the allocation of regulated flows in Australian irrigation areas and to allocate more water for environmental flows and river health, particularly in low-flow periods.

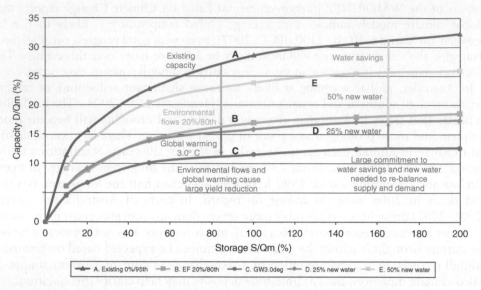

Figure 14.1 **Adapting to climate change and other impacts (See also colour plate 7)**

Curve B in Figure 14.1 shows the estimated reduction in system capacity for an environmental flow allowance of the 80th percentile flow plus 20% of all flows in excess of the 80th percentile flow. It can be seen that environmental flow allocations may absorb a considerable proportion of existing system capacity necessitating major capital investment to provide new system capacity.

14.3.4 Climate change impacts

Climate change resulting from global warming is forecast to bring about lower rainfalls, higher evaporation, lower streamflows and more frequent droughts over much of Australia.

Projected changes in average temperatures and rainfalls in eastern Australia have been presented by Hennessey et al. (2004). It is projected that winter and spring rainfalls will be lower. Coupled with a projected increase in evaporation, streamflows are likely to be reduced, particularly in inland areas. In coastal areas, there is some chance of increases in summer rainfalls, so that reductions in annual streamflows will be less than in inland areas.

Figure 14.1 shows the additional impact of global warming of 3.0°C on the capacity of the water supply, assuming rainfall reductions corresponding to the mid-range of the forecasts of temperature and rainfall changes in Hennessey et al. (2004) and 20%/80th percentile environmental flows apply (Curve C).

The large range of global warming projections leaves a wide range of uncertainty about the impacts of global warming on streamflows. Water strategies should include contingency plans to bring on additional sources of supply if impacts are greater than projected, or if greenhouse gas abatement strategies are not as effective as planned.

It can also be seen in Figure 14.1 that for systems which already have a large amount of storage (e.g., S/Qm >100%), there are only limited benefits in adding more storage. Such systems have long critical drought and considerable evaporation losses. Impacts are likely to be greater in the inland river catchments in the Murray–Darling Basin.

14.3.5 Adapting with water savings and water reuse

It may be possible to offset the impacts of environmental flow allocations and climate change by a combination of water savings (reductions in demands) and by introducing new water sources to achieve a new balance between supply and demand; e.g., Curves D and E in Figure 14.1.

New water sources may include additional river and groundwater sources if not already allocated. New water sources may also include: harvesting of urban rainwater and stormwater; water recycling (reuse of treated wastewater); or desalination of sea-water or saline groundwater sources.

Given the importance of water supply to the economy and to social wellbeing, and the uncertainties attached to climate change impacts, it is essential that the new water strategies are more diverse, flexible and adaptable. Water reuse and desalination are attractive for an adaptable strategy because they are not affected by drought.

14.4 ADAPTATION CASE STUDY

14.4.1 The Sydney water system

The principal source of supply for the Australian city of Sydney is the Hawkesbury–Nepean river system. The Hawkesbury–Nepean river basin has a catchment area of 22,000 km², an average annual rainfall of 890 mm per year and an average annual discharge of about 3,000 M m³ per year. Like all Australian catchments, the rainfall and the streamflow are highly variable. The Hawkesbury River enters the Pacific Ocean about 30 km north of Sydney. Since the Nepean River was first tapped for Sydney's water supply in 1886, the population of Sydney has grown from less than 1 million to more than 4 million. The population is expected to grow by another million over the next 20 years.

The major Sydney water storages command about 44% of the Hawkesbury–Nepean catchment and have average annual inflows of about 1,300 M m³. In addition, there is a diversion system to pump water from the Shoalhaven River, 160 km south of Sydney. The Sydney catchments have experienced a prolonged drought period since 1999, and storage levels were about 36% in January 2007.

Environmental flow allocations below the storages are being increased over an eight-year period to improve river health in the Hawkesbury and Nepean rivers. The yield of the current Sydney water supply system has recently been revised downwards to about 570 M m³ per year, because of increased environmental flow allocations and other factors. The current unrestricted annual water consumption would be about 610 M m³, but has been reduced to about 520 M m³ by planned water restrictions as part of Sydney's drought management plan. Sydney currently reuses about 22 M m³ of recycled water per year.

14.4.2 The Sydney Metropolitan Water Plan 2006

A new Sydney Metropolitan Water Plan (DIPNR, 2004b; NSW, 2006) is aiming to achieve by 2015:

- Water savings of 40% in new, detached, single residential dwellings and 25% in new, medium-density dwellings, through a mandatory Building Sustainability Index (BASIX) scheme for water efficiency in new dwelling (DIPNR, 2004a). The BASIX requirements are also being extended to houses undergoing major extensions or renovations.
- Annual water savings of $120 \, M \, m^3$ to $140 \, M \, m^3$ by existing consumers through a variety of programmes including: *Water Efficiency Plans* for large users; the Sydney Water *Leakage Reduction* programme; *Water Efficiency Labelling* for new household appliances; a programme of *Rebates for Rainwater Tank and Water Efficient Washing Machines*; and a household *Water Audit and Retrofit* programme for existing dwellings.
- Up to $70 \, M \, m^3$ per year of recycled water per year through new water reuse projects for urban, industrial, agricultural and environmental flow supplementation in western Sydney. It is proposed to supply recycled water to 160,000 new houses for residential garden watering and toilet flushing in the new development areas in western Sydney. Initially, $18 \, M \, m^3$ per year of recycled water will be supplied for environmental flow allocations with possible later expansion to $30 \, M \, m^3$ per year or more.
- Reuse of $30 \, M \, m^3$ per year of recycled water per year in existing urban areas through development of a recycled water network.
- Significant levels of new supply through rainwater and stormwater harvesting.

Also the secure yield of existing water supply system is being increased by various measures including:

- *Deep Storage:* Access to the deep storage below the previous minimum operating levels.
- *Shoalhaven Transfers:* Extra transfers from Shoalhaven River using the existing pumping system more frequently.
- *Groundwater:* Use of groundwater aquifers as a drought contingency measure.
- *Desalination:* If the Sydney storages fall below 30%, a desalination plant would be built as a drought contingency measure.

In combination these measures are likely to cater for growth and environmental flow allocations for about the next 20 years.

14.4.3 Managing drought risks

White and Campbell (2006) have demonstrated that it is possible to reduce the severity of drought restrictions and the probable cost of infrastructure development for Sydney by adopting a drought contingency plan in which construction of a costly desalination plant would only be triggered if the Sydney storages fall below 30%. This 'virtual' infrastructure contingency plan, which has a low probability of being triggered, provides an insurance policy against supply failure, and has been adopted as part of the Sydney Metropolitan Water Plan 2006.

14.4.4 Enhanced stochastic analyses

Enhanced stochastic analyses of supply and demand scenarios can be used to identify appropriate strategy responses to drought and climate change (Anderson, 2007). In the enhanced analyses, the operation of the water supply system is modelled for a period of 20 to 50 years, starting with the storages at the current levels. The analyses incorporate projected increases in consumer demands offset by projected water savings through the implementation of water saving and water reuse measures. Demands are varied seasonally, taking account of variations in rainfall and evaporation. Planned water reuse measures and other new water sources are switched on to match the nominated construction and commissioning schedules.

The projected reductions in rainfall and streamflows due to global warming and the corresponding increases in demands are incorporated into the system analyses month-by-month through the period of analysis.

For each supply scenario, the system is run for at least 1,000 data replicates, synthesized from the historical data sets. The results from each replicate are ranked to identify the probability of the storages falling to the levels that trigger water restrictions and the drought contingency works. These probability analyses provide measures of both the reliability and security of supply year-by-year, and clearly identify any decline in security coming from climate change.

14.4.5 Economic analyses

Cost functions can be added to the stochastic analysis model to record in each replicate:

- The investment and operating costs for drought contingency works.
- The cost to consumers of water restrictions during droughts, measured using a loss of consumer utility function.

The analysis also incorporates discounted net present value costs for each replicate. The net present value costs can be ranked to identify the probability of incurring high drought contingency and loss of customer utility costs. By varying starting levels, it is possible to identify appropriate storage trigger levels for drought contingency works and time triggers for new supply works.

By comparing supply scenarios, it is possible to identify the probable economic benefits by way of reduced (avoided) drought contingency and water restriction costs through implementing water reuse or other new supply measures.

Figure 14.2 shows an example of the probable savings in restriction and desalination contingency costs through the implementation of the water conservation and water reuse measures in the 2006 Sydney Metropolitan Water Plan. The system has been analyzed for 1,000 replicates, and assumes a starting storage level of 35% in January 2007. The net present value analysis assumes a 7% discount rate.

The economic analysis indicates that the probable average net present value savings in water restriction and desalination contingency costs will be about US$100 per person across all replicates, or about US$400 M total, from the implementation of water saving and water reuse projects. If conditions continue to be drier than the long-term

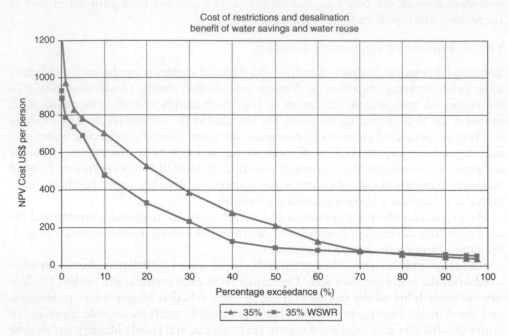

Figure 14.2 **Probable benefits from implementing water savings and water reuse in Sydney (See also colour plate 8)**

historical average, then the savings would probably be about double, about US$800 M. These savings are sufficiently large to justify substantial investment in water saving and water reuse projects.

It can also be demonstrated that the water saving and water reuse projects will increase the effective yield of the system expressed in terms of the future population served, and make the system more robust in future drought events.

14.4.6 Another example

The stochastic analysis approach can also be used to identify the size and timing of water reuse and other projects needed to offset projected reductions in the reliability of supply due to global warming. For example, the Central Coast system serves about 300,000 people in an area about 100 km north of Sydney. The system relies on small coastal catchments, and is vulnerable to drought and climate change impacts. The rising demand stochastic analysis indicates that use of about 20 ML/d of recycled water to meet environmental flow allocations below the main water supply weir may be sufficient to offset projected reductions in streamflows due to climate change.

14.5 ADDITIONAL DROUGHT SECURITY ISSUES

14.5.1 Drought severity

In Australia, the current drought has been sufficiently severe to alter the statistics used in the generation of streamflow replicates for stochastic analyses. System analyses need to be revised to include the current drought. If not already done, the statistics should also be extended to the maximum extent possible, using historical rainfall records.

14.5.2 Hindcasting

It is reported that average temperatures in Australia have risen by about 0.8°C in the last 100 years. A number of Australian water authorities have recorded reductions in rainfall and streamflows over the last 30 years, outside previous historical experience. As well as climate change forecasts, there is a case for climate change hindcasts to identify the changes in rainfall, evaporation and streamflows that have already occurred during the period of historical records. If the mid-range rainfall changes per degree of global warming from Hennessey et al. (2004) are applied to normalize streamflows in Australian coastal rivers to a year 2000 base year, then mean annual flows would be about 3% less than the recorded figures. This is a small reduction, but nonetheless significant in any assessment of future system capacity.

14.5.3 Starting storage

As demands approach system capacity, storages are seldom likely to be full. Estimates of system yield, based on analyses that commence with the storages starting full, overestimate the reliability and security in some systems. This factor will become more important if climate change causes more frequent droughts.

14.5.4 Demand variability

Drought security and reliability calculations are sensitive to demand variations during droughts. Unrestricted demands are often higher during droughts because of lower rainfall and higher irrigation needs. System analyses assuming constant demands may over-estimate system capacity.

14.5.5 Demand hardening

The success of water efficiency and water-saving measures over the last two decades has resulted in demand hardening. The old drought contingency of water restrictions is rapidly becoming less effective. Calculations of future system capacity and drought security need to account for further demand hardening, which will occur as houses become more water efficient and greater use is made of rainwater and recycled water for outdoor use.

14.5.6 Building diverse water portfolios

Beuhler (2006) has proposed that the modern financial portfolio theory developed for financial markets can be applied to water resource portfolios. The financial portfolio theory was developed by the economist Harry Markowitz in the 1950s to address risks in portfolios of financial assets such stocks, bonds and bank bills. The financial assets with high returns, like stocks, often have high volatility and high risk. Exposure to financial risks can be reduced by building diversified financial portfolios which contain a proportion of low volatility assets such as bonds.

Many water utilities have water resource portfolios which rely heavily on river water sources (low cost but high volatility). Volatility is reduced by providing surface water reservoirs and, in some cases, by conjunctive use of groundwater sources.

There is an opportunity to build more robust water systems (lower risk of not meeting needs) by building water portfolios that include a proportion of water from sources such as recycled water and desalination (higher cost but low volatility) that are not reduced by drought.

14.6 CONCLUSIONS

Climate change and environmental flow allocations will have large impacts on the reliability and security of urban and rural water systems in Australia. Australian water utilities are implementing a range of adaptation strategies including large commitments to water efficiency for new consumers, water savings for existing consumers. Adaptation will also involve extensive development of new water sources. As many river and groundwater sources are already heavily allocated, there will also be extensive development of new water sources, including water reuse, desalination, and urban rainwater and stormwater harvesting, to restore and maintain the balance between supply and demand.

Improved stochastic analysis tools can be applied to identify:

- Drought risks and trigger levels for drought contingency works.
- Emerging supply risks due to climate change.
- The economic benefits of water reuse in reducing drought and climate change risks.
- The timing for implementation of water reuse projects and other new supply works.

Given the importance of water supply to the economy and to social wellbeing, and the uncertainties attached to climate change impacts, it is essential that new water strategies are more diverse, flexible and adaptable. Water reuse is a particularly attractive new supply component for diversified, adaptable water supply strategies because, unlike conventional river sources, it provides a continuous supply which is not vulnerable to climate change and droughts.

Water utilities in Australia and elsewhere find themselves propelled into a new era by global warming. The old ways of slow evolution and the old certainties have been swept away by the new imperatives of drought and climate change. Every water utility needs to overhaul its water supply strategy to be able to meet the future water needs of consumers.

REFERENCES

Anderson, J.M. 2007. *Adapting to Climate Change with Water Savings and Water Reuse*. 3rd AWA Water Reuse and Recycling Conference, UNSW Sydney, 16–18 July 2007.

Beuhler, M. 2006. Application of Modern Financial Portfolio Theory to Water Resource Portfolios. *Proc. IWA 5th World Water Congress*, Beijing, September 2006.

DIPNR. 2004a. *BASIX – Building Sustainability Index*. NSW Department of Infrastructure. Planning and Natural Resources, May 2004. (www.basix.nsw.gov.au).

DIPNR. 2004b. *Meeting the challenges – Securing Sydney's Water Future: The Metropolitan Water Plan 2004*. NSW Department of Infrastructure, Planning and Natural Resources, October 2004. (www.dipnr.nsw.gov.au).

Erlanger, P. and Neal, B. 2005. *Framework for Urban Water Resource Planning*. Water Services Assoc of Australia, Occasional Paper No.14, June 2005.

Hennessey, K.J., Page, C., McInnes, K., Jones, R., Barthols, J., Collins, D. and Jones, D. 2004. *Climate Change in New South Wales*. Consultancy Report for the NSW Greenhouse Office by CSIRO and the Australian Bureau of Meteorology, July 2004.

IPCC. 2007. *Climate Change 2007: The Physical Science Basis – Summary for Policymakers*. Contribution of Working Group 1 to the Fourth Assessment Report of the Intergovernmental Panel on Climate Change, February 2007.

NSW. 2006. *Metropolitan Water Plan: Securing Sydney's Water Supply*. New South Wales Government, April 2006, ISBN 0 7313 5460 5. (www.waterforlife.nsw.gov.au).

White, S. and Campbell, D. 2006. *Review of the Metropolitan Water Plan: Final Report*. ISF, ACIL Tasman and SMEC for NSW Government, April 2006.

Index

Global and continental temperature change

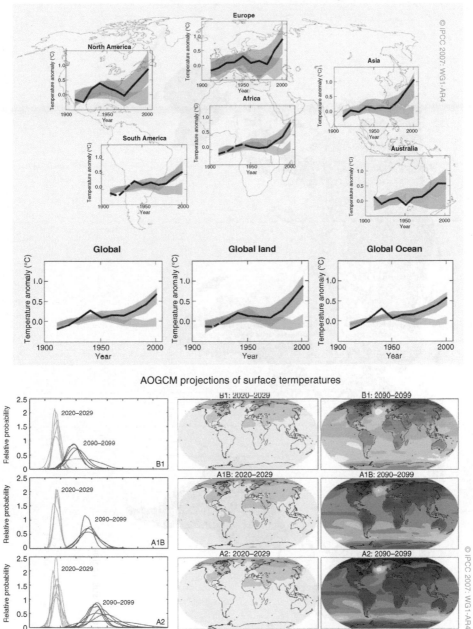

AOGCM projections of surface termperatures

Plate I **Comparison of observed continental- and global-scale changes in surface temperature with results simulated by climate models using natural and anthropogenic forcings**

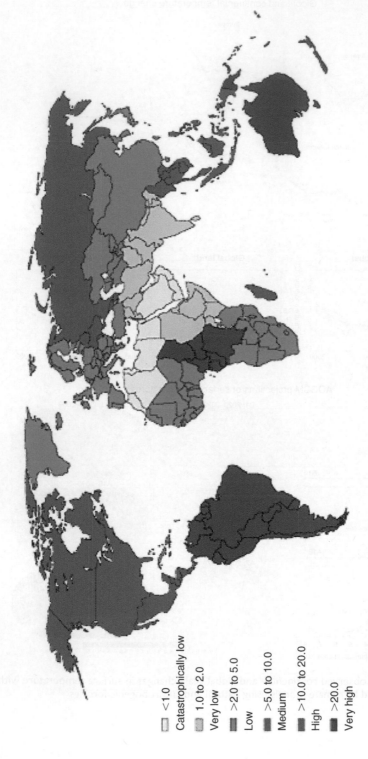

Plate 2 Water availability by sub-region

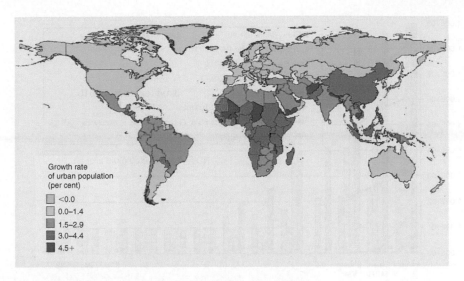

Plate 3 Urban population in major areas according to UN (2005)

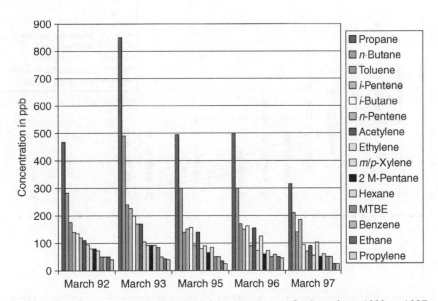

Plate 4 Pollutants (volatile organic compounds) found in Mexico City's air from 1992 to 1997

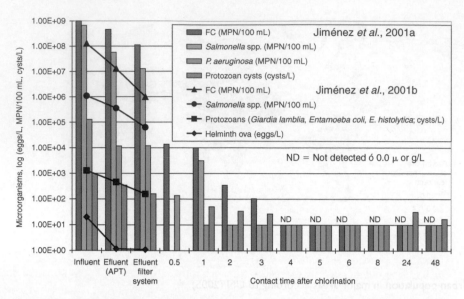

Plate 5 Removal of microorganisms from wastewater using different treatment processes

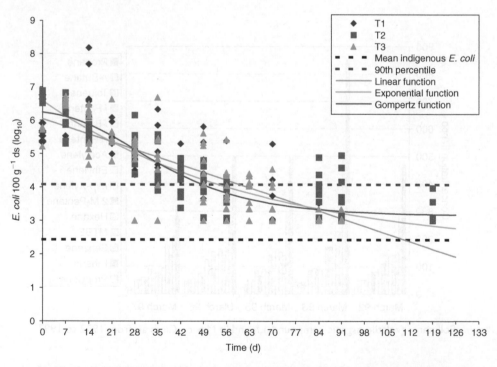

Plate 6 Decline in *E. coli* numbers in relation to time following application of dewatered mesophilic, anaerobically digested biosolids to a sandy loam soil at a rate of 10 t DS ha^{-1} and incorporation depth of 10 cm (Lang et al., 2003)

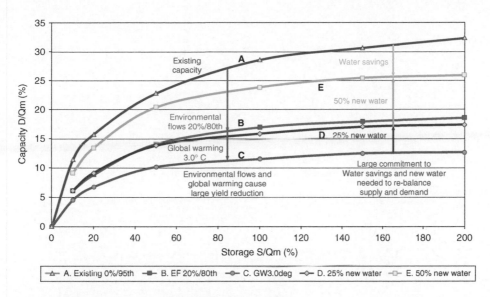

Plate 7 Adapting to climate change and other impacts

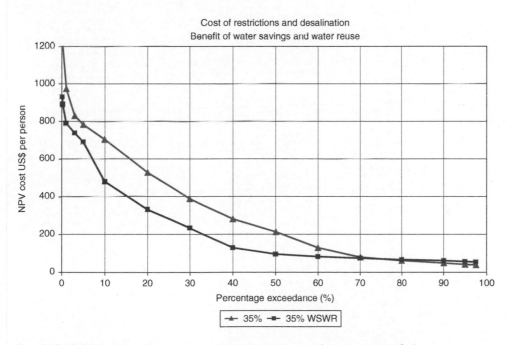

Plate 8 Probable benefits from implementing water savings and water reuse in Sydney

T - #0075 - 071024 - C6 - 246/174/19 - PB - 9780415485678 - Gloss Lamination